STUDIES IN APPLIED MECHANICS 7

Mechanics of Granular Materials:
New Models and Constitutive Relations

STUDIES IN APPLIED MECHANICS

1. Mechanics and Strength of Materials (Skalmierski)

2. Nonlinear Differential Equations (Fučík and Kufner)

3. Mathematical Theory of Elastic and Elastico-Plastic Bodies
   An Introduction (Nečas and Hlaváček)

4. Variational, Incremental and Energy Methods
   in Solid Mechanics and Shell Theory (Mason)

5. Mechanics of Structured Media, Parts A and B (Selvadurai, Editor)

6. Mechanics of Material Behaviour (Dvorak and Shield, Editors)

7. Mechanics of Granular Materials: New Models and Constitutive
   Relations (Jenkins and Satake, Editors)

STUDIES IN APPLIED MECHANICS 7

Mechanics of Granular Materials

New Models and Constitutive Relations

Proceedings of the U.S./Japan Seminar on New Models and Constitutive Relations in the Mechanics of Granular Materials, Ithaca, New York, August 23–27, 1982

Edited by

James T. Jenkins

Department of Theoretical and Applied Mechanics, Cornell University, Ithaca, New York, U.S.A.

and

Masao Satake

Department of Civil Engineering, Tohoku University, Aramaki, Sendai, Japan

Sponsored by the United States National Science Foundation and the Japan Society for Promotion of Science

ELSEVIER – Amsterdam – Oxford – New York 1983

ELSEVIER SCIENCE PUBLISHERS B.V.
Molenwerf 1
P.O. Box 211, 1000 AE Amsterdam, The Netherlands

Distributors for the United States and Canada:

ELSEVIER SCIENCE PUBLISHING COMPANY INC.
52, Vanderbilt Avenue
New York, NY 10017

Library of Congress Cataloging in Publication Data

U.S./Japan Seminar on New Models and Constitutive
Relations in the Mechanics of Granular Materials
(1982 : Ithaca, N.Y.)
Mechanics of granular materials.

(Studies in applied mechanics ; 7)
"Sponsored by the United States National Science
Foundation and the Japan Society for Promotion of
Science."
Bibliography: p.
1. Granular materials--Congresses. I. Jenkins,
James T., 1942- . II. Satake, Masao. III. National
Science Foundation (U.S.) IV. Nippon Gakujutsu
Shinkōkai. V. Title. VI. Series.
TA418.78.U17 1982 624.1'9 83-5616

ISBN 0-444-42192-0 (Vol.7)
ISBN 0-444-41758-3 (Series)

Printed in The Netherlands

PREFACE

The papers contained in this volume are the proceedings of the United States-Japan Seminar on New Models and Constitutive Relations in the Mechanics of Granular Materials held in Ithaca, New York, at Cornell University from August 23 through August 27, 1982.

The purpose of the Seminar was to bring together selected Japanese, United States, and Third Nation scientists who had made fundamental contributions to the understanding of the mechanics of granular materials. In particular, the Seminar was designed to facilitate an exchange of ideas between researchers working with discrete models of granular materials, especially statistical theories and computer simulations, and those theoreticians and experimentalists employing continuum models. The participants and observers represented a wide variety of fields in which an understanding of the mechanics of granular materials is crucial. The papers collected here reflect their broad range of interests.

Two-thirds of the papers are devoted to experimental and theoretical studies of quasi-static deformations of granular materials. Novel features of the irreversible deformation of granular materials are reported, the statistical micromechanics of interparticle contacts is further developed, computer simulations of contacting granular arrays are described, and phenomenological theories are proposed and discussed. The remaining third of the papers treat rapid deformations of granular materials. Here interparticle collisions rather than enduring contacts govern the mechanics. Computer simulations of colliding grains in a rapidly sheared granular mass are elaborated on and the predictions of continuum theories, some of which are derived by considering the statistics of particle collisions, are reviewed or presented for the first time.

The Seminar was sponsored by the Japan Society for the Promotion of Science and the United States National Science Foundation under the United States-Japan Cooperative Science Program. The publication of the Proceedings was made possible by a grant from the Geotechnical Engineering Division of the United States National Science Foundation. The technical sessions were enriched by the Bendix Corporation and Exxon Incorporated and the social program owed its existence to the generosity of Clarkson College and Cornell University. We are grateful for the finanical support that made the Seminar possible.

Finally, we would like to thank the contributors to this volume for participating in the Seminar and for their care in the preparation of their manuscripts.

James T. Jenkins and Masao Satake.

TABLE OF CONTENTS

PARTICIPANTS

Norbert L. Ackermann
Professor
Dept. of Civil and Environmental Engineering
Clarkson College
Potsdam, New York 13676

Christopher Brennen
Professor
California Institute of Technology
Pasadena, California 91125

Stephen C. Cowin
Professor
Department of Biomedical Engineering
Tulane University
New Orleans, Louisiana 70118

Peter A. Cundall
Associate Professor
Department of Civil Engineering
University of Minnesota
Minneapolis, Minnesota 55455

Jamshid Ghaboussi
Associate Professor
Department of Civil Engineering
University of Illinois
Urbana, Illinois 61801

Kolumban Hutter
Laboratory of Hydraulics, Hydrology, and Glaciology
Federal Institute of Technology
CH-8092 Zurich, Switzerland

Isao Ishibashi
Associate Professor
Dept. of Civil and Environmental Engineering
Cornell University
Ithaca, New York 14853

Kenji Ishihara
Professor
Department of Civil Engineering
University of Tokyo
Bunkyo-ku, Tokyo 113

James T. Jenkins
Associate Professor
Department of Theoretical and Applied Mechanics
Cornell University
Ithaca, New York 14853

Ken-ichi Kanatani
Associate Professor
Department of Computer Science
Gunma University
Kiryu, Gunma 376

Tadahiko Kawai
Professor
Institute of Industrial Science
University of Tokyo
Minato-ku, Tokyo 106

Yuji Kishino
Lecturer
Department of Civil Engineering
Tohoku University
Aramaki, Sendai 980

Jun-ichi Konishi
Associate Professor
Department of Civil Engineering
Shinshu University
Nagano, Nagano 380

Joseph T.C. Liu
Professor
Division of Engineering
Brown University
Providence, Rhode Island 02912

Kazutaka Makino
Associate Professor
Department of Chemical Engineering
Kyoto University
Sakyo-ku, Kyoto 606

Hajime Matsuoka
Associate Professor
Department of Civil Engineering
Nagoya Institute of Technology
Showa-ku, Nagoya 466

Siavouche Nemat-Nasser
Professor
Department of Civil Engineering
Northwestern University
Evanston, Illinois 60201

Jace W. Nunziato
Fluid Mechanics and Heat Transfer Division
Sandia National Laboratories
Albuquerque, New Mexico 87185

Masanobu Oda
Associate Professor
Department of Foundation Engineering
Saitama University
Urawa, Saitama 338

Nobunori Oshima
Professor
College of Industrial Technology
Nihon University
Narashino, Chiba 275

Stephen L. Passman
Computational Physics and
Mechanics Division
Sandia National Laboratories
Albuquerque, New Mexico 87185

Masao Satake
Professor
Department of Civil Engineering
Tohoku University
Aramaki, Sendai 980

Stuart B. Savage
Professor
Department of Civil Engineering
and Applied Mechanics
McGill University
853 Sherbrooke Street West
Montreal, PQ Canada H3A 2T6

Anthony J.M. Spencer
Professor
Department of Theoretical Mechanics
The University of Nottingham
Nottingham NG7 2RD, England

Otto D.L. Strack
Associate Professor
Department of Civil Engineering
University of Minnesota
Minneapolis, Minnesota 55455

OBSERVERS

Koichi Hashiguchi
Associate Professor
Department of Agricultural Engineering
Kyusyu University
Higashi-ku, Fukuoka 812

George W. Hawkins
Bendix Advanced Technology Center
9140 Old Annapolis Road
Columbia, Maryland 21045

Morteza M. Mehrabadi
Assistant Professor
Department of Mechanical Engineering
Tulane University
New Orleans, Louisiana 70118

Nobuchika Moroto
Associate Professor
Department of Civil Engineering
Hachinohe Institute of Technology
Hachinohe, Aomori 031

Behzad Rohani
Geomechanics Division
U.S. Army Engineer Waterways Experiment Station
P.O. Box 631
Vicksburg, Mississippi 39180

Mohsen Shahinpoor
Professor
Department of Mechanical and Industrial Engineering
Clarkson College
Potsdam, New York 13676

Hayley Shen
Assistant professor
Department of Civil and Environmental Engineering
Clarkson College
Potsdam, New York 13676

S.L. Soo
Professor
Department of Mechanical and Industrial Engineering
University of Illinois
Urbana, Illinois 61801

Fumio Tatsuoka
Associate Professor
Institute of Industrial Science
Minato-ku, Tokyo 106

Yoshio Tobita
Research Associate
Department of Civil Engineering
Tohoku University
Aramaki, Sendai 980

Otis R. Walton
Earth Sciences Division L-200
Lawrence Livermore Laboratory
Livermore, California 94550

Yasuo Yamada
Research Associate
Department of Civil Engineering
University of Tokyo
Bunkyo-ku, Tokyo 113

Mechanics of Granular Materials: New Models and Constitutive Relations, edited by
J.T. Jenkins and M. Satake, 1983
Elsevier Science Publishers B.V., Amsterdam — Printed in The Netherlands

STRESS AND FABRIC IN GRANULAR MASSES

S. NEMAT-NASSER[1] and M. MEHRABADI[2]

[1]Civil Engineering Dept., Northwestern University, Evanston, Ill. (USA)

[2]Mechanical Engineering Dept., Tulane University, New Orleans, La. (USA)

ABSTRACT

Recent results on the description of a macroscopic stress measure and measures of the fabric of granular masses which support external loads through frictional contact are briefly reviewed. Relations between the overall stress and fabric measures are developed.

INTRODUCTION

The term fabric refers to the microstructure of the granular mass, namely the relative arrangement of the particles within the overall assembly. To quantify the fabric of a granular mass, various measures may be used. One such measure is the void ratio which is closely related to the coordination number, i.e. the average number of contacts per particle (refs. 1-6). Experiments show that the void ratio is not sufficient to completely describe the microstructure of a granular mass, since two samples of the same granular material of the same void ratio may have very different mechanical responses (refs. 7-14). Thus, other measures of fabric have been introduced and used in the constitutive characterization of granular materials (refs. 5, 6, 13-24). In this paper some of these measures are examined in relation to the overall stress measure for the granular mass.

Since stress is a macroscopic continuum concept, its use as a dynamical measure for a particulate medium requires careful micromechanical considerations (refs. 24-26). Earlier works (refs. 19, 25) sought to define stress in terms of the average contact forces transmitted across an imagined plane. Recently, it has been shown (ref. 26) that the overall stress may be defined in terms of the volume average of the tensor product of the contact forces and the associated branch vectors which are vectors connecting the centroids of the corresponding contacting granules. Furthermore, it has been established that these two notions of stress are indeed identical under rather mild assumptions (refs. 24,27).

In this paper a brief review of some fabric measures and their relation to the overall stress measure is presented.

FABRIC MEASURES

Consider a typical pair of contacting granules. There are two unit vectors which may be included in the characterization of the relative arrangement of these granules: (1) the contact unit normal, $\underset{\sim}{n}$, and (2) the unit branch, $\underset{\sim}{m}$, i.e. a unit vector in the direction of the branch vector $\underset{\sim}{\ell}$ which connects the centroid of the first granule to that of the second; see Fig. 1. Note that while there are two $\underset{\sim}{n}$'s (i.e., $\underset{\sim}{n}$ and $-\underset{\sim}{n}$), and two $\underset{\sim}{m}$'s, the tensor product $\underset{\sim}{n} \otimes \underset{\sim}{m}$ can be uniquely defined for each contact. In addition, one has the branch length, ℓ, which is the length of branch $\underset{\sim}{\ell}$.

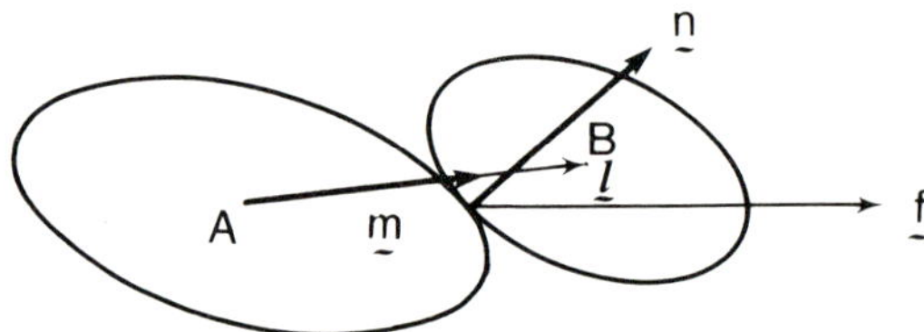

Figure 1.

Suppose it is required to generate volume averages of tensorial measures which involve $\underset{\sim}{n}$ and $\underset{\sim}{m}$. Since for each $\underset{\sim}{n}$ there is a $-\underset{\sim}{n}$ and for each $\underset{\sim}{m}$, a $-\underset{\sim}{m}$, it is clear that only tensors of even order emerge.

The simplest quantity of this kind is a tensor of order zero, i.e. a scalar, which is commonly identified with the void ratio, or porosity, or equivalently, the coordination number.

For a second measure, one may use*

$$J_{ij} = \langle m_i m_j \rangle, \quad J_{ij}^{(1)} = \langle n_i n_j \rangle, \quad H_{ij} = \langle n_i m_j \rangle. \tag{1}$$

The particular choice will depend on the particular usage.

Naturally, more information is obtained about the microstructure if higher order tensors are involved, e.g.

$$\langle n_i n_j n_k n_\ell \rangle, \quad \langle m_i m_j m_k m_\ell \rangle, \quad \langle n_i n_j m_k m_\ell \rangle, \tag{2}$$

---

*Rectangular Cartesian components are used, and $\langle \cdots \rangle$ represents the average over all contacts within a representative sample, i.e., $\langle n_i n_j \rangle \equiv \frac{1}{M} \sum_{\alpha=1}^{M} n_i^{(\alpha)} n_j^{(\alpha)}$, where there are M contacts in the sample, and $\underset{\sim}{n}^{(\alpha)}$ is the unit contact normal at the αth contact.

and others. Quantities of this kind do emerge in a micromechanical modeling of overall stress (ref. 24).

If a size effect is to be included in the fabric measures, then one may modify, for example, $(1)_3$ to read

$$\bar{H}_{ij} = \langle \ell n_i m_j \rangle / \hat{\ell}, \tag{3}$$

where $\hat{\ell}$ is the average branch length used here to render the fabric tensor physically dimensionless; note that all tensors in (1) and (2) are dimensionless.

Additional information on the microstructure is included in the fabric tensor if one also involves a measure of "contact area." Let at a typical contact, a be a measure of this kind with physical dimensions of length squared (i.e. area) and define

$$\hat{H}_{ij} = \frac{M}{V} \langle a\ell n_i m_j \rangle = \langle \varepsilon\, n_i m_j \rangle \tag{4}$$

where M is the total number of contacts in volume V, and

$$\varepsilon = M\,\frac{a\ell}{V} = M\,\frac{v}{V}, \qquad v = a\ell. \tag{5}$$

The average $\bar{\varepsilon}$ of ε itself is a fabric measure which includes some information on the average coordination number, the average particle size, and the average contact area.

Of course, the first two fabric tensors in (1) may also be modified to yield

$$\hat{J}_{ij} = \langle \varepsilon m_i m_j \rangle, \qquad \hat{J}^{(1)}_{ij} = \langle \varepsilon n_i n_j \rangle, \tag{6}$$

but except for spherical (in three-dimensions) or circular (in two dimensions) granules, $\hat{H}_{ij} = \langle \varepsilon n_i m_j \rangle$ contains more information about microstructure.

OVERALL STRESS

An imaginary plane with unit normal $\underset{\sim}{\nu}$ intersects a set of granules which may be divided into two classes: class A with the centroid on the negative side of the $\underset{\sim}{\nu}$-plane, and class B with the centroid on the positive side. Let $\underset{\sim}{f}^{(\alpha)}$ be the set of contact forces representing the action of class A granules upon those in class B. Then the traction transmitted across the $\underset{\sim}{\nu}$-plane may be defined by (refs. 6, 19)

$$\underset{\sim}{T}^{(\nu)} = \sum_{a=1}^{N^{(\nu)}} \underset{\sim}{f}^{(a)}, \tag{7}$$

where $N^{(\nu)}$ is the number of contact forces per unit area of the $\underset{\sim}{\nu}$-plane. The stress tensor at a point in a continuum equivalent to the granular mass may then be characterized by definition (7) applied to three mutually orthogonal planes at that point.

A different definition of the overall Cauchy stress emerges from the principle of virtual work (refs. 24, 26) and the representation of the average stress, $\bar{\underset{\sim}{\sigma}}$, by its natural definition (refs. 28, 29)

$$\bar{\underset{\sim}{\sigma}} = \frac{1}{V}\int_V \underset{\sim}{\sigma}(\underset{\sim}{x})dV, \tag{8}$$

where $\underset{\sim}{\sigma}(\underset{\sim}{x})$ with components σ_{ij} is a variable stress field in equilibrium with the overall applied boundary tractions, i.e.

$$\sigma_{ij,j} = 0 \quad \text{in } V, \qquad \sigma_{ij}\nu_j = T_i^0 \quad \text{on } S, \qquad i,j = 1,2,3, \tag{9}$$

where T_i^0 are tractions prescribed on the boundary S of the sample of volume V; a comma followed by an index denotes partial differentiation with respect to the corresponding coordinate, and repeated subscripts are summed. Indeed, if v_i is a virtual overall velocity field which produces a virtual relative separation velocity $\dot{\Delta}_i$ at a typical contact, one has

$$M\langle f_i\dot{\Delta}_i\rangle = \int_V T_i^0 \, v_i \, dV. \tag{10}$$

Now, assume $v_i = \phi_{ij}x_j$, where ϕ_{ij} is an arbitrary constant matrix, and obtain (ref. 26)

$$\bar{\sigma}_{ij} = N\langle f_i\ell_j\rangle, \qquad N = \frac{M}{V}. \tag{11}$$

It appears that this definition of stress was first given by Christoffersen, Mehrabadi, and Nemat-Nasser (refs. 26, 30). A different definition, but with a similar appearance, has been discussed in the literature (refs. 25, 31). It involves an average over the sample *surface* S; note that $\langle f_i\ell_j\rangle$ is the average of the tensor product $f_i\ell_j$ taken over all contacts within the *volume* V. Indeed, from (8), (9), and the Gauss theorem, it follows that

$$\bar{\sigma}_{ij} = \frac{1}{V}\int_V \sigma_{ik}\delta_{jk}dV = \frac{1}{V}\int_V (\sigma_{ik}x_j)_{,k}\, dV$$

$$= \frac{1}{V}\int_S \sigma_{ik}\nu_k x_j dS = \frac{1}{V}\int_S T_i^0(\underset{\sim}{x})x_j \, dS,$$

where δ_{jk} is the Kronecker delta. The surface integral may be replaced by a sum over points on the exterior surface S, at which external concentrated forces are applied to the sample (possibly through contact with other bodies). Then

$$\bar{\sigma}_{ij} = \frac{1}{V} \sum_{\alpha=1}^{M_S} F_i^{(\alpha)} x_j^{(\alpha)}. \tag{12}$$

Here, M_S is the total number of points on S at which concentrated forces $F_i^{(\alpha)} = T_i^{(0)}(\underset{\sim}{x}^{(\alpha)})\Delta S^{(\alpha)}$ are prescribed over the areas $\Delta S^{(\alpha)}$ of the corresponding granules, and $x_j^{(\alpha)}$ are the corresponding position vectors. While both definitions (11) and (12) are useful, they are quite different. Here, definition (11) will be used.

It is convenient to require that the overall stress be symmetric,

$$\bar{\sigma}_{ij} = \frac{1}{2} N \langle f_i \ell_j + f_j \ell_i \rangle, \qquad \langle f_i \ell_j \rangle = \langle f_j \ell_i \rangle. \tag{13}$$

STRESS AND FABRIC

Let every contact within the sample be surrounded by an elementary volume in such a manner that the sum of these volumes equals V. Introduce within a typical elementary volume, a stress field τ_{ij} such that

$$f_i = a\tau_{ij} n_j, \tag{14}$$

where a is the effective contact area. The stress field τ_{ij} varies from elementary volume to elementary volume, and within each volume it is independent of the particular orientation of a plane on which local tractions $\tau_{ij} n_j$ are acting. Equation (11) then becomes

$$\begin{aligned} \bar{\sigma}_{ij} &= \langle \varepsilon \tau_{ik} n_k m_j \rangle = \langle \tau_{ik} \rangle \langle \varepsilon n_k m_j \rangle \\ &= T_{ik} \hat{H}_{kj} = \frac{1}{2}(T_{ik}\hat{H}_{kj} + T_{jk}\hat{H}_{ki}), \end{aligned} \tag{15}$$

where T_{ik} is the volume average of τ_{ij}, and (13) is also imposed.

Note that the local stress τ_{ij} is not identified with the smoothly varying stress field $\sigma_{ij}(\underset{\sim}{x})$ which enters Eqs. (8) and (9). The stress τ_{ij} is of a more discrete nature, with possibly sharper variations from neighborhood to neighborhood. Thus the volume average T_{ik} is not equal to $\bar{\sigma}_{ik}$. On the other hand, both T_{ik} and $\bar{\sigma}_{ik}$ depend on the overall measure of fabric, $\hat{H}_{kj}$. It may therefore be reasonable to take T_{ik} as an isotropic function of $\hat{H}_{ij}$,

$$T_{ik} = A_0 \delta_{ik} + B_0 \hat{H}_{ik} + C_0 \hat{H}_{i\ell} \hat{H}_{\ell k}, \tag{16}$$

where parameters A_0, B_0, and C_0, in general, will depend on the basic invariants of $\bar{\sigma}$, e.g.

$$I = \bar{\sigma}_{ii}, \qquad II = \frac{1}{2!}\bar{\sigma}_{ij}\bar{\sigma}_{ji}, \qquad III = \frac{1}{3!}\bar{\sigma}_{ij}\bar{\sigma}_{jk}\bar{\sigma}_{ki}, \tag{17}$$

as well as those of $\hat{H}_{ij}$. From (15) and (16) it follows that

$$\bar{\sigma}_{ij} = [A_0\hat{H}_{ij} + B_0\hat{H}_{ik}\hat{H}_{kj} + C_0\hat{H}_{i\ell}\hat{H}_{\ell k}\hat{H}_{kj}]_S, \tag{18}$$

where index S stands for the symmetric part of the expression. If the fabric tensor $\hat{H}_{ij}$ turns out to be symmetric, then with the aid of the Hamilton-Cayley theorem, (18) reduces to

$$\bar{\sigma}_{ij} = [A_1\delta_{ij} + B_1\hat{H}_{ij} + C_1\hat{H}_{ik}\hat{H}_{kj}] . \tag{19}$$

For the spherical granules, the unit vectors $\underset{\sim}{n}$ and $\underset{\sim}{m}$ coincide and hence

$$\hat{J}_{ij} = \hat{J}^{(1)}_{ij} = \hat{H}_{ij} = \langle \varepsilon n_i n_j \rangle, \tag{20}$$

a common symmetric tensor.

An alternative approach (ref. 32) is to represent the typical contact force f_i by

$$f_i = \frac{a}{N\ell}\, a_{ik} m_k, \tag{21}$$

and, instead of (15), obtain

$$\bar{\sigma}_{ij} = \langle a_{ik} m_k m_j \rangle = A_{ik} J_{kj}. \tag{22}$$

If it is assumed that A_{ik} is an isotropic function of J_{ij}, it then follows that

$$\bar{\sigma}_{ij} = A_2\delta_{ij} + B_2 J_{ij} + C_2 J_{ik} J_{kj}; \tag{23}$$

this follows exactly if $\bar{\sigma}_{ij}$ and A_{ij} are both symmetric tensors. For two-dimensional granules, an expression of this kind seems to accord well with experimental results obtained from biaxial loading of photoelastic granules (ref. 33).

It may appear that two major assumptions are used to obtain Eq. (23), i.e.: (1) the local stress a_{ij} is not correlated with $\varepsilon m_i m_j$, and (2) the average quantity A_{ij} can be expressed as an isotropic function of the corresponding fabric tensor. We now show that the second assumption follows from the symmetry of $\bar{\sigma}_{ij}$ and A_{ij}.

From the symmetry of $\bar{\sigma}_{ij}$, one has

$$A_{ik}J_{kj} - A_{jk}J_{ki} = 0, \tag{24}$$

and hence A_{ij} is coaxial with J_{ij}. Let $\underset{\sim}{N}^{\alpha}$, $\alpha = 1,2,3$, be the principal directions of the coaxial tensors A_{ij} and J_{ij}, with A_{α} and J_{α} as the corresponding principal values. Then

$$A_{ik} = \sum_{\alpha=1}^{3} A_{\alpha}N_i^{\alpha}N_k^{\alpha}, \quad J_{kj} = \sum_{\alpha=1}^{3} J_{\alpha}N_k^{\alpha}N_j^{\alpha}. \tag{25}$$

From (22) it follows that

$$\bar{\sigma}_{ij} = \sum_{\alpha=1}^{3} A_{\alpha}J_{\alpha}N_i^{\alpha}N_j^{\alpha} \tag{26}$$

so that $\bar{\sigma}_{ij}$, A_{ij}, and J_{ij} are coaxial. Hence the representation (23) is exact. It is therefore seen that the only major assumption for obtaining (23) is that the symmetric local stress a_{ij} is not correlated with $\varepsilon m_i m_j$. To a first order of approximation, this assumption may be reasonable.

Note that if $\hat{H}_{ij}$ is symmetric, then (19) also follows directly from assumption that the symmetric local stress τ_{ij} is not correlated with $\varepsilon n_i m_j$. However, in general, the fabric tensor $\hat{H}_{ij}$ may not be symmetric for nonspherical granules (noncircular, in two dimensions). When $\hat{H}_{ij}$ is symmetric and since the same overall stress must result from representations (15) and (22), then it follows that all five tensors $\bar{\sigma}_{ij}$, T_{ij}, $\hat{H}_{ij}$, A_{ij}, and J_{ij} must be coaxial. For nonspherical (in three dimensions) or noncircular (in two dimensions) granules, there seems to be no compelling theoretical reason to suggest a priori that $\hat{H}_{ij}$ and J_{ij} should be coaxial. It appears that the representation (15) includes more information about the microstructure and therefore may be preferred. The on-going experiments on biaxial deformation of oval cross-sectional photoelastic rods (in the Earthquake Research and Engineering Laboratory at Northwestern University) should provide some guidance for a more informed selection of a suitable fabric measure and the corresponding stress-fabric relations.

ACKNOWLEDGMENT

This work has been supported by the National Science Foundation under Grant No. CEE-8007764 to Northwestern University.

REFERENCES

1 W.O. Smith, P.D. Foote and P.F. Busang, Phys. Rev., Ser. 2, 34 (1929) 1271-1274.
2 J.D. Bernal and J. Mason, Nature, 188 (1960) 910-911.
3 W.G. Field, in Proc. 4th A. and N.Z. Conf. on Soil Mech., 1963, 143-148.
4 W.A. Gray, The Packing of Solid Particles, Chapman & Hall Ltd., London, 1968.
5 M. Oda, J. Konishi and S. Nemat-Nasser, Earthquake Research and Engineering Laboratory Technical Report No. 80-3-26, Dept. of Civil Engrg., Northwestern University (1980); Géotechnique, 30, 4 (1980) 479-495.
6 M. Oda, S. Nemat-Nasser and M.M. Mehrabadi, EREL Tech. Rept. No. 80-4-28, Dept. of Civil Engrg., Northwestern University (1980); Int. J. Numer. Anal. Methods Geomech., 6 (1982) 77-94.
7 D. Lafeber, Engineering Geology, 1 (1966) 261-290.
8 J.R.F. Arthur and B.K. Menzies, Géotechnique, 22, 1 (1972) 115-128.
9 M. Oda, Soils and Foundations, 12, 4 (1972) 45-63.
10 J.R.F. Arthur and A.B. Phillips, Géotechnique, 25, 4 (1975) 799-815.
11 A. Mahmood and J.K. Mitchell, Clays and Clay Minerals, 22, 5/6 (1974) 397-408.
12 J.P. Mulilis, C.K. Chan and H.B. Seed, The Effects of Method of Sample Preparation on the Cyclic Stress-Strain Behavior of Sands, Report No. EERC 75-18, Earthquake Engrg. Research Center, U. of California, Berkeley (1975).
13 S. Nemat-Nasser, EREL Tech. Rept. No. 79-6-19, Dept. of Civil Engrg., Northwestern University (1979); Soils and Foundations, 20, 3 (1980) 59-73.
14 S. Nemat-Nasser and Y. Tobita, EREL Tech. Rept. No. 81-2-40, Dept. of Civil Engrg., Northwestern University (1981); Mechanics of Materials, 1, 1 (1982) 43-62.
15 M.R. Horne, Proc. Roy. Soc. London, A286 (1965) 62-78, 79-97.
16 M.R. Horne, Proc. Roy. Soc. London, A310 (1969) 21-34.
17 T. Mogami, Soils and Foundations, 5, 2 (1965) 26-36.
18 J.K. Wilkins, Rock Mech., 2 (1970) 205-222.
19 M. Oda, Soils and Foundations, 14, 1 (1974) 13-27.
20 H. Matsuoka, Soils and Foundations, 14, 1 (1974) 29-43.
21 S.K. Sadasivan and V.S. Raju, J. Geotech. Engrg. Div., ASCE, 103, GT8 (1977) 851-861.
22 R.A. Davis and H. Deresiewicz, Acta Mech., 27 (1977) 69-89.
23 M. Satake, in Theoretical and Applied Mechanics, 26, University of Tokyo Press, 1978, 257-266.
24 M.M. Mehrabadi, S. Nemat-Nasser and M. Oda, EREL Tech. Rept. No. 80-4-29, Dept. of Civil Engrg., Northwestern University (1980); Int. J. Numer. Anal. Methods Geomech., 6 (1982) 95-108.
25 A. Drescher and G. de Josselin de Jong, J. Mech. Phys. Solids, 20 (1972) 337-351.
26 J. Christoffersen, M.M. Mehrabadi and S. Nemat-Nasser, J. Appl. Mech., 48 (1981) 339-344.
27 S. Nemat-Nasser, in P.A. Vermeer and H.J. Luger (Eds.), Deformation and Failure of Granular Materials, A.A. Balkema, Rotterdam, 1982, 37-42.
28 R. Hill, J. Mech. Phys. Solids, 11 (1963) 357-372.
29 R. Hill, J. Mech. Phys. Solids, 15 (1967) 79-95.
30 J. Christoffersen, M.M. Mehrabadi and S. Nemat-Nasser, A Micromechanical Description of Granular Material Behavior, EREL Tech. Rept. No. 80-1-22, Dept. of Civil Engrg., Northwestern University (1980).
31 O.D.L. Strack and P.A. Cundall, The Distinct Element Method as a Tool for Research in Granular Media, Part I, Dept. of Civil and Mineral Engrg., U. of Minnesota (1978).
32 M.M. Mehrabadi and S. Nemat-Nasser, Stress, Dilatancy, and Fabric in Granular Materials, 1982, to be submitted for publication.
33 M. Oda, J. Konishi and S. Nemat-Nasser, EREL Tech. Rept. No. 82-7-47, Dept. of Civil Engrg., Northwestern University (1982); Mechanics of Materials, 1, 4 (1982) in press.

Mechanics of Granular Materials: New Models and Constitutive Relations, edited by
J.T. Jenkins and M. Satake, 1983
Elsevier Science Publishers B.V., Amsterdam — Printed in The Netherlands

FUNDAMENTAL QUANTITIES IN THE GRAPH APPROACH TO GRANULAR MATERIALS

M. SATAKE
Civil Engineering Dept., Tohoku University, Sendai (Japan)

ABSTRACT

Some new quantities are introduced for the macroscopic description of graphical and mechanical characteristics of granular materials. These quantities are defined statistically by averaging the microscopic quantities, such as contact force, relative displacement etc., and the inverse relationships expressing microscopic quantities in terms of macroscopic ones are also discussed. It is seen that the fabric tensor plays the most fundamental role in such analysis.

INTRODUCTION

It has been recognized that the graphical approach (ref. 1, 2) is a precise and useful method for the investigation of microscopic structure of granular materials. Satake (ref. 2) proposed some fundamental quantities in the graphical approach with respect to a single grain or void in a granular materials. In this paper, by considering a mesodomain, which includes a sufficient number of particles to define local average measures, such as stress and strain, these quantities are extended statistically to more general and suitable forms. The inverse relationships expressing microscopic quantities in terms of macroscopic ones and mutual relationships of these quantities are also discussed. It is seen that three kinds of tensors describing the anisotropic characteristics of graphs obtained from a granular assembly may be introduced and that the fabric tensor is the most fundamental among them. Recently Oda, Nemat-Nasser and Mehrabadi (ref. 3) introduced the fabric tensor for a statistical measure of a granular assembly and Christoffersen, Mehrabadi and Nemat-Nasser (ref. 4) considered the representation of stress in a more general case in which individual granules are not necessarily spherical. However, as it is considered that the anisotropic property is the most essential in the granular materials a simplified definition is used for the fabric tensor and other concerned quantities in this paper and discussions are given from such a view point. Throughout this paper symbolic notations are used for vector and tensor.

FABRIC TENSOR

To simplify the analysis, we assume that the concerned granular material is simulated by a 2-dimensional assembly of discs. Let R be a region in the assembly which includes a sufficient number of particles to apply a statistical analysis and yet be small enough to give a uniform value for macroscopic measures, such as stress and strain. We call such a region R a *mesodomain* of granular materials (ref. 5).

Firstly, we define the *fabric tensor* $\underset{\sim}{\phi}$ in a mesodomain R by the form.

$$\underset{\sim}{\phi} = \frac{1}{2n} \sum_{R} \underset{\sim}{n}\underset{\sim}{n} \tag{1}$$

where $\underset{\sim}{n}$ denotes a contact normal and n is the number of contact points in R. As we have two contact normals with mutually opposite directions for each contact point, 2n becomes the number of contact normals. Thus, $\underset{\sim}{\phi}$ is regarded as an average of $\underset{\sim}{n}\underset{\sim}{n}$ in R.

As we assume that R contains a relatively large number of particles, we can also write $\underset{\sim}{\phi}$ as

$$\underset{\sim}{\phi} = \int_{0}^{2\pi} f(\theta)\underset{\sim}{n}\underset{\sim}{n}d\theta \tag{2}$$

where f(θ) is the *probability density function of contact normals*, or briefly the *contact normal density* in R and θ denotes the inclination angle of $\underset{\sim}{n}$ to a fixed axis in the space. From the prescribed condition

$$\int_{0}^{2\pi} f(\theta)d\theta = 1 \tag{3}$$

we have

$$\mathrm{tr}\ \underset{\sim}{\phi} = 1 \tag{4}$$

Next, we shall discuss the fabric tensor through the graph representation of a granular assembly. The replaced graph

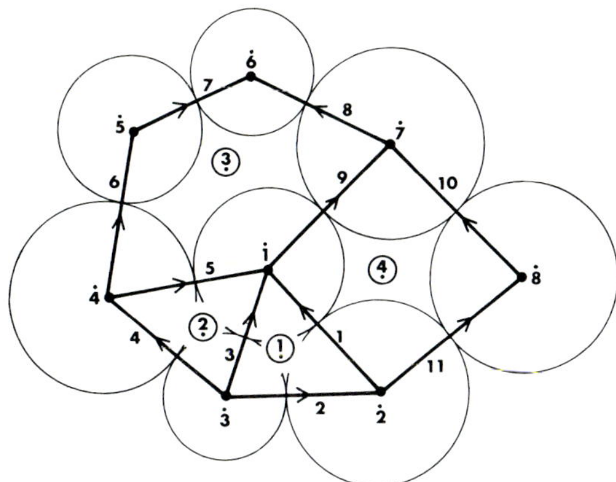

(a) Replaced graph (n = 8, n = 11, n = 4)

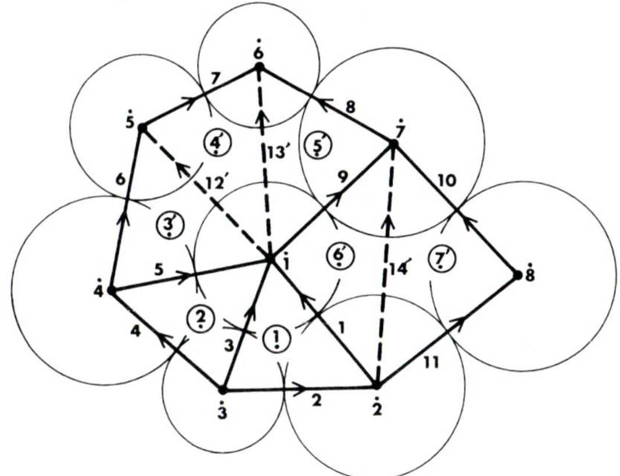

(b) Modified graph (n' = 8, n' = 14, n' = 7)

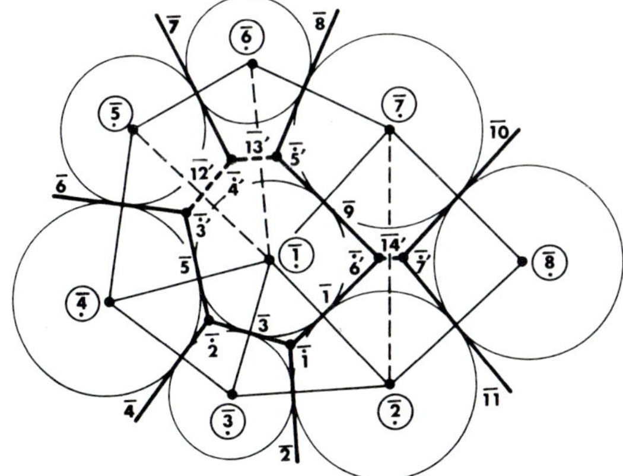

(c) Dual graph (n = 7, n = 14, n = 8 orientations are omitted)

Fig. 1 Graph representation of granular assembly

(Fig. 1(a)) is an oriented graph and each *branch vector* corresponds to one of two contact normals at the corresponding contact point. Thus, Eq. (1) is transformed as

$$\underset{\sim}{\phi} = \frac{1}{n} \sum_{R} \frac{\underset{\sim}{\ell}\,\underset{\sim}{\ell}}{\ell^2} \tag{5}$$

where $\underset{\sim}{\ell}$ denotes the branch vector, $\ell = \sqrt{\underset{\sim}{\ell} \cdot \underset{\sim}{\ell}}$ and n is the number of branches of the replaced graph in R.

In a quite similar manner, we next define the fabric tensor $\underset{\sim}{\phi}^M$ of the modified graph (Fig. 1(b)) by the form

$$\underset{\sim}{\phi}^M = \frac{1}{n'} \sum_{R}^{M} \frac{\underset{\sim}{\ell}\,\underset{\sim}{\ell}}{\ell^2} \tag{6}$$

where n' is the number of branches of the modified graph and Σ^M means that the summation should be applied according to the modified graph, namely the virtual branches should be taken into account. Using the branch direction $\underset{\sim}{n} = \frac{\underset{\sim}{\ell}}{\ell}$ we have

$$\underset{\sim}{\phi}^M = \frac{1}{n'} \sum_{R}^{M} \underset{\sim}{n}\,\underset{\sim}{n} \tag{7}$$

It may be supposed that $\underset{\sim}{\phi}^M$ tends to become more isotropic than $\underset{\sim}{\phi}$ by the addition of virtual branches.

For the dual graph (Fig. 1(c)), we can also define the fabric tensor $\underset{\sim}{\phi}^D$ expressed as

$$\underset{\sim}{\phi}^D = \frac{1}{n'} \sum_{R}^{M} \hat{\underset{\sim}{n}}\,\hat{\underset{\sim}{n}} \tag{8}$$

where $\hat{\underset{\sim}{n}}$ denotes the dual vector (see the Appendix) of $\underset{\sim}{n}$. From Eq.(8), it follows that

$$\underset{\sim}{\phi}^D = \hat{\hat{\underset{\sim}{\phi}}}{}^M \tag{9}$$

Thus $\underset{\sim}{\phi}^D$ is the dual tensor (see the Appendix) of $\underset{\sim}{\phi}^M$.

WEIGHTED FABRIC TENSOR

In this section, we shall consider two kinds of weighted fabric tensors. The first one is a tensor expressed as

$$\underset{\sim}{B} = \frac{\sum\limits_{R} \ell \underset{\sim}{n}\underset{\sim}{n}}{\sum\limits_{R} \ell} \tag{10}$$

where ℓ is the length of branch, $\underset{\sim}{n}$ denotes the branch direction, and the summation is applied according to the replaced graph. If the length of branch ℓ is randomly distributed independently of $\underset{\sim}{n}$, $\underset{\sim}{B}$ naturally reduces to $\underset{\sim}{\phi}$. However if the mean value of ℓ for a direction $\underset{\sim}{n}$ depends on θ, we write it as $\bar{\ell}(\theta)$.† Then we have

$$\underset{\sim}{B} = \frac{2}{\bar{\ell}} \int_0^{\pi} \bar{\ell}(\theta) f(\theta) \underset{\sim}{n}\underset{\sim}{n} d\theta \tag{11}$$

where

$$\bar{\ell} = 2\int_0^{\pi} \bar{\ell}(\theta) f(\theta) d\theta = \frac{1}{n} \sum_{R} \ell \tag{12}$$

We shall call $\underset{\sim}{B}$ the *branch tensor*. It is easy to see that

$$\mathrm{tr}\, \underset{\sim}{B} = 1 \tag{13}$$

and if $\bar{\ell}(\theta)$ is assumed to have the same symmetry about $\frac{\pi}{2}$ as $f(\theta)$, $\underset{\sim}{B}$ has the same principal directions as $\underset{\sim}{\phi}$. Further, denoting

$$\underset{\sim}{B}^* = \frac{1}{2} \underset{\sim}{\phi}^{-1} \cdot \underset{\sim}{B} \tag{14}$$

we can give the inverse relation of Eq. (10) expressed as

$$\underset{\sim}{\ell} = 2\bar{\ell} \underset{\sim}{n} \cdot \underset{\sim}{B}^* + \delta \underset{\sim}{\ell} \tag{15}$$

where $\delta \underset{\sim}{\ell}$ is the *residual branch vector* having the property

$$\sum_{R} \underset{\sim}{n} \delta \underset{\sim}{\ell} = 0 \tag{16}$$

B^* is called the *co-branch tensor*.

The second weighted fabric tensor is expressed as

---

† Without loss of generality, we assume that $0 \leq \theta \leq \pi$ where θ is the inclination angle of $\underset{\sim}{n}$.

$$\underset{\sim}{C} = \frac{\sum_R s\underset{\sim}{n}\underset{\sim}{n}}{\sum_R s} \tag{17}$$

where s is the length of the corresponding branch of dual graph which denotes the contact length. Note that the summation is applied only to the real branches in the replaced graph. It is easy to see that $\underset{\sim}{C}$ also reduces to $\underset{\sim}{\phi}$ if s is randomly distributed independently of $\underset{\sim}{n}$ but $\underset{\sim}{C}$ becomes different from $\underset{\sim}{\phi}$ when the mean value of s for a direction $\underset{\sim}{n}$ depends on θ. Denoting this mean value by $\overline{s}(\theta)$, we have

$$\underset{\sim}{C} = \frac{2}{\overline{s}}\int_0^{\pi} \overline{s}(\theta)f(\theta)\underset{\sim}{n}\underset{\sim}{n}d\theta \tag{18}$$

where

$$\overline{s} = 2\int_0^{\pi} \overline{s}(\theta)f(\theta)d\theta = \frac{1}{n}\sum_R s \tag{19}$$

We shall call $\underset{\sim}{C}$ the *contact tensor*. We have also

$$\text{tr}\,\underset{\sim}{C} = 1 \tag{20}$$

and if $\overline{s}(\theta)$ is assumed to have the same symmetry about $\frac{\pi}{2}$ as $f(\theta)$, $\underset{\sim}{C}$ has the same principal directions as $\underset{\sim}{\phi}$. Further, denoting

$$\underset{\sim}{C}^* = \frac{1}{2}\underset{\sim}{\phi}^{-1}\cdot\underset{\sim}{C} \tag{21}$$

we can give the inverse relationship of Eq. (17) expressed as

$$s\underset{\sim}{n} = 2\overline{s}\underset{\sim}{n}\,C^* + \delta(s\underset{\sim}{n}) \tag{22}$$

where $\delta(s\underset{\sim}{n})$ is the *residual contact vector* having the property

$$\sum_R \underset{\sim}{n}\delta(s\underset{\sim}{n}) = 0 \tag{23}$$

C* is called the *co-contact tensor*.

It may be easily seen that $\underset{\sim}{B}$ and $\underset{\sim}{C}$ are regarded as natural extensions of $\underset{\sim}{K}_k$ and $\underset{\sim}{H}_i$ respectively which were introduced as anisotropy tensors of loop k and grain i respectively in a previous paper (ref.2). It is noteworthy that all of three tensors $\underset{\sim}{\phi}$, $\underset{\sim}{B}$, $\underset{\sim}{C}$ introduced here are regarded as expressions of anisotropy of an granular assembly, and that if functions $\bar{\ell}(\theta)$ and $\bar{s}(\theta)$ are given explicitly, the mutual relations must be clarified.

STRESS AND STRAIN

In the graph analysis of granular assembly, the *stress tensor* is defined in the form

$$\underset{\sim}{\sigma} = \frac{2 \sum\limits_{R} \underset{\sim}{n}\underset{\sim}{F}}{\sum\limits_{R} s} = \frac{2}{n\bar{s}} \sum_{R} \underset{\sim}{n}\underset{\sim}{F} \tag{24}$$

where $\underset{\sim}{F}$ denotes the *contact force* applying to a contact plane corresponding to $\underset{\sim}{n}$.† It is noted that Eq (24) is quite analogous to the form

$$\underset{\sim}{\sigma} = \lim_{a \to 0} \frac{2 \oint \underset{\sim}{n}\underset{\sim}{f} ds}{\oint ds} \tag{25}$$

which is the definition of stress in a continuum, where $\underset{\sim}{f}$ denotes the traction vector and the integral is applied to a circumference with radius a. The contact force is given conversely by

$$\underset{\sim}{F} = \bar{s}(\theta)\underset{\sim}{n} \cdot \underset{\sim}{\sigma}^* + \delta\underset{\sim}{F} \tag{26}$$

where $\underset{\sim}{\sigma}^*$ is the *co-stress tensor* defined, by virtue of the contact tensor $\underset{\sim}{C}$, as

$$\underset{\sim}{\sigma}^* = \frac{1}{2} \underset{\sim}{C}^{-1} \cdot \underset{\sim}{\sigma} \tag{27}$$

and $\delta\underset{\sim}{F}$ denotes the *residual contact force*. By use of Eqs.(17), (24), (26) and (27), it is easy to show the property

$$\sum_{R} \underset{\sim}{n}\delta\underset{\sim}{F} = 0 \tag{28}$$

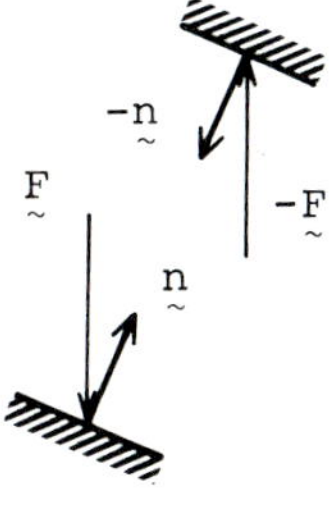

† As we have two contact forces with mutually opposite directions for a contact point, we must use the one applying to the contact plane with outwards normal $\underset{\sim}{n}$.

It is also noted that Eq (26) is analogous to the relation in a continuum

$$\underset{\sim}{f} = \underset{\sim}{n} \cdot \underset{\sim}{\sigma} \tag{29}$$

and it may be recognized that in the graph analysis of granular assembly, two kinds of stress tensor, i.e. $\underset{\sim}{\sigma}$ and $\underset{\sim}{\sigma}^*$ are necessary to give the relationship between the contact force and the stress tensor. In the case where $\bar{s}(\theta)$ does not depend on θ, namely $\bar{s}(\theta) = \bar{s}$, Eqs (26) and (27) reduce to

$$\underset{\sim}{F} = \bar{s}\underset{\sim}{n} \cdot \underset{\sim}{\sigma}^* + \delta\underset{\sim}{F} \tag{30}$$

$$\underset{\sim}{\sigma}^* = \frac{1}{2} \underset{\sim}{\phi}^{-1} \cdot \underset{\sim}{\sigma} \tag{31}$$

In a quite similar manner as above, we define the *displacement gradient tensor* in a granular assembly by the form

$$\underset{\sim}{\gamma} = \frac{2 \sum_R \underset{\sim}{n} \Delta \underset{\sim}{u}}{\sum_R \ell} = \frac{2}{n\bar{\ell}} \sum_R \underset{\sim}{n} \Delta \underset{\sim}{u} \tag{32}$$

where $\Delta\underset{\sim}{u}$ denotes the *relative displacement* of contact points of two particles after deformation. It is seen that Eq. (32) is also analogous to the form in a continuum

$$\underset{\sim}{\gamma} = \lim_{a \to 0} \frac{2 \oint \underset{\sim}{n} d\underset{\sim}{u}}{\oint d\ell} \tag{33}$$

The relative displacement is given conversely by

$$\Delta\underset{\sim}{u} = \bar{\ell}(\theta) \underset{\sim}{n} \cdot \underset{\sim}{\gamma}^* + \delta\Delta\underset{\sim}{u} \tag{34}$$

where $\underset{\sim}{\gamma}^*$ is the *co-displacement-gradient tensor* defined, by virtue of the branch tensor $\underset{\sim}{B}$, as

$$\underset{\sim}{\gamma}^* = \frac{1}{2} \underset{\sim}{B}^{-1} \cdot \underset{\sim}{\gamma} \tag{35}$$

and $\delta\Delta\underset{\sim}{u}$ denotes the *residual relative displacement*. By use of Eqs (10), (32), (34) and (35), it is easy to show the property

$$\sum_R \underset{\sim}{n}\delta\Delta\underset{\sim}{u} = 0 \qquad (36)$$

and it is seen that Eq (34) is analogolous to the relation in a continuum

$$d\underset{\sim}{u} = d\ell\underset{\sim}{n}\cdot\underset{\sim}{\gamma} \qquad (37)$$

In the case where $\bar{\ell}(\theta)$ does not depend on θ, namely $\bar{\ell}(\theta) = \bar{\ell}$, Eqs (34) and (35) reduce to

$$\Delta\underset{\sim}{u} = \bar{\ell}\underset{\sim}{n}\cdot\underset{\sim}{\gamma}^* + \delta\Delta\underset{\sim}{u} \qquad (38)$$

where

$$\underset{\sim}{\gamma}^* = \frac{1}{2}\underset{\sim}{\phi}^{-1}\cdot\underset{\sim}{\gamma} \qquad (39)$$

The work done by contact forces $\underset{\sim}{F}$ against the relative displacements $\Delta\underset{\sim}{u}$ in R is given approximately by

$$\sum_R \underset{\sim}{F}\cdot\Delta\underset{\sim}{u} = \sum_R \bar{s}(\theta)\bar{\ell}(\theta)(\underset{\sim}{n}\cdot\underset{\sim}{\sigma}^*)\cdot(\underset{\sim}{n}\cdot\underset{\sim}{\gamma}^*)$$

$$= 2(\sum_R \bar{A}(\theta)\underset{\sim}{n}\underset{\sim}{n})\cdot\cdot(\underset{\sim}{\sigma}^*\cdot\underset{\sim}{\gamma}^{*T}) \qquad (40)$$

where $\bar{A}(\theta)$ denotes the mean value of the *accompanied areas* with branches in R (Fig. 2) given by

$$\bar{A}(\theta) = \frac{1}{2}\bar{s}(\theta)\bar{\ell}(\theta) \qquad (41)$$

and terms concerning with $\delta\underset{\sim}{F}$ and $\delta\Delta\underset{\sim}{u}$ are neglected in Eq. (40). In the case where $\bar{A}(\theta)$ is regarded not to depend on θ, we can put as

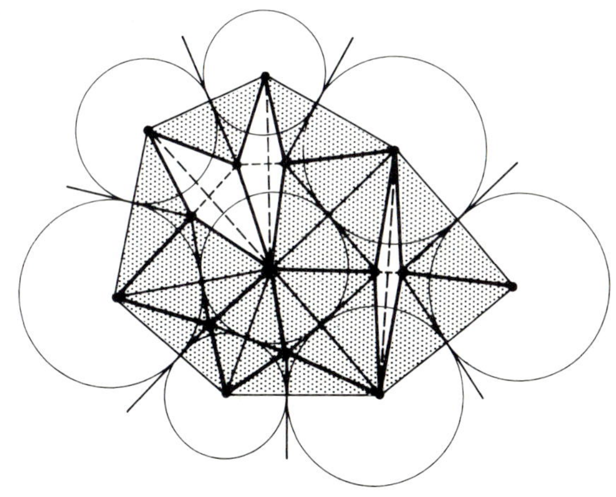

Fig. 2 Accompanied area with branches

$$\bar{A}(\theta) = \bar{A} = \frac{1}{2}\bar{s}\bar{\ell} \tag{42}$$

and we have

$$\sum_R \underset{\sim}{F}\cdot\Delta\underset{\sim}{u} = 2n\bar{A}\underset{\sim}{\phi}\cdot\cdot(\underset{\sim}{\sigma}^*\cdot\underset{\sim}{\gamma}^{*T})$$

$$= n\bar{A}\underset{\sim}{\sigma}\cdot\cdot\underset{\sim}{\gamma}^*$$

$$= \nu_R A_R \underset{\sim}{\sigma}\cdot\cdot\underset{\sim}{\varepsilon} \tag{43}$$

where A_R denotes the total area of R,

$$\nu_B = \frac{n\bar{A}}{A_R} \tag{44}$$

which will be called the *graph porosity* in R, and $\underset{\sim}{\varepsilon}$ is the *strain tensor* given as the symmetric part of $\underset{\sim}{\gamma}^*$. It is noted here that, as is seen in Eq. (43), the factor of porosity becomes necessary for the expression of internal work or energy in granular materials (ref. 6).

CONCLUDING REMARKS

Along the line of the graph approach proposed by the author previously, some quantities which seem to be fundamentally important in the analysis are defined and explained. These quantities are defined with respect to the so-called mesodomain in a granular assembly, so it is considered that they are more reasonable and useful for theoretical consideration and practical application than the quantities formerly introduced (ref. 2).

To extend the graph analysis to more general cases, such as an assembly of ellipses, a lot of modifications may become necessary. (For example, the branch direction should be distinguished from the contact normal. ref. 4) It is expected, however, that the development of the graph approach to more general applications may bring plentiful information to the analysis of granular materials.

REFERENCES

1 M. Satake, Constitution of Mechanics of Granular Materials through Graph Representation, Theor. Appl. Mech. 26, University of Tokyo Press, Tokyo, 1978, 257-266.
2 M. Satake, Constitution of Mechanics of Granular Materials through the Graph Theory, Proc. U.S.-Japan Seminar on Continuum Mechanical and Statistical Approaches in the Mechanics of Granular Materials, Gakujutsu Bunken Fukyukai, Tokyo, 1978, 47-62.
3 M. Oda, S. Nemat-Nasser and M. M. Mehrabadi, A Statistical Study of Fabric in a Random Assembly of Spherical Granules, Int. J. Numer. Anal. Methods Geomech. 6 (1982) 77-94.
4 J. Christoffersen, M. M. Mehrabadi and S. Nemat-Naaser, A Micromechanical Description of Granular Material Behavior, J. Appl. Mech. 48 (1981) 339-344.
5 D. R. Axelrad, Micromechanics of Solids, Elsevier Scientific Publishing Company, Amsterdam, 1978, 49.
6 M. Satake, On Fabric Tensor in Granular Materials (in Japanese), Mechanical Constitution of Granular Materials (Edit. by M. Satake), 1982, 1-10.

APPENDIX

For a vector $\underset{\sim}{v} = (v_1\ v_2)$, we define the *dual vector*

$$\hat{\underset{\sim}{v}} = (v_2, -v_1) \qquad \text{(a 1)}$$

$\hat{\underset{\sim}{v}}$ is expressed symbolically as

$$\hat{\underset{\sim}{v}} = \underset{\sim}{I} \times \underset{\sim}{v} \qquad \text{(a 2)}$$

where $\underset{\sim}{I}$ is the unit tensor,and has a property

$$\underset{\sim}{v} \cdot \hat{\underset{\sim}{v}} = 0 \qquad \text{(a 3)}$$

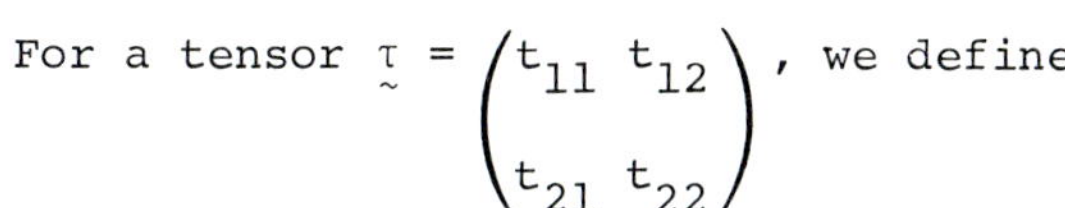

For a tensor $\underset{\sim}{\tau} = \begin{pmatrix} t_{11} & t_{12} \\ t_{21} & t_{22} \end{pmatrix}$, we define

the *dual tensor*

$$\hat{\hat{\underset{\sim}{\tau}}} = \begin{pmatrix} t_{22} & -t_{21} \\ -t_{12} & t_{11} \end{pmatrix} \qquad \text{(a 4)}$$

$\hat{\hat{\underset{\sim}{\tau}}}$ is expressed symbolically as

$$\hat{\hat{\underset{\sim}{\tau}}} = -(\underset{\sim}{I} \times \underset{\sim}{\tau}) \times \underset{\sim}{I} \qquad \text{(a 5)}$$

and has a property

$$\tau \cdot\cdot \hat{\hat{\tau}} = 2\,|\tau| \qquad \text{(a 6)}$$

Mechanics of Granular Materials: New Models and Constitutive Relations, edited by
J.T. Jenkins and M. Satake, 1983
Elsevier Science Publishers B.V., Amsterdam — Printed in The Netherlands

EXPERIMENTAL MICROMECHANICAL EVALUATION OF THE STRENGTH OF GRANULAR MATERIALS: EFFECTS OF PARTICLE ROLLING

MASANOBU ODA[1], JUNICHI KONISHI[2] and SIAVOUCHE NEMAT-NASSER[3]

[1]Dept. of Foundation Engineering, Saitama University, Urawa, Japan

[2]Dept. of Civil Engineering, Shinshu University, Nagano, Japan

[3]Dept. of Civil Engineering, Northwestern Univ., Evanston, Ill. 60201, U.S.A.

ABSTRACT

Particle rolling appears to be a major microscopic deformation mechanism, especially when interparticle friction is large. There are relatively few contacts at which relative sliding is dominant, and this seems to be true even when the assembly reaches the overall failure state; this observation is in contradiction to the common assumption that particle sliding is the major microscopic deformation mode.

INTRODUCTION

Cohesionless granular materials support overall shear stresses through frictional contacts. The overall shear resistance of a material of this kind decreases with decreasing interparticular friction. In the absence of any friction, the overall material response would be more like that of an inviscid fluid.

In the past, it has been commonly assumed that frictional sliding is the major component of the microscopic deformation mechanism, with particle rolling being negligibly small. On the other hand, it has been commented on in the literature that particle rolling does take place during the flow of cohesionless granular materials; see, for example, Roscoe and Schofield (1969), Oda (1972), and Ochiai (1972). The main purpose of the present work is to examine the extent of particle rolling and its effect on the overall mechanical response of granular materials.

EXPERIMENT

The material, the sample preparation technique, and the test procedure are reported elsewhere in this volume; see Konishi, Oda, and Nemat-Nasser (1982). Two kinds of cross-sectional shape--referred to as oval I and Oval II in the sequel-- are employed. For each shape three different size particles (i.e., large, medium, and small) are prepared. The ratio of the maximum, r_1, to the minimum, r_2, of the principal radii of oval I is 1.1, whereas that of oval II is 1.4.

To assess the influence of friction, two sets of experiments are performed, one with non-lubricated particles of overall friction angle 52°, the other with particles which are lubricated with talcum powder that results in an overall friction angle of 26°.

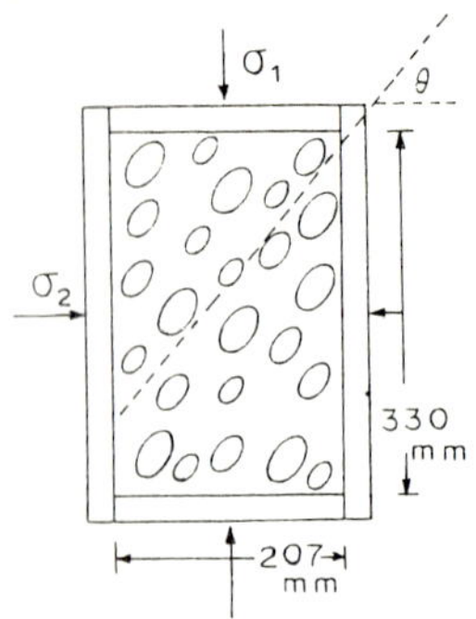

Fig.1 Definition of bedding angle θ.

The testing apparatus consists of an overall frame within which a biaxial loading frame (330 mm high and 207 mm high) is mounted. The overall frame can be tilted in its plane and held at a desired angle; it can also be tilted and held out of its plane. This flexibility permits sample preparation with various bedding angles θ, Fig.1, and hence, various inherent anisotropies. Particles are stacked by hand randomly in the tilted frame with their longer axis essentially parallel to the bedding plane. The sample is confined laterally by a constant force of 4.41 N (0.45 kg), and it is compressed vertically by incremental displacement downward of the upper part of the frame.

TEST RESULTS

In what follows, the axial stress and strain are denoted by σ_1 and ε_1, respectively, and the lateral ones by σ_2 and ε_2. The interparticle friction angle is denoted by ϕ_μ and is assumed to be constant.

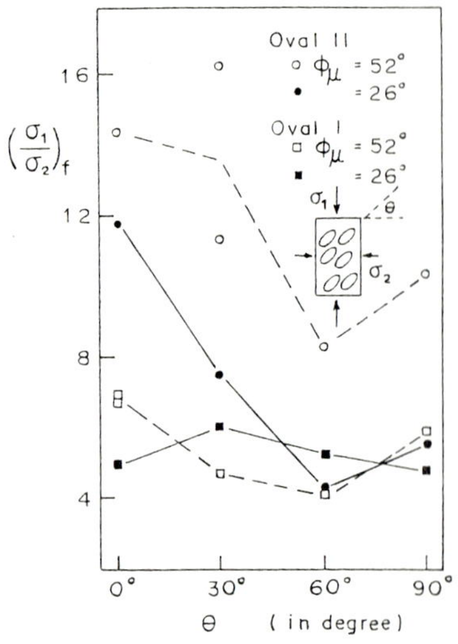

Fig.2 Effect of bedding angle θ on stress ratio at failure for oval I and oval II.

In Fig.2, stress ratio at failure, (σ_1/σ_2), is plotted against the bedding angle, θ. The following results are revealed: (1) For oval II (flat) particles the value of the stress ratio at failure depends strongly on the bedding angle θ. For both 26° and 52° interparticle friction angles the peak value of the stress ratio is greatest for θ=0°, becomes a minimum at θ=60°, and then again increases, as shown in Fig.2. For the oval I (more round) particles, on the other hand, no such distinct variation of the peak stress ratio with bedding angle is observed.

2. The assembly of oval II particles exhibits considerable dependence of the magnitude of its peak stress ratio on the value of the internal friction, whereas no such dependence is observed for oval I particles. Similar results have been reported by Skinner (1969) who sheared wet and dry assemblies of 1 mm glass ballotini in a shear box. Although the wet glass ballotini interparticle friction angle (27°-38°) is considerably higher than that of dry ones (2°-5°), both exhibit essentially the same overall shear resistance, i.e. the same overall effective friction coefficient. Skinner attributes this to the dominance of particle rolling during shear deformation. Our experimental results seem to support Skinner's conclusion.

PARTICLE ROLLING AND FABRIC EVOLUTION

According to our experimental results the value of the stress ratio at failure, $(\sigma_1/\sigma_2)_f$ for the rounder oval I particles remains essentially the same when the interparticle friction is changed from ϕ_μ=26° to ϕ_μ=52°, whereas considerable dependence of this quantity on the interparticle friction is displayed by the flatter oval II particles. This suggests the presence of particle rolling during the course of deformation, as pointed out by Skinner (1969). We have analyzed sequences of photoelastic pictures, in an effort to see if particle rolling takes place.

Figure 3 shows two particles, G_1 and G_2, initially in contact at point c_1. During the overall shearing of a granular mass which contains these particles, particle G_1 is displaced relative to G_2 to a new position denoted by G_1', with the new contact point c_2. The old contact point has moved to a new position denoted by c_1'.

We may identify three possible mechanisms which produce the relative displacement of the two particles. These are pure sliding, pure rolling, and a combined sliding and rolling. Including the contacts of mutually immobile particles, we may classify all contacts into the following four groups.

(1) Pure sliding contacts (Fig.3(a)):

Let ω be the angular rotation of G_1 relative to G_2. For pure sliding contact, ω=0. Moreover, points c_1 and c_1' remain on opposite sides of point c_2. Let the arc $\widehat{c_1c_2}$=a be regarded positive. Then $\widehat{c_1'c_2}$=a will be negative. With

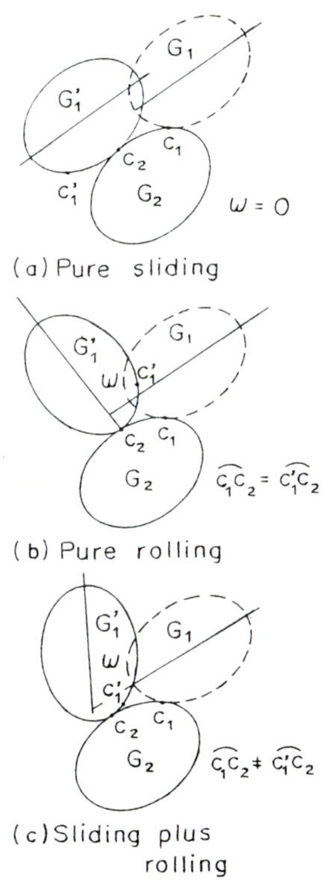

Fig.3 Microscopic deformation mechanism.

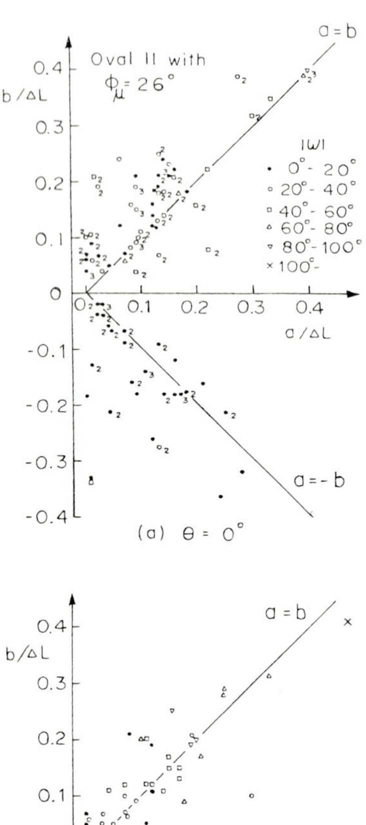

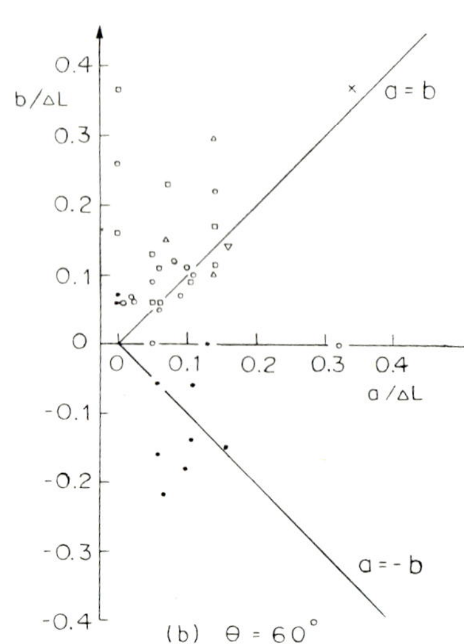

Fig.4 Division of contacts into two groups with ϕ_μ=26°. (Data points falling on the upper half of the plane correspond to contacts at which rolling is dominant; the opposite is the case for points in the lower half of the plane.)

particles of circular cross-sections and equal radii, a=-b is satisfied.

(2) Pure rolling contacts (Fig.3(b)):

The angular rotation ω is nonzero and points c_1 and c_1' remain on the same side of point c_2, in this case. The length $\widehat{c_1c_2}$ equals the length $\widehat{c_1'c_2}$, and the a=b relation identifies this group.

(3) Sliding and Rolling Contacts (Fig.3(c)):

These are contacts at which both sliding and rolling take place. Most active contacts belong to this group. It is possible to further subdivide this group, depending on whether sliding or rolling is dominant. This may

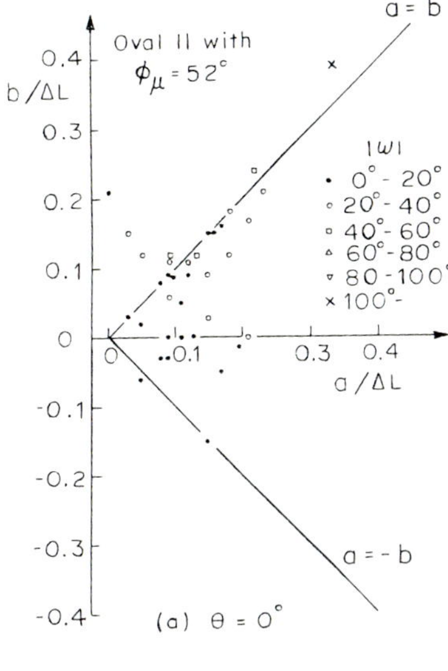

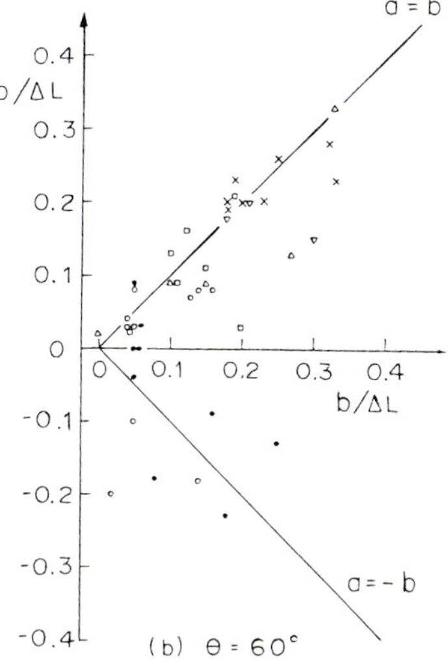

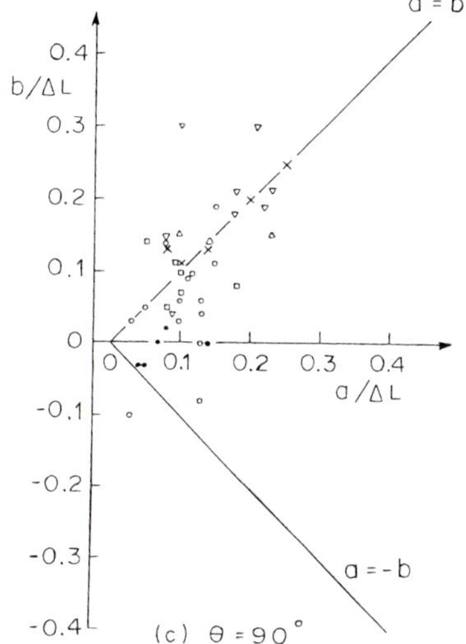

Fig.5 Division of contacts into two groups in oval II with ϕ_μ=52°.

be done by considering the relative magnitudes of a,b, and ω.

(4) Rigid Contacts:

There is no relative movement between adjacent particles at these contacts.

A number of photoelastic pictures have been carefully analyzed with a view toward identifying the relative importance of rolling and sliding. Contacts are selected if relatively large contact forces are transmitted across them, and if the corresponding relative displacements are large enough to allow reasonably accurate measurement. It is observed that a relatively large number of contacts that transmit large contact forces, remain rigid (no relative motion). Furthermore, a contact which begins sliding or rolling continues to be active (slide or roll), even after large deformations, up to the residual state.

Figures 4(a), (b), (c) represent measured results for initial bedding angles of $\theta=0°$, 60°, and 90°, respectively, for oval II particles with $\phi_{\mu}=26°$. In these figures ΔL is the maximum vertical displacement of the upper frame. Each data point gives the values of a and b, as well as the range for the corresponding absolute value of ω. Marks which do not carry any suffix are data for overall axial strain $\varepsilon_1=3.55\%$, those with suffix 2 are for $\varepsilon_1=5.45\%$, and those with suffix 3 are for $\varepsilon_1=6.25\%$. Analysis of other photoelastic pictures yields essentially the same results.

Figures 5(a), (b) and (c) are for the oval II particles with $\phi_{\mu}=52°$, and the same corresponding bedding angles.

From these figures the following results may be inferred:

(1) Data points that fall close to the a=-b line correspond to rolling angles $|\omega|<40°$, whereas those which fall close to a=b may possess rolling angles up to 120°. Since a=-b characterizes the sliding contacts, data points falling on the upper half of the plane correspond to contacts at which rolling is dominant; the opposite is the case for points in the lower half of the plane.

(2) Let R be the ratio of the number of contacts at which sliding is dominant over the number of those at which rolling is dominant. Measurements reveal that R ranges from 0.12 to 0.51, depending on the value of interparticle friction and the initial bedding angle (for oval II assemblies). Hence, particle rolling rather than sliding seems to be a major mechanism of microdeformation.

(3) In the oval II assemblies, R ranges from 0.20 to 0.51 for $\phi_{\mu}=26°$, whereas it ranges only from 0.12 to 0.23 at the higher interparticle friction, $\phi_{\mu}=52°$. Hence as suggested by Skinner (1969), rolling becomes more important as the interparticle friction increases.

(4) For the flatter oval II particles the higher interparticle friction ($\phi_{\mu}=52°$) produces about a 40% higher value of the stress ratio at failure than the lower interparticle friction ($\phi_{\mu}=26°$), whereas no such differential result is produced by friction for the more round oval I particles. This may be attributed to the fact that the more round particles roll with greater ease than the flatter ones.

Figure 6 show the frequency histograms of the rolling angle (in radians) for oval II assembly with $\phi_{\mu}=26°$. In these figures, ω is normalized by dividing it by the corresponding axial strain. Hence, the value of $\omega/\varepsilon_1=20$ means that a particle rolls 20 radians around its neighboring particle, when the axial strain is changed by $\varepsilon_1=1$. The figure clearly show considerable rolling of particles.

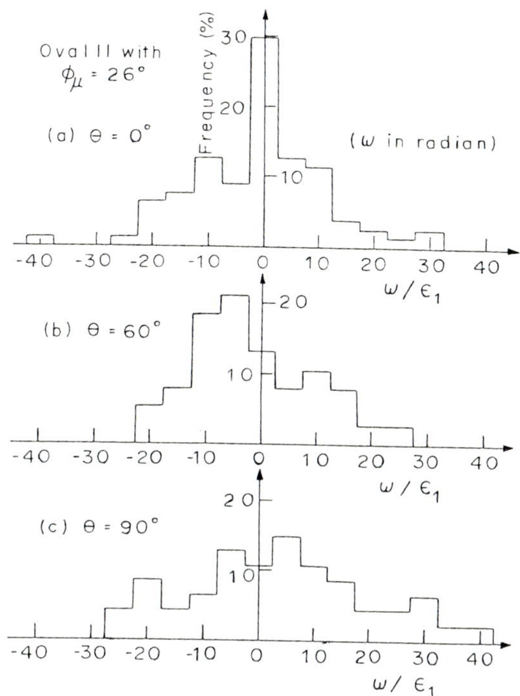

Fig.6 Histogram of particle rolling angle ω

CONTACT FORCES

Figure 7 shows a typical contact with contact force $\underset{\sim}{f}$ inclined at an angle δ (positive clockwise) with respect to the contact unit normal $\underset{\sim}{n}$. Sliding cannot occur at a contact for which $|\delta|$ is less than ϕ_μ. The angle δ may therefore be called the mobilized angle of friction. It can be measured from the fringe patterns in the photoelastic pictures; Konishi (1978). Figures 8 and 9 are the corresponding histograms for oval I and II assemblies with interparticle friction of 26° and 52°, respectively, and the indicated initial bedding angles. These histograms are obtained from the analysis of each assembly sheared up to the peak stress state. They reveal the following facts:

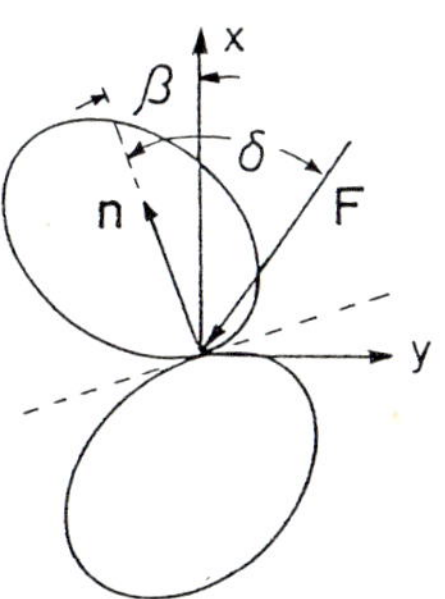

Fig.7 Mobilized angle of friction

(1) The mobilized angle of friction, δ, falls in the range ±40° when ϕ_μ=26°, and in the range ±50° when ϕ_μ=52°. The histograms display similar features

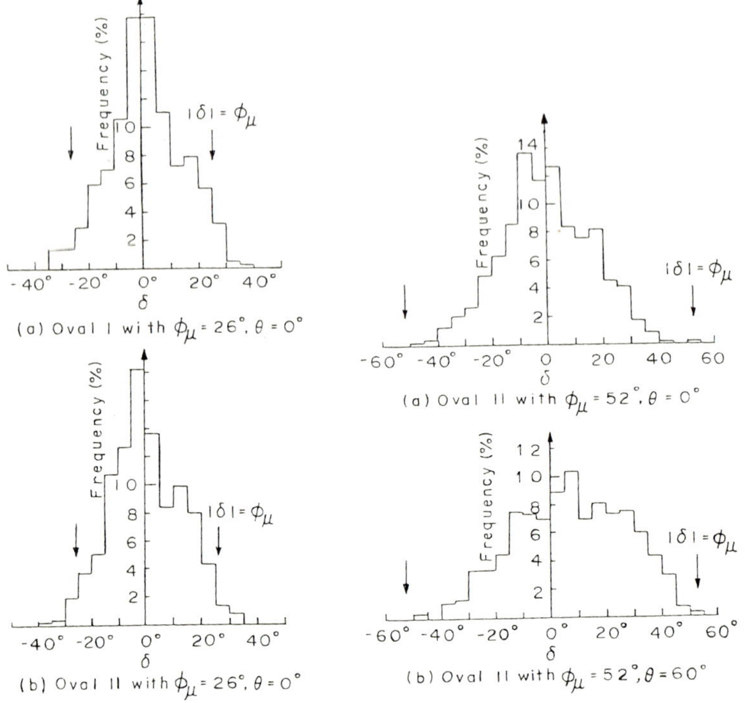

Fig.8 Histogram of mobilized angle of friction for oval I and II with ϕ_μ=26°.

Fig.9 Histogram of mobilized angle of friction for oval II with ϕ_μ=52°.

with peaks at δ=0° for both particle shapes.

(2) The distribution range is wider for the higher interparticle friction ϕ_μ=52° than for interparticle friction ϕ_μ=26°.

(3) There are few contacts at which $|\delta|=\phi_\mu$, even in assemblies that are just at failure. Close examination of Figs. 8 and 9 show that the condition $|\delta|=\phi_\mu$ tends to be satisfied at more contacts in assemblies with the lower interparticle friction than with the higher one. This observation is in harmony with the result that rolling is more dominant at the higher interparticle friction.

Similar results have been reported by Oda and Konishi (1974) and Konishi (1978) for simple shearing of two-dimensional granules with circular cross-sections. These authors, however, explain their results in terms of Horne's hypothesis (1965), which suggests that sliding does not occur at the majority of contacts, and that overall deformation takes place by relative motion between instantaneously rigid groups of particles. These groups continually reform by the division and coalescence of contacting points. Our experimental observations do not seem to support such an explanation which places too much emphasis on the role of interparticle sliding as a major microdeformation mechanism. The experimental results presented in Figs. 7 and 9 require explanation which must give due account to the role of particle rolling in shear deformation of granular masses.

CONCLUSIONS

Biaxial compression tests have been performed on assemblies of oval cross-sectional rods, in an effort of evaluate the effects of interparticle friction, particle shape, and initial fabric on the overall strength of granular materials. The variation in the spatial arrangement of the particles (fabric) and particle rolling and sliding are monitored by taking photoelastic pictures at various stages during the course of deformation. Based on this, the following conclusions are obtained: (1) Particle rolling appears to be a major microscopic deformation mechanism, especially when interparticle friction is large. (2) There are relatively few contacts at which relative sliding is dominant, and this seems to be true even when the assembly reaches the overall failure state; this observation is in contradiction to the common assumption that particle sliding is the major microscopic deformation mode.

ACKNOWLEDGMENTS

This work has been supported in part by the United States Air Force Office of Scientific Research under Grant No. AFOSR-80-0017, and in part by the National Science Foundation under Grant No. CME-8007764 to Northwestern University. The testing apparatus has been designed and assembled by Mr. John Schmidt, who has been very helpful in providing technical assistance during the entire course of this research. The assistance of Messrs. Z.L. Qiu and K. Takahashi is also gratefully acknowledged.

REFERENCES

Horne, M.R. (1965). The behavior of an assembly of rotund, rigid, cohesionless particles (I and II), Proc. Roy. Soc. A., Vol. 286, pp. 62-97.

Konishi, J. (1978). Microscopic model studies on the mechanical behaviour of granular materials, Proc. of US-Japan Seminar on Continuum-Mechanical and statistical Approaches in the Mechanics of Granular Materials, Edited by Cowin and Satake, Gakujutsu Bunken Fukyukai, Tokyo, Japan, pp. 27-45.

Konishi, J., Oda, M. and Nemat-Nasser, S. (1982). Published in U.S.-Japan Seminar (in preparation).

Ochiai, T. and Yamanouchi, T. (1972). Movement and rotation of particle during shear in two dimensional model assemblies, Interial Meeting on Soil Mech. Found. Engrg., JSSMFE, pp. 181-184. (in Japanese).

Oda, M. (1972). The mechanism of fabric changes during compressional deformation of sand, Soil and Foundations, Vol. 12, No. 2, pp. 1-18.

Oda, M. and Konishi, J. (1974). Microscopic deformation mechanism of granular material in simple shear, Soils and Foundations, Vol. 14, No. 4, pp. 25-38.

Skinner, A.E. (1969). A note on the influence of inparparticle friction on the shearing strength of a random assembly of spherical particles, Géotechnique, 19, pp. 150-157.

Roscoe, K.H. and Schofield, A.N. (1964). Discussion on Rowe's paper, Proc. of ASCE, Vol. 90, SM. 1, pp. 136-150.

Mechanics of Granular Materials: New Models and Constitutive Relations, edited by
J.T. Jenkins and M. Satake, 1983
Elsevier Science Publishers B.V., Amsterdam — Printed in The Netherlands

INDUCED ANISOTROPY IN ASSEMBLIES OF OVAL CROSS-SECTIONAL RODS IN BIAXIAL COMPRESSION

J. KONISHI[1], M. ODA[2] and S. NEMAT-NASSER[3]

[1]Dept. of Civil Engineering, Shinshu University, Nagano 380, Japan

[2]Dept. of Foundation Engineering, Saitama University, Urawa 338, Japan

[3]Dept. of Civil Engineering, Northwestern Univ., Evanston, Ill. 60201, U.S.A.

ABSTRACT

Stress-induced anisotropy of granular materials which have initially strong anisotropic fabric is discussed based on experimental observations. Major principal axis of fabric tends to rotate toward the major principal axis of stress during the course of deformation. The generation of new contacts in the direction of maximum principal compression is observed up to peak stress ratio. This trend seems to be closely related to the formation of column-like load paths and to the induced anisotropy.

INTRODUCTION

The microstructure (or the fabric) of granular materials changes during their plastic flow in response to applied loads, resulting in load-induced anisotropy. Such fabric changes have been reported by Oda (ref. 1), for natural sands, and by Biarez and Wiendieck (ref. 2), Konishi (ref. 3) and others for two-dimensional model assemblies. It has been concluded that the evolution of fabric is closely related to the variation in the distribution of the contact normals, and that this distribution seems to have a close relation to the overall applied stress: the distribution of contact normals changes in such a manner as to produce a greater concentration of contact normals along an orientation which parallels the direction of maximum principal compression.

How does an initially strong anisotropic fabric change in response to an applied load? In order to investigate this problem, we have performed a series of biaxial compression tests on two-dimensional assemblies of photoelastic rods with oval cross sections, where each assembly is formed by stacking the rods within a tilted loading frame at a desired angle, producing a strong initial anisotropy.

We have already reported the overall results of the experiments elsewhere (ref. 4). Here, the induced anisotropy during deformation is discussed in some detail.

TABLE 1 Dimension of particles

CROSS-SECTION	AXIAL RATIO d_1/d_2	DIAMETERS d_1/d_2 (mm)	
OVAL I	1.1	LARGE : 14.8 / 13.4 MEDIUM : 9.9 / 8.9 SMALL : 6.3 / 5.7	
OVAL II	1.4	LARGE : 16.0 / 11.3 MEDIUM : 10.7 / 7.4 SMALL : 7.1 / 4.9	

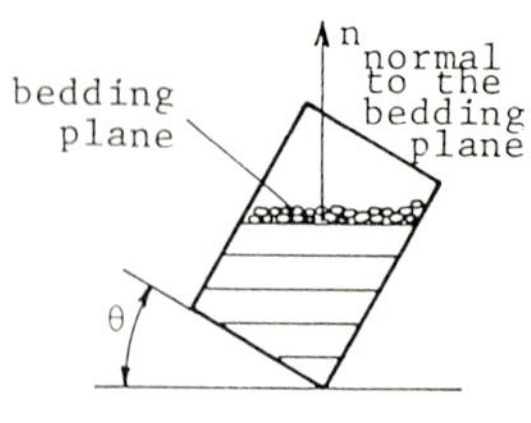

(a) during stacking

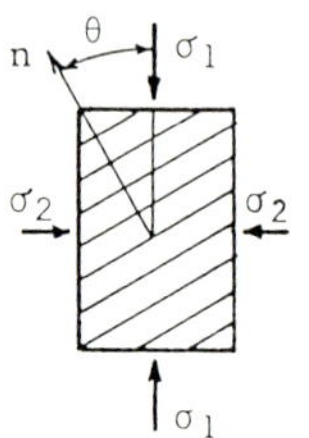

(b) while testing

Fig. 2 Definition of bedding angle

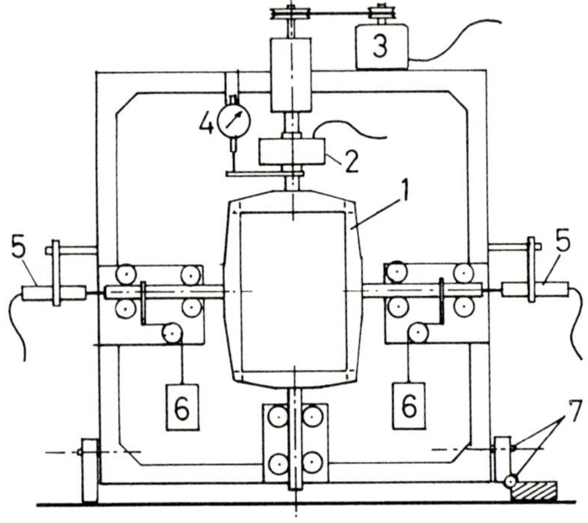

1 loading frame
2 load transducer
3 motor for vertical loading
4 dial gauge
5 LVDT's
6 weights for lateral loading
7 tilting axes

Fig. 1 A schematic diagram of the biaxial testing apparatus (Northwestern University)

OUTLINE OF THE EXPERIMENT

Particles

Rod-shaped particles of common length, 19 mm, with oval cross sections are cast from polyurethane rubber (photoelastic constant=82.5 mm/kgf). Two kinds of cross-sectional shapes are employed, each in three different sizes; the dimensions of particles are summarized in Table 1.

To assess the influence of friction, two sets of experiments are performed, one with non-lubricated particles of average friction angle of 52°, the other with particles which are lubricated with talcum powder that results in an average friction angle of 26°.

Testing Apparatus and Procedure

A schematic diagram of the biaxial compression apparatus is shown in Fig. 1. It consists of an overall frame within which a biaxial loading frame is mounted. The overall frame can be tilted in its plane and held at a desired angle. This flexibility permits sample preparation with various bedding angles θ, Fig. 2, and hence, various inherent anisotropies. The sample is formed within the loading frame of initial dimensions of about 330 mm high and 207 mm wide. Each sample consists of equal numbers of particles of the three sizes. Particles are stacked by hand randomly in the tilted frame with their longer axis essentially parallel to the bedding plane. After the completion of the assembly, the loading frame is set back to its normal position.

The number of particles and the initial void ratio of each sample are shown in Table 2.

The sample is confined laterally by a constant force of 4.41 N (0.45 kgf), and it is compressed vertically by incremental displacement downward of the upper part of the frame. The vertical load is measured by a load transducer and the vertical displacement by a dial gauge; the lateral displacements are measured by LVDT's. Photoelastic pictures are taken at appropriate stages during the course of deformation.

TEST RESULTS

Data for sixteen tests are summarized in Table 2.

The stress-strain and stress-dilatancy relations obtained in biaxial compression for two-dimensional assemblies of oval cross-sectional rods are quite similar to those of natural sands.

The stress ratio at failure $(\sigma_1/\sigma_2)_f$ varies with particle shape, interparticle friction angle ϕ_μ and the bedding angle θ. Though the effects of these parameters are coupled, the following results are revealed:

(1) For oval II (flat) particles the value of the stress ratio at failure depends strongly on the bedding angle θ. For both 26° and 52° interparticle friction angles the peak value of the stress ratio is greatest for $\theta=0°$, becomes a minimum at $\theta=60°$, and then again increases. For the oval I (more round) particles, on the other hand, no such distinct variation of the peak stress ratio with bedding angle is observed. This suggests that the inherent anisotropy to a large extent may stem from the parallel alignment of flat grains.

(2) The assembly of oval II particles exhibits considerable dependence of the magnitude of its peak stress ratio on the value of the interparticle friction, whereas no such dependence is observed

TABLE 2 Properties of assemblies

CROSS-SECTION	INTERPAR. FRICTION ANGLE ϕ_μ	BEDDING ANGLE θ	VOID RATIO e	TOTAL NO. OF PARTICLES	PEAK STRESS RATIO $(\sigma_1/\sigma_2)_{max}$	INTERNAL FRICTION ANGLE ϕ_{max}
OVAL I	52°	0°	0.190	674	7.0	49°
		30°	0.183	679	4.7	40°
		60°	0.185	676	4.1	37°
		90°	0.192	676	5.9	45°
OVAL II	52°	0°	0.177	719	14.4	60°
		30°	0.154	724	11.4	57°
		60°	0.159	721	8.3	52°
		90°	0.155	729	10.4	55°
OVAL I	26°	0°			4.9	42°
		30°	0.169	689	6.0	46°
		60°	0.174	686	5.4	43°
		90°	0.168	693	4.8	41°
OVAL II	26°	0°	0.158	726	11.8	58°
		30°	0.160	720	7.5	50°
		60°	0.155	726	4.3	39°
		90°		727	5.5	44°

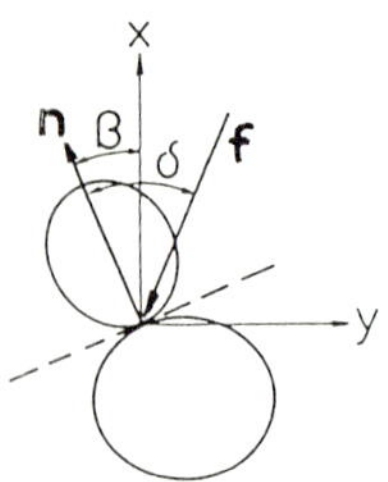

Fig. 3 Direction of contact normal $\underset{\sim}{n}$

for oval I particles. Similar results have been reported by Skinner (ref. 5) who sheared wet and dry assemblies of 1 mm glass ballotini in a shear box. Although the interparticle friction angle of the wet glass ballotini (27°-38°) is considerably higher than that of dry ones (2°-5°), both exhibit essentially the same overall shear resistance, i.e. the same overall effective friction coefficient. Skinner attributes this to the dominance of particle rolling during shear deformation. Our experimental results seem to support Skinner's conclusion; see Oda, Konishi, and Nemat-Nasser (ref. 6).

EVOLUTION OF FABRIC: INDUCED ANISOTROPY

Fabric

The concept of fabric for a macroscopically homogeneous sample of granular material should include the following (ref. 7): a measure of the orientation of individual particles (i.e. the orientation fabric), and a measure which reflects the mutual relationship of individual particles (i.e. packing). The packing of granular materials is defined by the average number of contact points per particle (co-ordination number) as well as by a distribution function $E(\underset{\sim}{n})$ which characterizes the spatial distribution of contact normals. In a two-dimensional case $E(\underset{\sim}{n})$ reduces to $E(\beta)$, where β is the inclination angle of $\underset{\sim}{n}$ with respect to the fixed vertical x-axis shown in Fig. 3.

The stress ratio σ_1/σ_2 and the volumetric strain ε_v are plotted in Fig. 4. Figure 4(a) is the plot for oval I particles with $\phi_\mu = 52°$, $\theta = 0°$; the arrows indicate initial, near the peak, and residual stages appearing in Figs. 5 and 10. Figure 4(b) is a similar diagram for non-lubricated oval II particles with $\phi_\mu =$ 52°.

Distribution of Contact Normals

Figures 5 to 8 are examples of the frequency distribution for angle β which measures the orientation of contact normals relative to the vertical x-axis; Fig. 3. The uppermost diagram in each figure shows the initial distribution of contact normals before loading, that is, the inherent anisotropy of the fabric.

For an assembly of the oval I particles with $\theta = 0°$, the fabric is rather isotropic with slight concentration of the contact normals close to $\beta = 0°$ as shown in Fig. 5a. On the other hand, assemblies of the oval II particles show distinct inherent anisotropy: contact normals concentrate in the direction normal to the bedding plane, i.e., in the direction of $\beta = \theta$; see Fig. 6a for $\theta = 0°$, Fig. 7a for $\theta = 60°$, and Fig. 8a for $\theta = 90°$.

These distributions change during the course of deformation, as the contact normals tend to concentrate in the orientation of the major principal stress, i.e. $\beta = 0°$. The resulting frequency diagrams for β at peak stress, however, do not necessarily have their peak in the direction of the major principal stress axis because strong inherent anisotropy may still remain at peak stress ratio or even in the residual state. By using the distribution of contact normals, we may define a tensor quantity

$$\phi_{ij} = \iint_{\Omega} E(\underset{\sim}{n}) n_i n_j d\Omega \qquad i,j = 1,\ 2,\ 3 \tag{1}$$

This tensor is called "fabric tensor" for an assembly of spheres by Satake (ref. 8). For two-dimensional assemblies Eq. (1) reduces to

$$\phi_{ij} = \int E(\beta) n_i n_j d\beta \qquad i,j = 1,\ 2 \tag{2}$$

We can easily calculate the deviation angle ξ of the major principal axis of ϕ_{ij} from the direction of the major principal axis of stress, i.e. the x-axis. Figure 9 shows the relation between the stress ratio σ_1 / σ_2 and $\cos \xi$. Each curve tends to $\xi = 0$ (or $\cos \xi = 1.0$) with increasing stress ratio.

In the case of $\theta = 0°$ (Fig. 5 for the oval I and Fig. 6 for the oval II) the initial fabric and the applied stress are coaxial ($\xi \simeq 0$). Therefore the frequency diagrams of β at peak stress have their maximum value at $\beta = 0°$. For an assembly of $\theta = 60°$, the orientation of the major axis of the initial fabric presumably is about -60° from the x-axis, whereas the major principal stress is

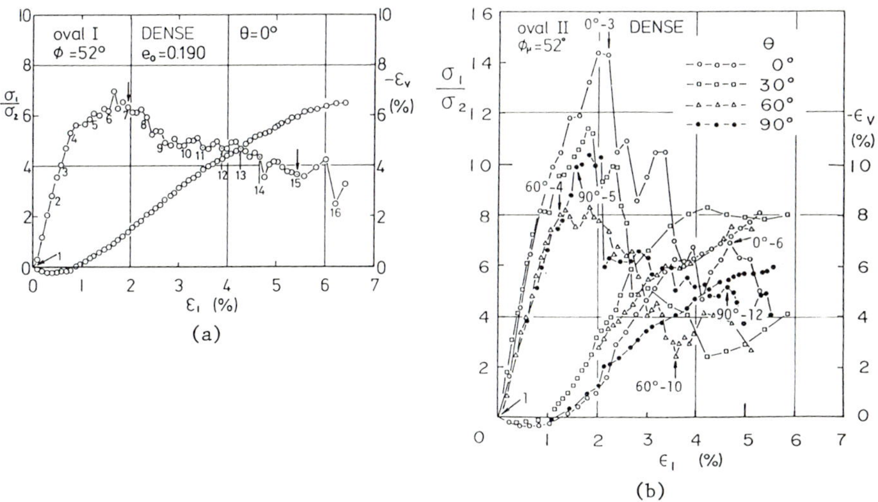

Fig. 4 Stress ratio, axial strain, volumetric strain relations:
(a) oval I, ϕ_μ=52°, θ=0°; (b) oval II, ϕ_μ=52°, θ=0°,30°,60° and 90°

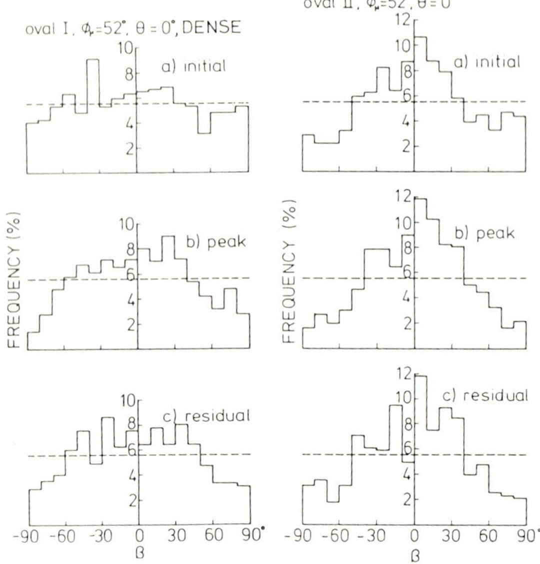

Fig.5 Change in fabric oval I, ϕ_μ=52°, θ=0°

Fig.6 Change in fabric oval II, ϕ_μ=52°, θ=0°

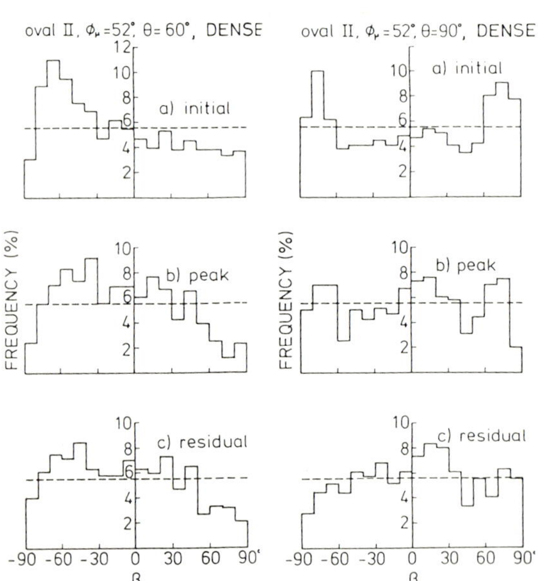

Fig.7 Change in fabric oval II, ϕ_μ=52°, θ=60°

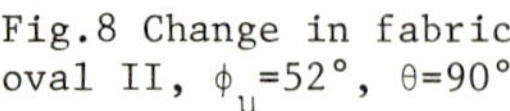

Fig.8 Change in fabric oval II, ϕ_μ=52°, θ=90°

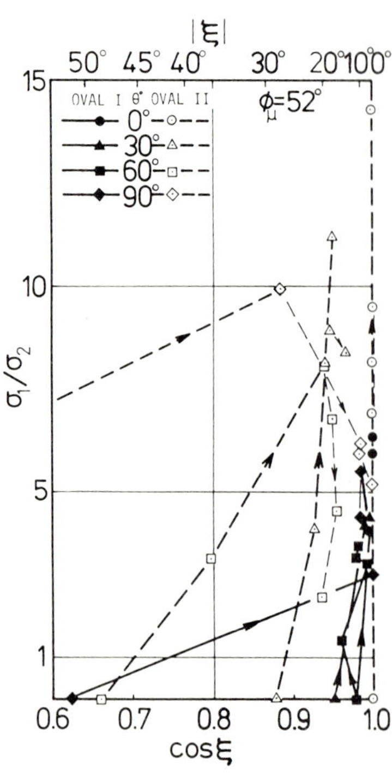

Fig.9 Change in angle ξ between principal directions of σ_{ij} and ϕ_{ij}

applied in the direction of the x-axis. As a result of this discrepancy, rotation of the axes of the fabric anisotropy (i.e., fabric tensor) occurs toward the axes of the applied stress. The distribution shown in Fig. 7b still exhibits a trace of the inherent anisotropy. A similar trace is also found at stress peak for the assembly with $\theta = 90°$, Fig. 8b.

The distributions of contact normals at the residual state are essentially similar to those at the peak; see diagram (c) in each figure.

Figure 10 shows the frequency diagrams of new contacts and of those which have disappeared during the course of deformation: a) from the initial state to the peak stress state, and b) from the peak to the residual stress state for an assembly of the oval I particles stacked at the bedding angle $\theta = 0°$. From the initial to the peak, as shown in Fig. 10a, most new contacts are generated close to the σ_1-direction, while among the contacts that have disappeared, most had orientations close to $\pm 90°$. The same tendency is seen in Fig. 11a for $\theta = 60°$ for the oval II particles, though the diagrams show some obliquity.

The generation of new contacts in the direction of maximum principal compression (the σ_1-direction) is closely related to the formation of column-like load paths,and the increasing axial stress necessitates progressive formation of new load-carrying axial columns. Figures 10b and 11b show the diagrams during the stage from the peak to the residual stress state. As is seen, new contacts do

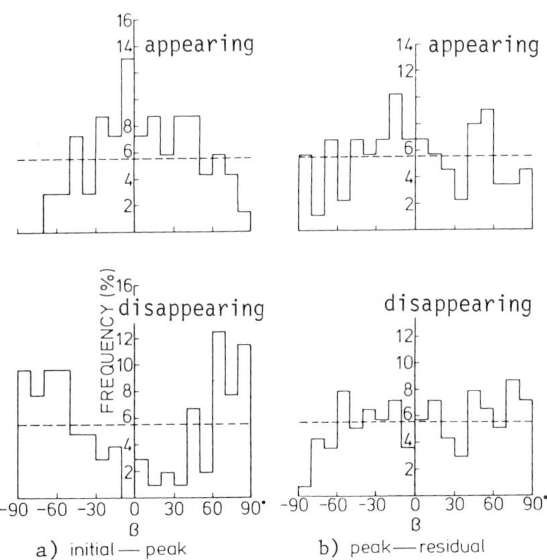

Fig. 10 Appearing new contacts and disappearing contacts: (a) up to peak stress, and (b) after peak stress; oval I, $\phi_\mu = 52°$, $\theta = 0°$

not concentrate in the σ_1-direction at the residual state anymore. Because of extensive dilatation, the column-like load paths lose stability during this process, leading to a fabric which should closely correspond to the critical state.

Fabric Tensors

In recent years, the concept of "fabric tensor" has been developed by several authors; for example, Satake (ref. 8), Oda, Nemat-Nasser, and Mehrabadi (ref. 9), and Mehrabadi, Nemat-Nasser, and Oda (ref. 10). A fabric tensor proposed by Oda, Nemat-Nasser, and Mehrabadi has the form

$$F_{ij} = N\hat{\ell}\iint_{\Omega} E(\Omega)n_i n_j d\Omega \tag{3}$$

where N is the density of contacts, and $\hat{\ell}$ is the average branch length. This fabric tensor was derived for spheres. For non-spherical particles, alternative expressions are given by Nemat-Nasser and Mehrabadi (ref. 11). A fabric tensor, Eqs. (1) and (2), in terms only of contact normals may be a crude approximation.

CONCLUSION

Biaxial compression tests have been performed for assemblies of oval cross-sectional rods made of photoelastically sensitive materials. Based on experimental results, the following conclusions are obtained: (1) Even when the initial

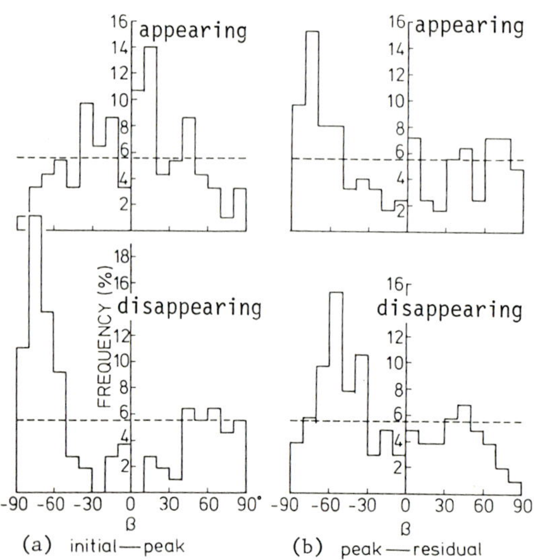

Fig. 11 Appearing new contacts and disappearing contacts: (a) up to peak stress, and (b) after peak stress; oval II, ϕ_μ=52°, θ=60°

fabric of the granular material and the applied stress are not coaxial, contact normals tend to concentrate in the orientation close to the σ_1-direction.

(2) During deformation up to the peak stress, new contacts with normals close to the σ_1-direction are formed, while many contacts with initial orientations close to the σ_3-direction, disappear.

ACKNOWLEDGMENTS

This work has been supported in part by the National Science Foundation under Grant No. CEE-8007764 and in part by the United States Air Force Office of Scientific Research under Grant No. AFOSR-80-0017 to Northwestern University.

REFERENCES

1 M. Oda, Soils and Foundations, 12, No. 2 (1972) 1-18.
2 J. Biarez and K. Wiendieck, C. R. Acad. Sci., 256 (1963) 1217-1220.
3 J. Konishi, in S.C. Cowin and M. Satake (Eds.), Proc. U.S.-Japan Seminar on Continuum Mechanical and Statistical Approaches in the Mechanics of Granular Materials, Gakujutsu Bunken Fukyukai, Tokyo, 1978, 27-45.
4 J. Konishi, M. Oda and S. Nemat-Nasser, in P.A. Vermeer and H.J. Luger (Eds.), Deformation and Failure of Granular Materials, A.A. Balkema, Rotterdam, 1982, 403-412.
5 A.E. Skinner, Géotechnique, 19 (1969) 150-157.
6 M. Oda, J. Konishi and S. Nemat-Nasser, EREL Tech. Rept. No. 82-7-47, Dept. of Civil Engrg., Northwestern University, Evanston, IL (1982); Mechanics of Materials, 1, No. 4 (1982) in press.
7 M. Oda, in S.C. Cowin and M. Satake (Eds.), Proc. U.S.-Japan Seminar on Continuum Mechanical and Statistical Approaches in the Mechanics of Granular Materials, Gakujutsu Bunken Fukyukai, Tokyo, 1978, 7-26.
8 M. Satake, in S.C. Cowin and M. Satake (Eds.), Proc. U.S.-Japan Seminar on Continuum Mechanical and Statistical Approaches in the Mechanics of Granular Materials, Gakujutsu Bunken Fukyukai, Tokyo, 1978, 47-62.
9 M. Oda, S. Nemat-Nasser and M.M. Mehrabadi, EREL Tech. Rept. No. 80-4-28, Dept. of Civil Engrg., Northwestern University, Evanston, IL (1980); Int. J. Numer. Anal. Methods Geomech., 6 (1982) 77-94.
10 M.M. Mehrabadi, S. Nemat-Nasser and M. Oda, EREL Tech. Rept. No. 80-4-29, Dept. of Civil Engng., Northwestern University, Evanston, IL (1980); Int. J. Numer. Anal. Methods Geomech., 6 (1982) 95-108.
11 S. Nemat-Nasser and M. Mehrabadi, Stress and fabric in granular masses, in these proceedings.

Mechanics of Granular Materials: New Models and Constitutive Relations, edited by
J.T. Jenkins and M. Satake, 1983
Elsevier Science Publishers B.V., Amsterdam — Printed in The Netherlands

THE SIGNIFICANCE OF THE FABRIC TENSOR IN THE DEFORMATION BEHAVIOR OF SATURATED SAND IN CYCLIC SHEARING

Y. TOBITA
Civil Engineering Dept., Tohoku University, Sendai, 980, (JAPAN)

ABSTRACT

The deformation behavior of saturated sand in the cyclic shearing apparatus is influenced to a great extent by the fabric state and its change during previous stress history as well as over consolidation ratio. In this paper, the effects of previous shear stress history on the deformation behavior of saturated sand are summarized and discussed based on the state of the microstructure in a granular assembly and its change, which are observed in the bi-axial compression test. The microscopic properties are incorporated into the fabric tensor defined by the spatial distribution of the contact areas between grain particles.

A modified stress tensor for granular materials is defined for the new modified plane , which is a transformed plane through the fabric tensor. The derivation of the well known stress dilatancy equation from continuum mechanical point of view is also presented.

INTRODUCTION

Liquefaction behavior of a saturated sand deposit in seismic excitation has been considered important for the earthquake resistant design of soil structures. A lot of experimental works have been performed in order to determine the cyclic number required to the initial liquefaction or the prescribed shear strain magnitude. The cyclic number is compared with the number determined by the magnitude of an earthquake, where the cyclic triaxial testing apparatus and the simple shear testing apparatus have been used. Recently, much attention has been paid to the stress-strain relationship of sand in a cyclic loading test in order to carry out the dynamic effective stress analysis for a saturated sand deposit. However, the proposed relationships up to now can't account for the whole aspects of the deformation behavior of saturated sand in the cyclic shearing apparatus.

The liquefaction potential of saturated sand to the cyclic loadings was believed to be determined by the void ratio. However, Mulilis et al(ref.1) and Ladd (ref.2) have shown the effects of the sample preparation method on the liquefaction potential in the cyclic triaxial testing apparatus. The effects of previous stress history on the deformation behavior of saturated sand in the succeeding cyclic shear loadings were shown by Finn et al. (ref.3), Ishihara (ref.4), and discussed from microscopic point of view by Nemat Nasser & Tobita (ref.5). These effects are believed to be attributable to the microstructure change in a saturated sand specimen during sample preparation, consolidation and shear deformation.

The microscopic change of granular materials like sand during monotonic shearing has been studied by Oda (ref.6), Oda & Konishi (ref.7), Konishi(ref.8), Matsuoka (ref.9) Konishi et al.(ref.10). It is very important for us to consider the deformation behavior of saturated sand in the cyclic shearing test based on the findings of the microstructure change in a granular assembly. The microscopic study will be helpful for the better understanding of the liquefaction behavior of a saturated sand deposit during earthquake excitation, if the differences between the condition in situ and the experimental condition are properly taken into consideration. In ref.11, the discrepancy between them is discussed.

Recently, in order to identify the microstructure in a granular assembly, several kinds of fabric tensors have been proposed and discussed by Oda (ref.12), Satake (ref.13), Kanatani (ref.14), Mehrabadhi & Nemat Nasser (ref.15). The concept of the fabric tensor is very helpful for the understanding of the deformation behavior of saturated sand in the cyclic shearing tests. It has been found that the increase of the first invariant of the fabric tensor is closely related to the hardening of the saturated sand specimen and the increase of the second invariant of it lead to the softening of the specimen.

In this paper, the deformation behavior of saturated sand in the cyclic shearing tests and the effects of previous stress history on the deformation behavior of sand are reviwed and summerized. Typical experimental results are included. The important features of the microstructure change in a granular assembly during monotonic shearing are summarized and incorporated into the fabric tensor defined in terms of the distribution of contact areas between grains.

Since the fabric tensor is a continuum mechanical description of the microstructure of granular material, It is possible for us to develop a constitutive equation reflecting the microscopic changes by taking the fabric tensor and its evolution into account, which is a continuum mechanical description of the behavior of granular material. A modified stress tensor is introduced in this paper in order to derive a stress dilatancy equation from continuum mechanical point of view, which is defined on the new isotropic space transformed by the fabric tensor.

REVIEW OF EXPERIMENTAL RESULTS

Fundamental behavior of saturated sand in undrained cyclic shearing

The fundamental features of the deformation characteristics of saturated sand in the undrained cyclic shearing test can be summarized as follows:

(1) The pore pressure increases gradually at each half cycle of the loadings. However, the development of shear strain is not so remarkable.(Loading Regime 1)

(2) After the effective stress path reaches to the line of phase transformation, the large increase of pore pressure occurs in unloading and results in the

state of initial liquefaction: complete loss of the effective confining pressure. On the other hand, the effective stress increases in loading due to the tendency of positive dilatancy. (Loading Regime 2)

(3) As the number of cyclic loadings increases, the magnitude of shear strain developed in the saturated sand specimen increases, and the stress strain relationship in this state shows an S shape.

The distinct different deformation behavior of saturated sand between loading Regime 1 and 2 was first clearified by Ishihara (ref.4). The importance of the line of phase transformation have been reexpressed by the term "Characteristic state" by Luong (ref.16) and by " Loading Regime" by Nemat Nasser & Tobita (ref.5).

During the cyclic loadings, the internal slippage between grains leads to a change of the microstructure in a sand specimen. In drain conditions, the microstructure change leads to a macroscopic volume change and the shear deformation. In undrained condition, the microstructure change results in the pore pressure change. The volume change in drained condition and the pore pressure change in undrained condition can be related with each other (see, Martin et.al (ref.17), Lindenberg et.al (ref.18)).

The effect of previous stress history on the deformation behavior of saturated sand in cyclic shearing

During the performance of liquefaction test, the effects of the sample preparation method and previous stress history on the deformation behavior of saturated sand in liquefaction test have been found to be important.

The effects of the sample preparation method were studied by the triaxial compression testing apparatus (refs.1&2). However, such a remarkable effect as these reports has not been found in the simple shearing tests. (refs.5&19) Thus, the effects of the sample preparation method depends on the type of the testing apparatus, so that this effect is not discussed in this paper.

The effect of previous stress history can be found for both testing apparatus and summarized as follows:

(1) The small preshearing leads to the increase of the resistancy of a saturated sand specimen to the liquefaction test.

(2) The previous large shearing beyond the line of phase transformation leads to decrease of the resistancy.

These effects can be observed whether the previous loadings are applied in undrained condition or in drained condition. The previous stress effects are considered to be attributed to the microstructure change of the saturated sand specimen during the previous loadings.

In order to see the fundamental characteristics of saturated sand in cyclic shearing, the quasi-static cyclic shearing tests are often performed .
Let us see two examples of the effects of previous stress history in the quasi-static cyclic shearing tests. Fig.1 shows the dilatancy behavior of a sand

specimen in a large cyclic shearing test. Large volume reduction can be observed in unloading and in the opposite direction of loading. This behavior is very important for the understanding of liquefaction behavior. Large volume reduction in the opposite direction of loading lead to the rapid increase of the pore pressure in a saturated sand specimen in undrained condition. (see Fig.2)
First, the sample was subjected to large shearing beyond the line of phase transformation in an extension direction and unloaded to the isotropic stress state, the sample showed an almost elastic behavior in the same direction of loading, however, the sample showed a rapid increase of pore pressure in the opposite direction of loading: compression in this case. The similar results in the simple shearing device were reported by Nemat Nasser & Tobita (ref.5). It seems essential for us to discuss the deformation behavior of saturated sand based on the microstructure change in a granular assembly during shearing.

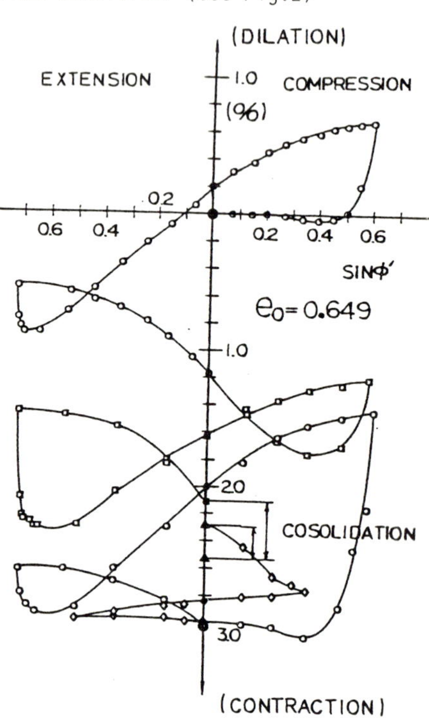

Fabric change in a granular assembly during monotonic shearing

The fabric change in a granular assembly during monotonic and cyclic shearing has been studied by Oda (ref.6), Oda & Konishi (ref.7), Konishi (ref.8) and Konishi et al.(ref.10). It seems to the author that we can summerize the fabric change in a granular assembly obsereved in these experimental works as follows:
(1) The coordination number (an average number of contacts a particle) increases during the first stage of shearing (L. R.1) and decreases in the following shearing. In a fairly large shear strain state, (critical state) the coordination number has its minimum.

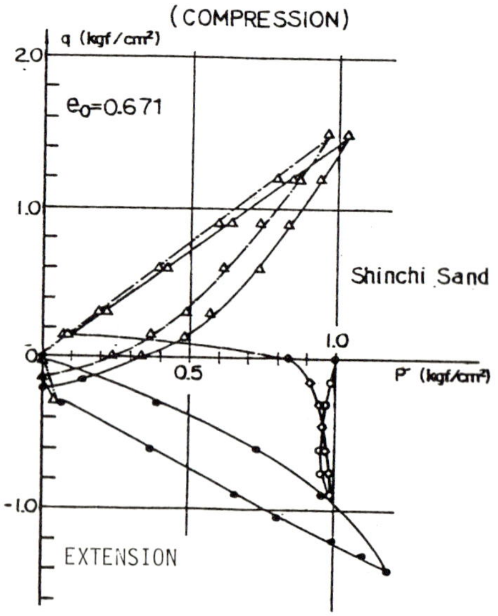

Fig.1 Dilatancy Behavior in Large Cyclic Shearing ↗
Fig.2 Effective Stress Path ⟶

(2) The anisotropy of the fabric state of granular materials in terms of the normal directions of the contact planes between grain particles increases slightly during the first stage of shearing (L.R.1) and increases until the maximum stress ratio is attained, and then it decreases during strain softening. It is naturally considered that these two microscopic changes are closely related to the deformation behavior of saturated sand in cyclic shearing.

It should be remembered that the sand sample subjected to a large shearing shows a remarkable heterogeneity, namely; the local void ratio varies from place to place within the sample. The heterogeneity of the sand sample may have a great effect on the stress deformation behavior of saturated sand in cyclic shearing. However, in what follows, we treat the sample as a homogeneous one because we don't have a sufficient data on the heterogeneity in the sand specimen which was subjected to previous stress history. This problem on the heterogeneity should be studied in the future .

In a next section, we first consider a fabric tensor which accounts for the microscopic change of granular materials during shearing. Succeedingly, we discuss the deformation behavior of saturated sand which was subjected to previous stress history in the cyclic shearing tests based on the fabric tensor.

FABRIC TENSOR

In order to describe the internal state of granular materials, some internal variables should be introduced. The fabric tensor, which identifies the spatial arrangement of grain particles and their associated voids, can be considered as a prominent one. Several definitions of the fabric tensors have been intorduced. (refs. 12,13,14,15)

In this paper, we define a fabric tensor in terms of the effective contact planes* between grain particles. The contact planes were used for the derivation of the stress dilatancy equation of sand in the triaxial compression test(ref.6) and the fabric ellipsoid possessing a tensor quality based on the distribution of the contact planes was proposed by Oda et al. (ref.12) We define the fabric tensor for a granular assembly occupying a volume V, which is large enough to permit the statistical consideration and small enough to be considered as a uniform assembly. It is possible for us to consider the volume fraction $V^{(k)}$ in the sand specimen $V = \Sigma V^{(k)}$. The volume fraction satisfying the conditions mentioned above may show a different behavior from the overall behavior if the fabric tensor is different from the overall fabric tensor. It is possible for us to consider the heterogeneity in a granular assembly by determining the fabric tensor for the volume fractions. In the following, however, we treat a uniform assembly.

* The contact between grain particles can be clasified into two categories: the effective contact through which the interparticle forces are transmitted, the dead contact which transmits little or no interparticle force.

If the distribution function of the contact area for a granular assembly occupying a volume V is given by $C(\Omega)$, we can define a following fabric tensor as

$$F_{ij} = \oint_{\Omega} C(\Omega)\, n_i n_j \, d\Omega \tag{1}$$

$$\oint_{\Omega} C(\Omega)\, d\Omega = \bar{C}, \quad tr(F_{ij}) = \bar{C} \tag{2}$$

where $\bar{C}$ means the summed contact area over the volume V , Ω is the solid angle, and n_i , n_j are direction cosines.

Let us imagine a reference state of the fabric for a granular assembly, where the fabric state is isotropic and the mean value of the trace is F_0. By referring to the isotropic fabric with a prescribed value F_0, we can discuss the effects of the void ratio and the anisotropy on the deformation and strength characteristics of granular material like sand. We consider the modified fabric tensor F^*_{ij}, which is normalized with respect to F_0 and defined as

$$F^*_{ij} = F_{ij} / F_0 \tag{3}$$

In what follows, however, we assume that $F_0 = 1$ and we denote the modified fabric tensor as F_{ij} and call it the fabric tensor for the simplicity.

Evolution of fabric tensor

The experimental works on the fabric change in a granular assembly are limited to the triaxial compression test (ref.6), two dimensional simple shear test (refs.7,8,9) and the bi-axial compression test (refs.8,10). Experimental works (refs.7,8) show that the fabric change in a granular assembly in the simple shear test occures in such a manner that the direction of the maximum intensity of the fabric orients toward the maximum principal stress direction. In other words, the rotation of the principal direction has a great effect on the fabric change, which lead to the inelastic strain and the associated volume change even if the applied stress ratio is kept constant. (see, refs.20,21) The effect of the rotation of the principal stress direction on the evolution of the fabric tensor should be taken into consideration in the future study, however, this effect is not considered in this paper because of the difficulty involved in this problem.

The fabric changes (1) and (2) described in the previous section: the coordination number and the induced anisotropy, are considered in this paper. This means that we will not treat the case where the drastic change of the direction of principal stress occurs. The first and second invariants of the fabric tensor are defined as

$$J_1 = tr\,(F_{ij}), \quad J_1 = \bar{C} = \oint_{\Omega} C(\Omega)\, d\Omega \tag{4}$$

$$J_2 = \frac{1}{2} F'_{ij} F'_{ij} {}^{*} \tag{5}$$

where a super-prime indicates the deviatoric component of the fabric tensor.

---

* This definition is not the same with the definition in tensor analysis, however, each has a close relation.

Since the first invariant J_1 is the summed contact area in the volume V, J_1 shows such an evolution as shown by the solid line in Fig.3. Remember that the evolution of J_1 has a similar characteristic to that of the coordination number. The second invariant J_2 is a quantity which expresses how the fabric state deviates from the isotropic state. Cosequently, J_2 has a similar trend to the intensity of the induced anisotopy and it will be expressed as shown by the dotted line in Fig.3.

Thus, J_1 and J_2 have close relations to the microscopic quantities of granular material observed in the experimental works. From the discussions in the preceeding sections, we can say that the increase of J_1 leads to the hardening for the resistance of a sand sample in the cyclic shearing test, and that increases of J_2 lead to the lower resistance of the sand specimen to the cyclic loading. The increase of J_2 means that the sand sample develops an anisotropic fabric: strong in an extension direction and weak in the compression in the case of Fig.2. Since the sand sample is subjected to a repeated reversal loading, the resistance of the sand is determined by the weak direction, so that the highly oriented fabric sand shows the lower resistance to the cyclic loadings. The increase of J_1, which is expected in a consolidation stage and in the small shearing, may be related to the disappearance of the unstable structures involved in the granular assembly and results in the hardening effects of the sand specimen.

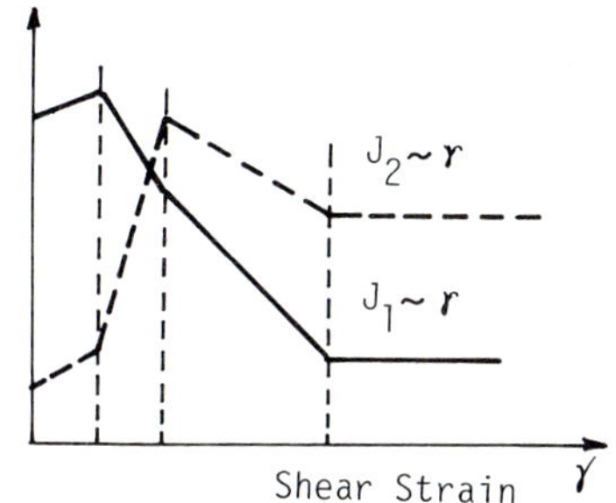

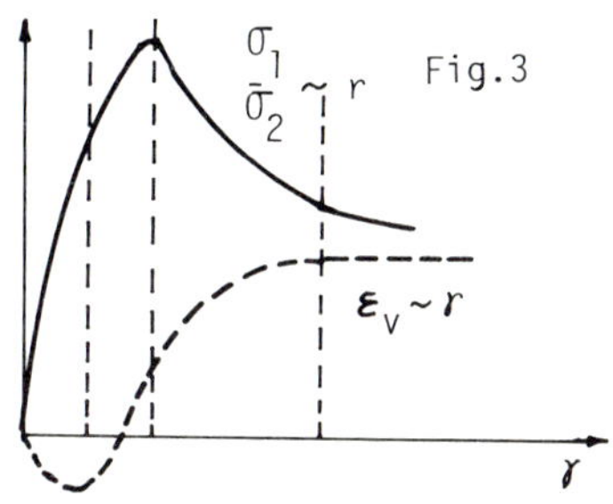

Fig.3

As mentioned above, the rotation of principal stress axes has a great effect on the deformation behavior of sand, which means that the effects of previous stress history can't be accounted for by the invariant quantities of the fabric tensor. In order to discuss this effect, we should consider the relation between the macroscopic compliance matrix C_{ijkl} and the fabric tensor F_{ij}, namely; we should consider how the change of the fabric tensor relate to the change of the compliance matrix which express the overall stress-strain relationship. This is a really tough problem, however, this problem seems to be essential for the constitution of granular mechanics as well as the better understandings of the behaviors of cohesionless soils.

New modified plane

We consider a new plane which is transformed by the fabric tensor as

$$S_i^* = F_{ik} S_k \tag{6}$$

where $S_k = S_{(n)}(\underset{\sim}{e}_k \cdot \underset{\sim}{n})$; $S_{(n)}$ is a plane the normal unit vector of which is $\underset{\sim}{n}$, $\underset{\sim}{e}_k$ is a base vector of the rectangular Cartesian coordinate system.

In what follows, the summation convention on the repeated indices in a term is used. In the new space formed by S_i^*: $S_i^* = S_{(\nu)}(\underset{\sim}{e}_i^* \cdot \underset{\sim}{\nu})$, the contact planes distribute equally in any directions and the contact area projected onto the plane S_i^* is equal to that projected onto the associated original plane S_i. In other words, we consider a new plane which has an isotropic configuration with respect to the contact area density and we make the original plane expand or shrinkage in order to satisfy the following condition,

$$F_0 S_i^* = F_{(i)} S_{(i)} \quad \text{... not summed} \tag{7}$$

where $F_{(i)}$, $S_{(j)}$ mean the principal value of the fabric tensor and the principal plane respectively.

We introduce a modified stress tensor A_{ij} and derive a well known stress dilatancy equation based on the modified stress tensor and the new modified plane transformed by the fabric tensor.

MODIFIED STRESS TENSOR AND STRESS DILATANCY EQUATION

We define a new stress tensor A_{ij} with respect to the new modified plane. Since the force vector $\underset{\sim}{f} = S_{(n)}\underset{\sim}{T}$ is the same with the force vector for the new modified plane : $\underset{\sim}{f} = S_{(\nu)}^* \underset{\sim}{T}^*$, we can get a new stress tensor as follows:

$$\underset{\sim}{f} = \underset{\sim}{\sigma}(S_{(n)} \cdot \underset{\sim}{n}) = \underset{\sim}{\sigma}(\underset{\sim}{F})^{-1} S_{(\nu)}^*$$

$$= \underset{\sim}{A}(S_{(\nu)}^* \cdot \underset{\sim}{\nu}) \tag{8}$$

$$\underset{\sim}{A} = (\underset{\sim}{F})^{-1} \underset{\sim}{\sigma} = \underset{\sim}{\varphi} \underset{\sim}{\sigma} \tag{9}$$

where super astrisk indicates the quantities with respect to the new modified plane, $\underset{\sim}{T}$: traction vector $\underset{\sim}{\sigma}$: Cauchy stress tensor for continous medium.

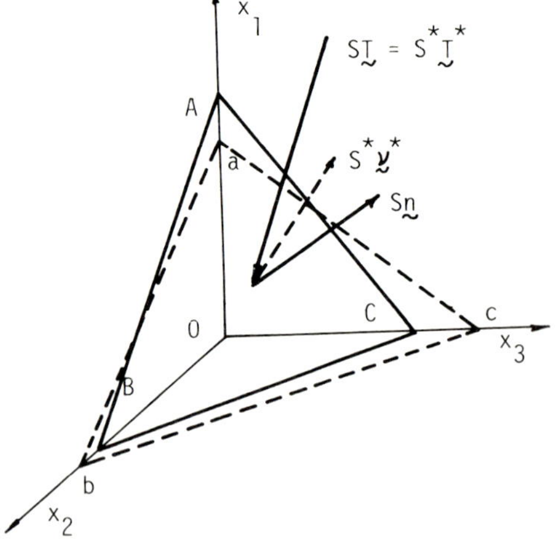

Fig.4 A new modified plane and A modified stress tensor

The new stress tensor defined by eq.(9) will be called "a modified stress tensor ". The similar stress tensors to eq.(9) have been obtained by Satake (ref.13) and Mehrabadhi & Nemat Nasser (ref.15). In thier papers, A_{ij} were called the sub-stress tensor and the local stress measures respectively. Eq.(15) indicates that the modified stress tensor has two important physical meanings as:

(1) If the density of the contact area in a volume fraction $V^{(k)}$ is small compared with the overall contact density defined for the whole volume V, the value of the modified stress tensor for the volume fraction $V^{(k)}$ becomes large, which inmplies a kind of stress concentration. Thus, by introducing the modified stress tensor for granular materials, the heterogeneity in a granular assembly can be accounted for, although it is a very difficult problem as discussed by Mehrabadhi & Nemat Nasser (ref.15);

(2) The anisotropic state of the fabric in a granular assembly lead to the diffrent stress obliquity between the modified stress tensor and the Cauchy stress tensor as discussed in the derivation of the stress dilatancy equation.

The introduction of the modified stress tensor was influenced to a great extent by the works due to Satake (ref.22), in which the transformation of a space due to a tensor is discussed, and due to Murakami (ref.23), in which the effective stress was developed for the solid materials including voids.

Stress dilatancy equation

The stress dilatancy equation for granular materials has been discussed by many reserchers and has been considered as a fundamental equation for granular mechanics. We derive a stress dilatancy equation from continuum mechanical point of view. The assumptions necessary for the derivation of this equation are summarized as follows:

(1) The sliding deformation of the frictional body in the new modified plane occurs toward the direction of the shear stress component. The initiation of the slippage is assumed to obey the Coulomb's frictional law:

$$\tau_{(\underset{\sim}{\nu})} / A_{(\underset{\sim}{\nu})} = \tan \phi_{\mu} = K \qquad (10)$$

(2) The granular packing is in a dense state and the granular assembly is already sheared up to the loading regime 2 (positive dilatancy), in which the coaxiality between the fabric tensor and the Cauchy stress tensor is expected.

(3) The granular assembly is in a plane strain condition and the initiation of the slippage will not be affected by the intermediate principal stress.

Let θ_1 and θ_2 denote the angles formed by the new plane and the original plane with the reference axis x_2 as shown in Fig.5, in which x_1 is a direction of maximum principal stress and x_2 is a direction of minimum principal stress.* According to the assumption (2), $\theta_2 > \theta_1$ is expected. The shear strain rate $d\gamma^*$ generates along the new plane the unit normal vector of which is denoted by $\underset{\sim}{\nu}$, and no dilatancy component is expected since the new plane has an isotropic fabric.

* Compressive is taken as positive as usually adopted in the branch of soil mechanics.

By considering the force balance on the new plane on which the maximum stress obliquity equal to K is working, we can express the principal stresses of the modified stress tensor in terms of $A^*_{(\nu)}$ and K as

$$A_1 = (1 + K\tan\theta_1)\, A_{(\nu)} \tag{11}$$

$$A_2 = (1 - K\cot\theta_1)\, A_{(\nu)}, \quad A_{(\nu)} = A_{ij}\nu_i\nu_j \tag{12}$$

Considering the relationship between the modified stress tensor and Cauchy stress tensor: eq.(9), and making its ratio, we can get the following equation as

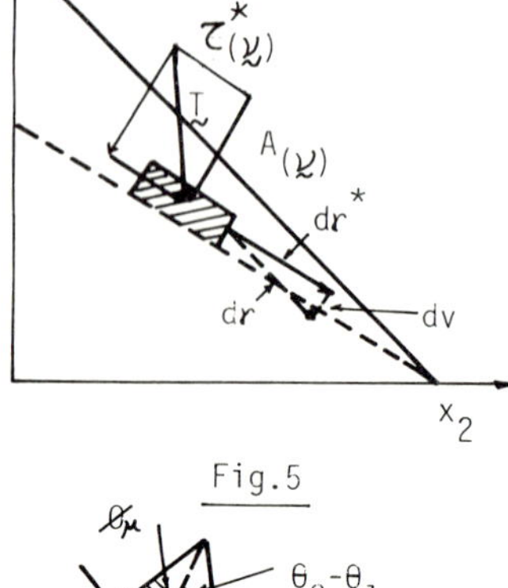

Fig.5

$$\sigma_1 / \sigma_2 = \cot\theta_1 (1 + K\tan\theta_1)/(1 - K\cot\theta_1) \tag{13}$$

In this equation, $\tan\theta_1 = F_2 / F_1$ expresses the effect of the fabric state of a granular assembly. Since $\cot\theta_1$ is larger than unity, it increases the stress ratio in terms of Cauchy stress tensor.

Now, we subdivide the shear deformation rate $d\gamma^*$ into the two parts, namely; the shear component $d\gamma$ and the dilatancy component dv as shown in Fig.5. Since $d\gamma = d\gamma^*\cos(\theta_2-\theta_1)$, $dv = d\gamma^*\sin(\theta_2-\theta_1)$, we can get the following equation by assuming that the plastic works dissipated on both planes must be the same,

$$\tau_{(\nu)} d\gamma^* = [\sigma_{(n)}\sin(\theta_2-\theta_1) + \tau_{(n)}\cos(\theta_2-\theta_1)]\, d\gamma^* \tag{14}$$

where $\tau^2_{(\nu)} = |T|^2 - A^2_{(\nu)}$, $\tau^2_{(n)} = |T|^2 - \sigma^2_{(n)}$

Dividing both sides of eq.(14) by $d\gamma^*$, $\sigma_{(n)}$ and $\cos(\theta_2-\theta_1)$,

$$\tau_{(\nu)} / \sigma_{(n)}\cos(\theta_2-\theta_1) = \tan(\theta_2-\theta_1) + \tau_{(n)} / \sigma_{(n)} \tag{15}$$

Since $\sigma_{(n)} = A_{(\nu)}\cos(\phi_\mu + \theta_2-\theta_1)/\cos\phi_\mu$, $\tau_{(\nu)}/\sigma_{(n)} = K\cos\phi_\mu\cos(\theta_2-\theta_1)/\cos(\phi_\mu+\theta_2-\theta_1)$

It is easily found that $\tau_{(\nu)}/\sigma_{(n)} > K$ from the last equation. The left hand side of eq.(15) is considered as the stress ratio mobilized in a critical state, which was found to be greater than K experimentally, and the first term in eq.(15) can be compared with the dilatancy ratio defined in the critical state soil mechanics. Eq.(15) implies that the dilatancy is a result of the anisotropic fabric of a granular assembly in the derivation of the stress dilatancy equation from continuum mechanical point of view.

This work can be compared with the microscopic approaches due to Newland & Alley (ref.24) and Nemat-Nasser & Tobita, in which the dilatancy angle: the angle between the actual slip direction and the macroscopic slip line, are considered as a paramount role. The derivation of the stress dilatancy equation based on the modified stress tensor and the new modified plane transformed by the fabric tensor seems to have shown the importance and the efficiency of the continuum mechanical description of the behaivor of granular material, although the sufficient interpretation on the quantities derived in this paper has not been made clear yet.

CONCLUDING REMARKS

The effects of previous stress history on the deformation characteristics of saturated sand in the cyclic shearing apparatus have been discussed based on the experimental results obtained by the bi-axial compression test and the two dimensional simple shear test. The microscopic state of granular material during the monotonic shearing test were incorporated into the fabric tensor proposed in this paper. It has been found that the effects of previous stress history on the deformation behavior of saturated sand in the succeeding cyclic shearing test can be accounted for by the invariants of the fabric tensor in a qualitative manner, if the direction of the principal stress axes will not rotate so drastically.

In order to account for the whole aspects of the deformation behavior of sand based on the microscopic mechanism, we should obtain an evolutionary equation for the fabric tensor in terms of the applied stress tensor. The relation between the overall compliance matrix and the fabric tensor is also necessary for the better understandings of the deformation behavior of granular material under general stress state. The continuum mechanical description of the behavior of granular material seems to be essential for the better understanding of the soil behavior, which is also important in the constitution of granular mechanics. As a first step of the continuum mechanical description, the stress dilatancy equation was derived based on the modified stress tensor and the new modified plane obtained by the fabric tensor.

ACKNOWLEDGEMENTS

The author wishes to express his sincere gratitude to Prof.E.Yanagisawa and Prof.M.Satake for their encouragements and suggestions. The critical discussions by Prof.Oda and Prof.Konishi are also appreciated. He also appreciates Prof. Nemat-Nasser for his kindness and meaningful suggestions to his study.

REFERENCES

1 Mulilis,j.p.,H.B.Seed,C.K.Chan,J.k.Mitchell and K.Arulanandan, Effect of Sample Preparation on Sand Liquefaction, ASCE, 103, GT2, 91 (1977)
2 Ladd, R.S., Specimen Preparation and Cyclic Stability of Sands, ASCE, 103, GT6, 535, (1977)
3 Finn,W.D.L.,P.L.Bransby, D.J.Pickering, Effect of train istory on iquefaction of Sand, ASCE, SM6, 1917, (1970)

4 Ihsihara,K., Stress History Effects on the Behavior of Sand, Proceedings of U.S. Japan Seminor on the Mechanics of Granular Materials, Gakujutsu Bunken Fukyukai, Tokyo, Japan, 71,(1978)

5 Nemat-Nasser,S. and Y.Tobita, Influence of Fabric on Liquefaction and Densification Potentail of Cohesionless Sand, Mechanics of Materials,1,1,43 (1982)

6 Oda,M. The Mechanism of Fabric Changes during Compressional Deformation of Sand, Soils & Foundations, 12, 2, 1 (1972)

7 Oda,M. and J.Konishi, Microscopic Deformation Mechanism of Granular Material in Simple Shear, Soils and Foundations, 14, 4, 25(1974)

8 Konishi,J., Microscopic Consideration on Deformation and Strength Behavior of Granular Material like Sand, Faculty of Enginnering Report, Shinshu University, (1978) - in Japanese

9 Matsuoka,H., A Microscopic Study on Shear Mechanism of Granular Materials, Soils and Foundations, 14, 1, 29 (1974)

10 Konishi,J., Oda M. and Nemat Nasser,S., Induced Anisotropy in Assemblies of Oval cross sectional rods in Bi-axial Compression, Proceedings of U.S.-Japan Seminor, Ithaca, NY, (1982)- Pre-Print

11 Tobita,Y., Effects of Fabric Change on Deformation Characteristics of Sand in Cyclic shearing, Advances in the Mechanics and the Flow of Granular Materials. Trans-Tech and McGraw Hill Publication, (1982)-in printing.

12 Oda,M.,J.Konishi and S.Nemat Nasser, Some Experimentally Based Fundamental Results on the Mechanical Behavior of Granular Materials, Geotechnique, 30, 4, 479 (1980)

13. Satake M., Fabric Tensor in Granular Materials, IUTAM Conference on Deformation and Failure of Granular Materials,Delft, 63 (1982)

14. Kanatani K., Mechanics of Particle Assemblage-Microscopic and Statistical Approach, Mechanical Consitution of Granular Materials, Satake,M.editor, 55, (1982)

15. Mehrabadhi,M.M. and S. Nemat Nasser, Stress Dilatancy, and Fabric in Granular Materials, U.S.-Japan Seminor, Ithaca, NY,(1982)-Pre print

16. Luong M.P., Stress Strain Aspects of Cohesionless Soils under Cyclic and Transient Loading, International Symposium on Soils under Cyclic and Transient Loading, Swansea, A.A.Balkema Pub. Rotterdam, Vol.1,315,(1980)

17 Martin,G.R.,Finn,W.D.L., and H.B.Seed, Fundamentals of Liquefaction under Cyclic Loadings, ASCE, 101, GT5,423, (1975)

18 Lindenberg,J.,and H.L.Koning, Critical Density of Sand, Geotechnique,31,2,351 (1981)

19 Silver,M.L.,F.Tatsuoka, A.Phukunhaphan and Anestis S.Avramidis, Cyclic Undrianed strength of Sand by Triaxial Test and Simple Shear Test, 7th World Conference on Earthquake Engineering, Vol.3,281 (1980)

20 Ishihara,K. and I.Towhata, Effects of Principal Axes Rotation on Undrained Deformation Characteristics of Loose Sand, Consitution of GRanular Mechanics, Satake,M. editor, 159, - in Japanese (1982)

21 Arthur,J.R.F.,Ken,S.Chua, Treve Dunston, and J.I. Rodringuez del C., Principal Stress Rotation-A Missing Parameter, ASCE, 106, GT4,419 (1980)

22 Satake,M., On Distortion of Tensor and Yield Criteria of Granular Materials International Jounal of Engineering Science, 19, 12, 1643 (1981)

23 Murakami S., Damage Mechanics - Continuum Mechanics Approach to Damage and Fracture of Materials, Review, Journal of the Society of Materials Science, Japan, Vol.31, 340, 1 - in Japanese (1982)

24 Newland,P.L. and B.H.Alley, Volume Changes in Drained Triaxial Tests on Granular Materials, Geotechnique 7,1,17 (1957)

Mechanics of Granular Materials: New Models and Constitutive Relations, edited by
J.T. Jenkins and M. Satake, 1983
Elsevier Science Publishers B.V., Amsterdam — Printed in The Netherlands

CYCLIC BEHAVIOR OF SAND DURING ROTATION OF PRINCIPAL STRESS AXES

K. ISHIHARA[1] and I. TOWHATA[2]

[1]Professor of Civil Engineering, University of Tokyo

[2]Research Associate of Civil Engineering, University of Tokyo, Bunkyo-ku
Tokyo, (Japan)

SYNOPSIS

By the use of a triaxial torsion shear test apparatus, statically operated cyclic loading tests were conducted on loose saturated specimens of the Japanese standard sand. The stress difference between the axial and horizontal stresses as well as the torsional shear stress was cyclically applied to the test specimens so that the continuous rotation of the principal stress directions could be achieved during the cyclic loading. The test results showed that, even when the amplitude of the combined shear stress (deviator stress) is maintained constant, the plastic deformation as represented by the pore water pressure build-up and progressive accumulation of volumetric strain could take place in the sand, if the rotation of the principal stress axes is executed during the cyclic loading.

INTRODUCTION

Laboratory cyclic loading tests by means of the triaxial, simple shear and torsion shear test apparatus have provided most of our understanding of the response of saturated sands under cyclic loading conditions. In these modes of cyclic deformation, the axes of the principal stresses are changed suddenly by 90 degrees at the instant of stress reversal. In many of the modes of stress application such as those encountered in the seismic loading, the change in the principal stress directions as above is deemed closely to simulate the manner of actual stress changes occurring in the soil deposits in the field. However, there exist several other instances of cyclic loading enviroments where the mode of load application is associated with continuous rotation of the principal stress directions. It was shown by Ishihara and Towhata (1983) that the change in shear stress induced in the seafloor deposits by waves travelling overhead involves a continuous rotation of the principal stress axes.

The effects of the principal stress axis rotation on the response of a clean sand have been extensively investigated by Arthur et al (1980, 1981) by means of a specially designed flexible boundary shear test apparatus. Some of their test results revealed the important features of the effects of rotation of the principal stress directions in understanding the deformation and strength characteristics of sand. Some studies have also been made by Ishihara and Towhata (1983) concerning the cyclic response of sand as affected by the continuous rotation of the principal stress directions using a triaxial torsion shear test apparatus.

The present study is a continuation of the previous investigation as cited above and mainly deals with the cyclic response of saturated sand under combined torsional and triaxial shear stress applications in which the rotation of the principal stress direction is executed only on the portion of the triaxial compression side.

TRIAXIAL TORSION SHEAR APPARATUS

The test apparatus used was a triaxial torsion shear apparatus which permits the torsional shear stress to be applied to the specimen independently of the vertical and horizontal stresses. A vertical cross section of the test device is shown in Fig.1. The hollow cylindrical test specimen 10cm in outer diameter, 6cm in inner diameter and 10.4cm in length encased in rubber membranes is put in place in the triaxial chamber as shown in Fig.1. The test apparatus was designed so that the pressure inside the hollow cylindrical specimen could be varied, if necessary, independently of the pressure outside the specimen. In the present study, however, the chamber pressures both inside and outside were held identical. Therefore, three components of stress, i.e., the vertical stress, lateral stress and torsional stress were varied to produce cyclic excursion of shear stresses in the test specimen.

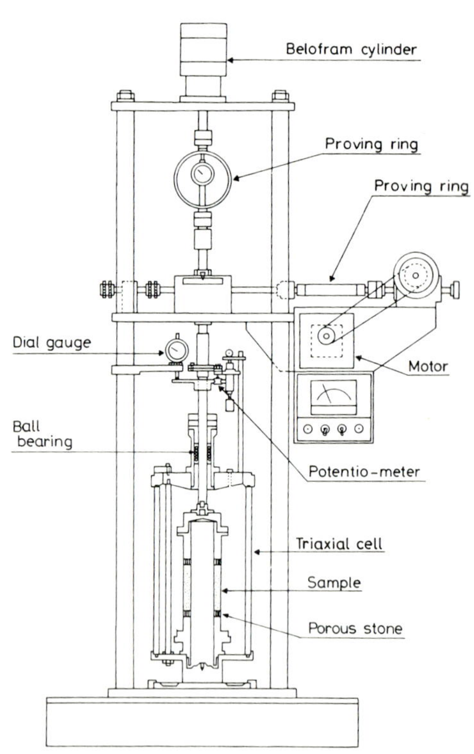

Fig. 1. Triaxial torsion shear test apparatus

The vertical load was applied through a vertical rod connected to a Belofram cylinder which was operated by air pressure. A proving ring was used to monitor the vertical load and a dial gage to measure the vertical displacement of the test specimen. The lateral stress was applied by the change in the chamber pressure. The torsional shear stress was applied by the rotational force transmitted to the vertical rod above the triaxial cell. The rotational force produced by a motor was monitored by a proving ring and the rotational angle measured by means of a potentiometer. Pore water pressures were monitored by an electric transducer installed at the base of the triaxial cell.

TEST MATERIAL AND PROCEDURES

The sand used in the present study was the Japanese standard sand called Toyoura sand. The mean particle size, D_{50}, and uniformity coefficient, U_C, are D_{50} = 0.17 mm and U_C = 2.0, respectively. Subrounded to subangular in grain shape, the sand has the maximum and minimum void ratios of e_{max} = 0.98 and e_{min} = 0.60, respectively. The specific gravity is G_s = 2.65.

Air-dried sand mixed with carbon dioxide gas, CO_2, was poured in the annular space between the outer and inner molds by means of a vinyl pipe. A height of fall of 10 to 13 cm was employed to spread the sand to a relative density of D_r = 40 ~ 50 %. To obtain a denser specimen with a relative density on the order of 80 %, the sand was poured from a height of approximately 30 cm. After forming the specimen, it was saturated under a back pressure of 29.4 kN/m^2. With the aid of the CO_2 gas, full saturation was achieved with a B-value exceeding 0.99. The specimen was consolidated under a confining stress of 294 kN/m^2. The cyclic shear stress application was made statically with a shear strain rate of 0.14 percent per minute. The tests reported in this paper were conducted either under undrained conditions with pore water pressure measurements or under drained conditions with measurements of volume change of the specimen.

LOADING SCHEME

The cyclic loading scheme involving changes in both the torsional shear stress and the axial stress difference was programmed so that the direction of the principal stress could be continuously rotated. The conventional types of the cyclic torsion shear test and of the cyclic triaxial shear test with abrupt change in the principal stress direction were also performed to provide a basis for comparison with the tests involving the continuous rotation of the principal stress axes.

Let the shear stress be denoted by τ_{vh}, and the vertical and horizontal effective normal stresses by σ_v and σ_h, respectively, as illustrated in Fig.2. Then, the major principal stress, σ_1, and the minor principal stress, σ_3, are given by,

$$\left.\begin{aligned} \sigma_1 &= \frac{\sigma_v + \sigma_h}{2} + \sqrt{\left(\frac{\sigma_v - \sigma_h}{2}\right)^2 + \tau^2_{vh}} \\ \sigma_3 &= \frac{\sigma_v + \sigma_h}{2} - \sqrt{\left(\frac{\sigma_v - \sigma_h}{2}\right)^2 + \tau^2_{vh}} \end{aligned}\right\} \qquad (1)$$

The angle between the major principal stress direction and the vertical, β, is given by

$$\tan 2\beta = \frac{2\tau_{vh}}{\sigma_v - \sigma_h} \tag{2}$$

Therefore, the deviator stress, defined as the difference between the major and minor principal stresses, is given by,

$$\frac{\sigma_1 - \sigma_3}{2} = \sqrt{\left(\frac{\sigma_v - \sigma_h}{2} \right)^2 + \tau^2_{vh}} \tag{3}$$

Corresponding to the stress components, it is possible to define the principal strain components as follows,

$$\left. \begin{aligned} \varepsilon_1 &= \frac{\varepsilon_v + \varepsilon_h}{2} + \frac{1}{2}\sqrt{(\varepsilon_v - \varepsilon_h)^2 + \gamma^2_{vh}} \\ \varepsilon_3 &= \frac{\varepsilon_v + \varepsilon_h}{2} - \frac{1}{2}\sqrt{(\varepsilon_v - \varepsilon_h)^2 + \gamma^2_{vh}} \end{aligned} \right\} \tag{4}$$

$$\tan 2\beta' = \frac{\gamma_{vh}}{\varepsilon_v - \varepsilon_h} \tag{5}$$

where ε_v and ε_h denote vertical and horizontal strains, respectively and γ_{vh} is the shear strain in the torsional mode of deformation. ε_1 and ε_3 are the major and minor principal strains, respectively.

Three types of cyclic loading tests were conducted by changing the two components of shear stress singly or in combination. The loading schemes of these tests are illustrated in a stress space shown in Fig.3, in which the shear stress and the stress difference are represented in a rectangular coordinate system.

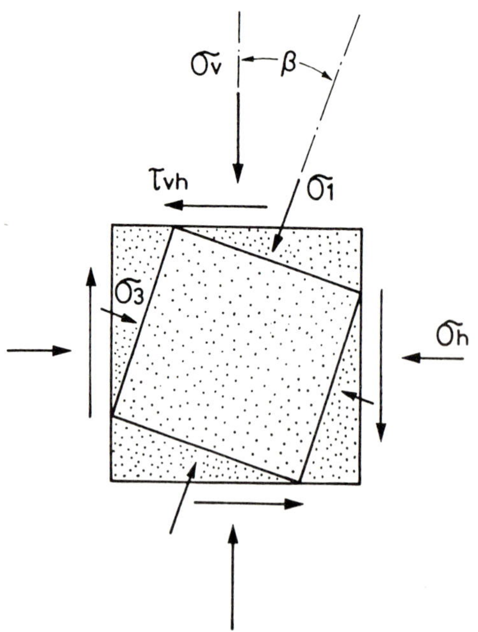

Fig. 2. Definition of the angle of the major principal stress direction

1. Cyclic Triaxial Shear Test

Only the stress difference, $(\sigma_v - \sigma_h)/2$, is cylically applied to the test specimen, while keeping the other component of shear stress,

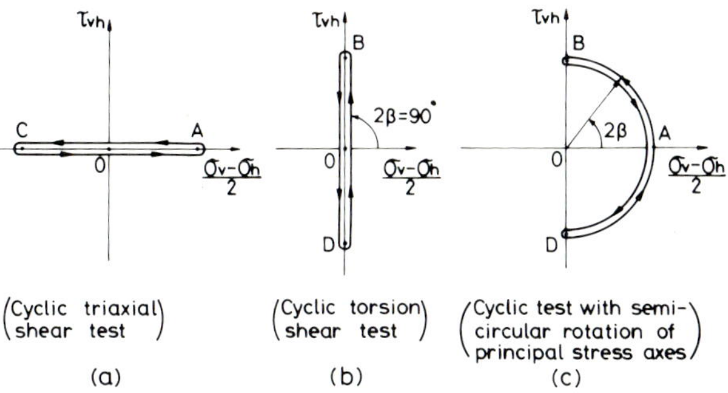

Fig. 3. Loading schemes adopted in the tests

τ_{vh}, equal to zero throughout the cyclic phase of loading. The test specimen was first consolidated isotropically and then subjected to cyclic changes in the vertical stress, σ_v, under undrained conditions. This type of test is identical with the conventional cyclic triaxial shear test. In this test, the axis of the major principal stress is directed vertically in the course of increasing the vertical stress, σ_v, but in the stage of decreasing the vertical stress, the major principal stress is oriented in the horizontal direction. Therefore, the axis of the major principal stress is abruptly rotated by 90 degrees at the instant of stress reversal as illustrated in Fig.3(a).

2. Cyclic Torsion Shear Test

In this type of test, only the shear stress, τ_{vh}, is applied cyclically to the specimen. The test specimen was isotropically consolidated and then subjected to the cyclic torsional shear stress under undrained conditions, while holding the vertical and horizontal stresses unchanged. This type of test is identical to the conventional cyclic torsion shear tests such as those conducted by Ishibashi and Sherif (1974), and Ishihara and Yasuda (1975). In this type of test, the axis of major principal stress is oriented in the direction 45 degrees from the vertical towards the left in the course of increasing the shear stress towards the left, but when the shear stress is reversely applied the axis of major principal stress is oriented towards the right 45 degrees from the vertical, as illustrated in Fig.3(b). Therefore, there occurs an abrupt rotation by 90 degree in the axis of the major principal stress, each time the direction of the shear stress is reversed during the cyclic loading.

3. Cyclic Test with Half-Circular Rotation of the Principal Stress Axes

In this type of test, both the shear stress, τ_{vh}, and the stress difference,

(σ_v - σ_h)/2, were changed so that the deviator stress, (σ_1 - σ_3)/2, as defined by Eq.(3), was held constant during the cyclic excursion of shear stress along the half-circle stress path as explained in Fig.3(c). The isotropically consolidated test specimen was first subjected to a stress difference by increasing the axial stress. This state of stress is represented by point A in Fig.3(c). Then, the shear stress, τ_{vh}, was gradually increased while reducing the axial stress, until the component of the stress difference was completely taken over by the shear stress component. During this phase of loading, the stress point moves along section of a cycle from point A to B in Fig.3(c). Upon reaching the state of purely torsional mode of shear, unloading in the torsional mode was executed simultaneously with reloading in the triaxial mode of stress application in such a way that the deviator stress was maintained constant. Therefore, the stress point in Fig.3(c) was taken back to the initial point A. Then, the torsional stress was increased in the opposite direction while reducing the stress difference, whereby the stress point moved to point D along the circle. The torsional stress was again reduced, while increasing the stress difference, to bring the test specimen back to the initial state of triaxial compression. The excursion of loads controlled as above was repeated until the test specimen deformed to a torsional shear strain of approximately 5 percent. Lastly, the specimen was brought to a state of zero deviator stress by removing the axial stress in the state of zero torsional stress. It is to be noted that in the course of a half circular excursion of the stress change as above, the direction of principal stress rotates continuously through an angle of 90 degrees according to Eq.(2), while the deviator stress as defined by Eq.(3) is kept at a constant level.

In the test conditions employed in the present study, the intermediate principal stress, σ_2, being applied to the hollow cylindrical sample is considered to be equal to the cell pressure which is kept constant throughout the cyclic loading. Therefore, the magnitude of major and minor principal stresses changes relative to the intermediate principal stress during the cyclic loading. However, since the cyclic changes in the deviator stress components, σ_1 - σ_2, and σ_2 - σ_3, take place on one side of loading under the conditions of $\sigma_1 - \sigma_2 \geqq 0$ and $\sigma_2 - \sigma_3 \geqq 0$, the plastic component of deformation caused by these complementary stress alteration can generally be deemed as small. Therefore, their effects on the pore water pressure build-up and residual strains will be neglected in the following presentation.

RESULTS OF TESTS

1. Cyclic Triaxial Shear Test

A test specimen was first consolidated isotropically under a confining

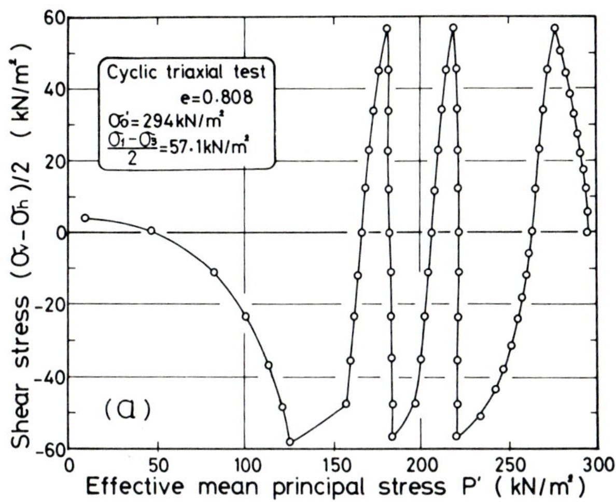

Fig. 4(a). Effective stress path in the cyclic triaxial test (Undrained test)

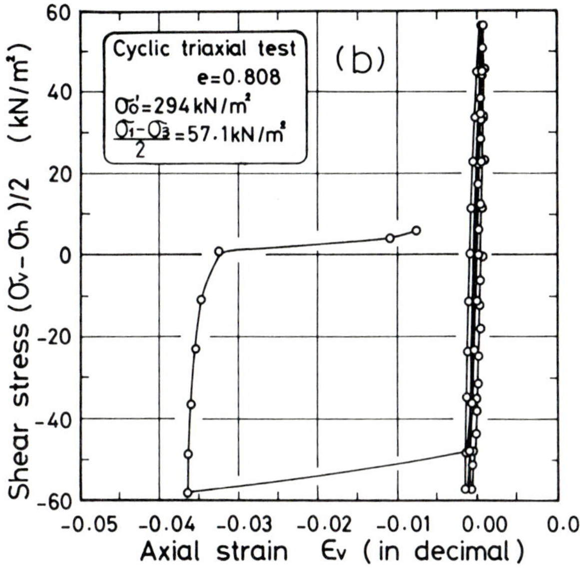

Fig. 4(b). Stress-strain relation in the cyclic triaxial test (Undrained test)

pressure of $\sigma_0' = 294$ kN/m^2 to a loose state of packing with a void ratio of e = 0.808. Then, the vertical stress was applied cyclically with a single amplitude of $\sigma_v = 114.2$ kN/m^2 under undrained conditions, while keeping the horizontal stress constant. In this test condition, the single amplitude of cyclic deviator stress was $(\sigma_1 - \sigma_3)/2 = 57.1$ kN/m^2 as defined by Eq.(3). The results of this test are shown in Fig.4. On the basis of the measured pore water pressures, the effective mean principal stress $p' = (\sigma_v' + 2\sigma_h')/3$ was obtained and used to construct the effective stress path during the cyclic stress application as shown in Fig.4(a), where the prime above the notations indicates the effective stress. The figure shows that the pore water pressure is generated significantly during the first cycle and much more remarkably during the last cycle. In the first cycle, the test specimen is in a virgin state and the loading involves development of plastic yielding all the way through. Consequently, the generation of the pore water pressure is greater in the first cycle as compared to the subsequent cycles. In the last cycle, the ratio of the stress difference to the existing effective mean principal stress exceeds, on the side of triaxial extension, a threshold value which is referred to as the angle of phase transformation. It has been shown in the triaxial tests on solid cylindrical samples (Ishihara et al, 1975) that, once a loose sand is brought to a state of stress above this threshold value in the stress space, the loose sand tends to develop a considerable amount of pore water pressure upon unloading and subsequent loading on the opposite side (triaxial compression or triaxial extension). Exactly the same characteristic behavior is observed, as shown in Fig.4(a), for the last cycle of the present test employing the hollow cylindrical specimen. In conformity to the sharp rise in the pore water pressure in the last cycle, the sand specimen reduces its stiffness and produces a large amount of axial strain of 3.7 percent on the triaxial extension side, as shown in the stress-strain plot of Fig.4(b).

2. Cyclic Torsion Shear Test

One of results of this type of tests is presented in Fig.5. In this particular test, a specimen was isotropically consolidated to a void ratio of $e = 0.809$ under the same confining pressure as before. The torsional shear stress with a single amplitude of $\tau_{vh} = 60.6$ kN/m^2 was applied cyclically under undrained conditions. Therefore, the single amplitude of cyclic deviator stress was $(\sigma_1 - \sigma_3)/2 = 60.6$ kN/m^2 as computed from Eq.(3). The stress path plotted in Fig.5(a) shows the same general characteristics as in the case of the cyclic triaxial shear test as demonstrated in Fig.3(a). The stress-strain plot in Fig.5(b) also shows the same characteristic trend as in the case of the cyclic triaxial shear test. The only difference in the cyclic behavior between these two types of tests is that as many as 8 cycles were necessary for the torsion type test to bring about a state of liquefaction, whereas only 3 cycles were sufficient to cause liquefaction in the sand specimen in the triaxial type of test, notwithstanding the larger deviator stress amplitude employed in the torsion shear test. One of the possible reasons for this difference would be anisotropic behavior of the specimen which tends to induce failure at a smaller shear stress ratio on the side of triaxial extension than that on the triaxial compression side.

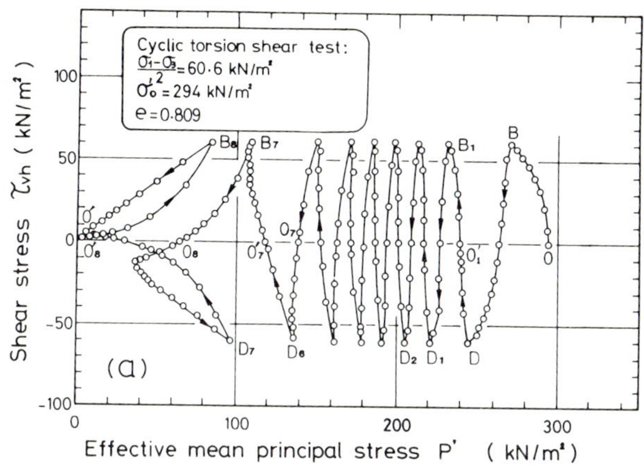

Fig. 5(a). Effective stress path in the cyclic torsion shear test (Undrained test)

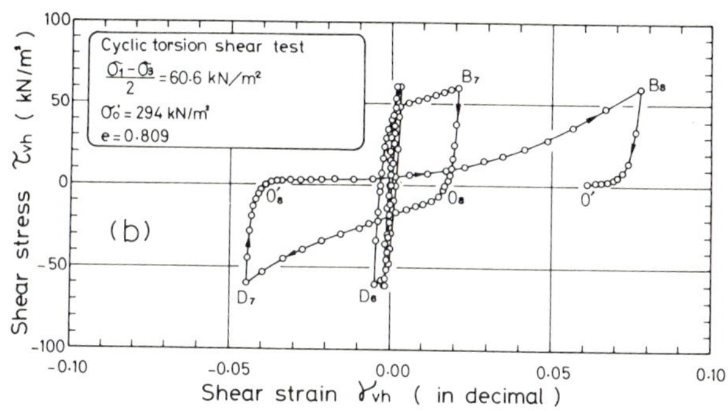

Fig. 5(b). Stress-strain relation in the cyclic torsion shear test (Undrained test)

3. Cyclic Test with Half-Circular Rotation of the Principal Stress Axes

A test specimen was first consolidated to a void ratio of e = 0.792 under a confining pressure of σ_0' = 294 kN/m^2. A vertical stress of σ_v = 120.8 kN/m^2 was then applied under undrained conditions to produce a deviator stress of 60.4 kN/m^2 in the specimen. The alternative interchange of the stress difference, (σ_v - σ_h)/2, and the shear stress, τ_{vh}, was continued cyclically, whereby producing the rotation of the principal stress directions between 45° and - 45° to the vertical, as illustrated in Fig.6. The result of such a test is demonstrated in Fig.7. The change in effective confining stress resulting from the pore water pressure build-up is shown in Fig.7(a) versus the angle of

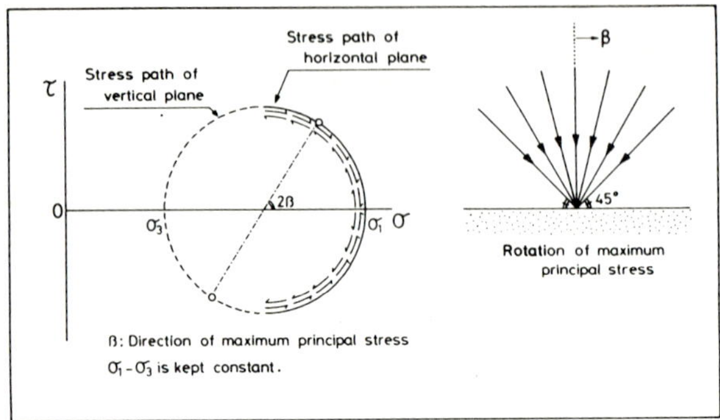

Fig. 6. Illustration for the semi-circular rotation of the principal stress directions

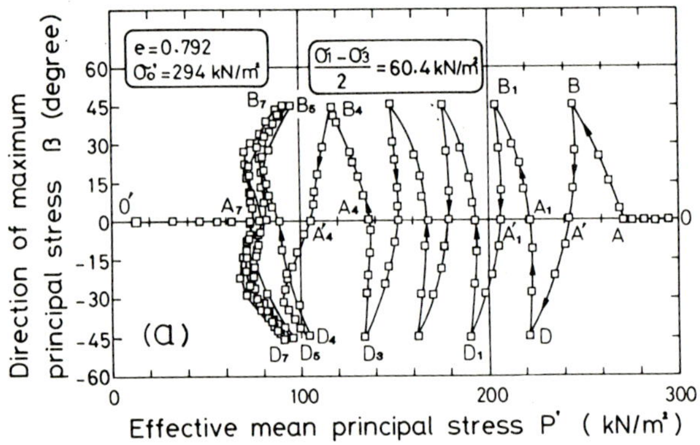

Fig. 7(a). Changes of effective mean principal stress due to rotation of the principal stress directions (Undrained test)

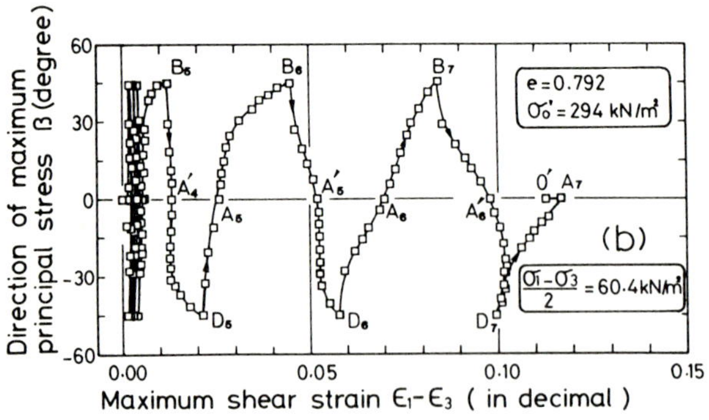

Fig. 7(b). Changes of deviator strain due to rotation of the principal stress directions (Undrained test)

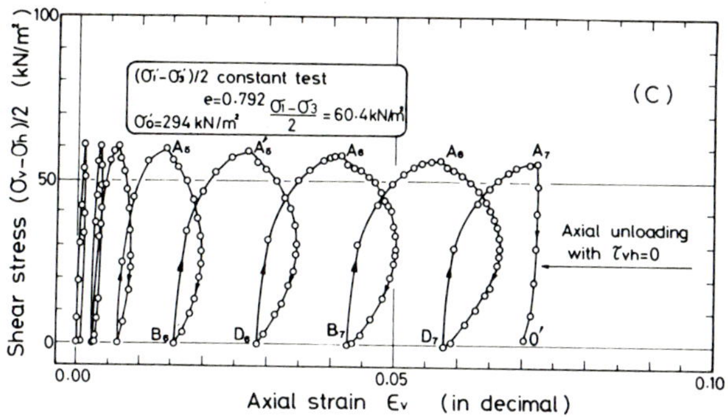

Fig. 7(c). Stress difference versus axial strain plot (Undrained test)

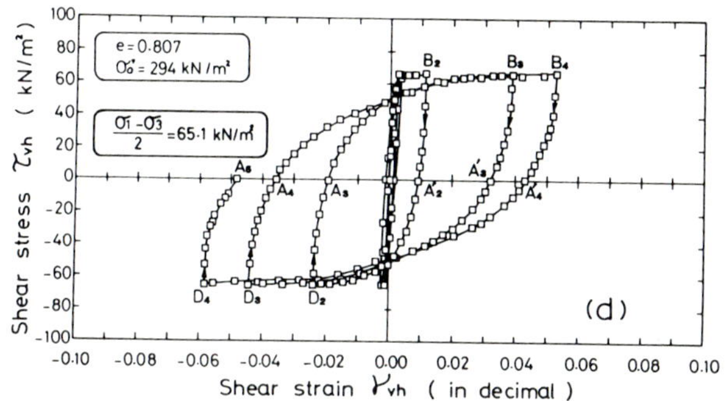

Fig. 7(d). Stress-strain relation in the torsional shear mode (Undrained test)

major principal stress, β, defined by Eq.(2). Fig.7(a) shows that, upon application of the initial stress difference, a pore water pressure of about 25 kN/m^2 was generated, as indicated by point 0 to A in Fig.7(a). In the course of the subsequent interchange of stresses from a state of triaxial compression to a state of simple shear, a pore water pressure of 25 kN/m^2 was also produced, as indicated by point A to B in Fig.7(a). During the process way back to the initial state of the triaxial compression, the pore water pressure stayed almost unchanged, but in the subsequent loading phase executed towards the opposite direction, about 20 kN/m^2 of pore water pressure is shown to have developed as represented by point A' to D in Fig.7(a). After several cycles of load application, the pore water pressure ceased to build up, staying at a value about 78 percent of the initial confining pressure. Then, as soon as the stress difference was removed, the pore water pressure built up to a value of the initial confining pressure and a state of initial liquefaction was induced in the specimen.

On the basis of the strain components measured during the cyclic tests, the deviator strain, $\varepsilon_1 - \varepsilon_3$, was computed using Eq.(4). By definition, the deviator strain, $\varepsilon_1 - \varepsilon_3$, may be visualized as representing an absolute value of the strain consisting of two components and with such an interpretation it may be of interest to observe how the deviator strain, $\varepsilon_1 - \varepsilon_2$, built up during the cyclic rotation of the principal stress axes. The variation of the deviator strain obtained in this test is presented in Fig.7(b) versus the rotational angle of the major principal stress axis. In comparison with the pore water pressure behavior shown in Fig.7(a), it can be seen that the deviator strain began to develop at a faster rate when the test specimen experienced the reduction in effective confining stress to about 30 percent of the initial confining pressure. The axial strains measured in the same test are plotted in Fig.7(c) versus the stress difference. It should be noted here that the axial strain under this type of loading condition is of the nature that can build up as the progression of cyclic load application. It may be seen in this figure that the rate of the axial strain build-up was accelerated after three and a half cycles of load application. This is consistent with the behavior of the deviator strain, $\varepsilon_1 - \varepsilon_3$, as discussed above. It is of interest to observe in this figure that the axial strain tends to increase at the early part of unloading from a state of triaxial compression. This could happen because the shear strain was simultaneously increased in this phase of unloading. The stress-strain plot in the torsional mode of deformation obtained in the same test is shown in Fig. 7(d). In contrast to the behavior of the axial strain, the shear strain, γ_{vh}, does not exhibit the build-up characteristics. It only changes cyclically around the zero point. It can be observed in Fig.7(d) that the amplitude of the shear strain began to increase considerably after three and a half cycle of load application. Such large shear strains, combined with the large axial strains, contributed for the development of the large deviator strain as demonstrated in Fig.7(b).

It has been shown in Fig.4 that, in the conventional cyclic triaxial shear test, the axial strain begins to increase sharply when the developed pore water pressure reaches a value equal to about 60 percent of the initial confining pressure. Exactly similar behavior was observed in the strain development for the case of the cyclic torsion shear test as shown in Fig.5. It should be noticed in Figs.4 and 5 that, when unloaded from a state of large shear strains at which the pore water pressure stays around 60 percent of the initial confining pressure, the test specimen generated a considerable amount of pore water pressure and a state of initial liquefaction could immediately occur. Similar mechanism in conformity to the above appears to be functioning also in the last stage of the present cyclic loading test in which the principal stress axes is rotated. At the end of the last cycle, the magnitude of the axial strain was as large as 7.5 % as indicated by point A_7 in Fig.7(c). Therefore, an amount of pore water pressure great enough to induce the initial liquefaction had been generated immediately after unloading from the state of triaxial compression, as indicated by point A_7 to O' in Fig.7(a).

It is of interest to notice in Fig.7(a) that the effective confining stress being retained during the last few cycles changes slightly depending upon the direction of the principal stresses. In view of the large shear strain both in the axial and in the torsional mode being imposed during the execution of the last few cycles, it may be appropriate to consider that the test specimen had been in a state of near-failure during these phases of cyclic loading. Viewed in this manner, the retained effective confining stress as mentioned above can be regarded as the effective confining stress required to mobilize the shear stress at failure which is in turn equal to the amplitude of the deviator stress being applied to the specimen in the present test. Since the shear stress at failure is a fixed value independent of the direction of the principal stresses, the slight variation of the retained effective confining stress as pointed out above would imply the fact that the angle of internal friction of the sand at failure could vary depending upon the direction of the principal stresses. As illustrated in Fig.8, the smaller the retained effective confining stress, the larger the corresponding angle of internal friction. The test results shown in Fig.7(a) indicates that the angle of internal friction takes its maximum value when the axis of the major principal stress is oriented approximately 25 degrees to the vertical, and its minimum value is encountered at $\beta = \pm 45°$ when the specimen is sheared in the purely torsional mode.

A similar test was conducted on another specimen of loose sand compacted to a void ratio of e = 0.807 with an amplitude of the deviator stress now increased to $(\sigma_1 - \sigma_3)/2 = 65.1$ kN/m^2. The results of the test are shown in Fig.9. Because the amplitude of the deviator stress was greater than employed in the previous test, the number of cycles required to cause failure strains was smaller as compared to the corresponding number of cycles in the previous test. The characteristic features of the

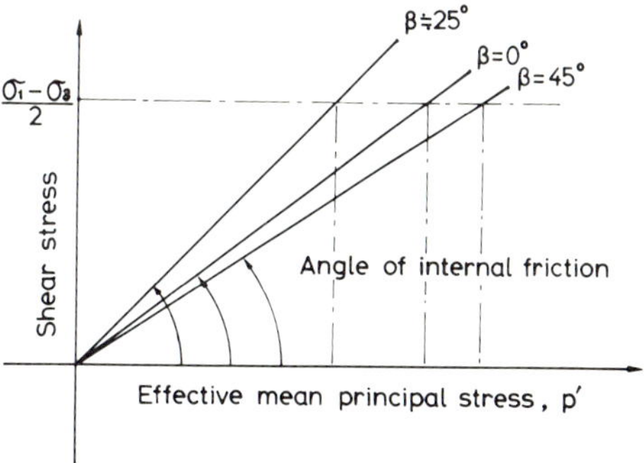

Fig. 8. Illustration for the angle of internal friction as influenced by the direction of the major principal stress

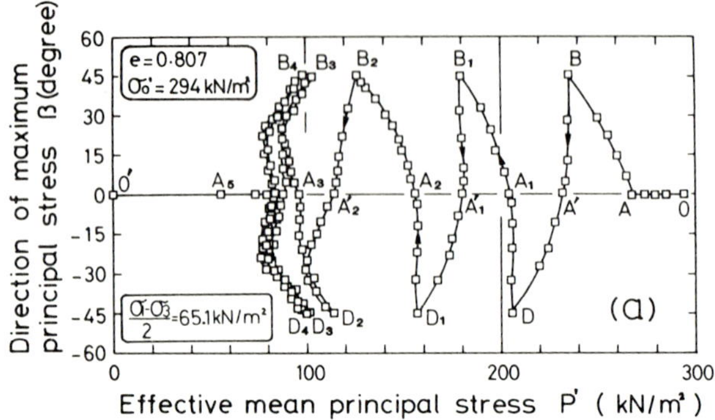

Fig. 9(a). Changes of effective mean principal stress due to rotation of the principal stress directions (Undrained test)

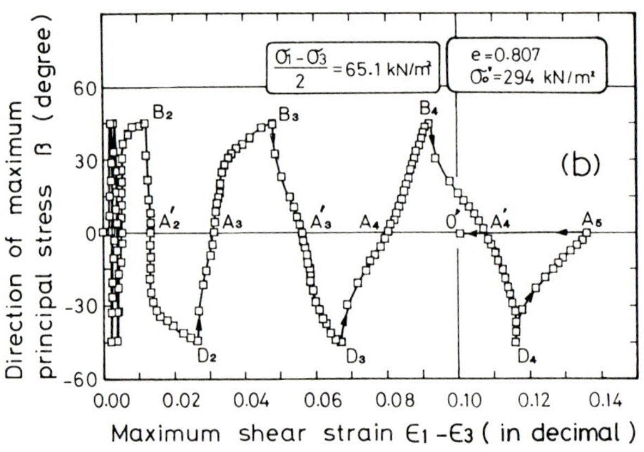

Fig. 9(b). Changes of deviator strain due to rotation of the principal stress directions (Undrained test)

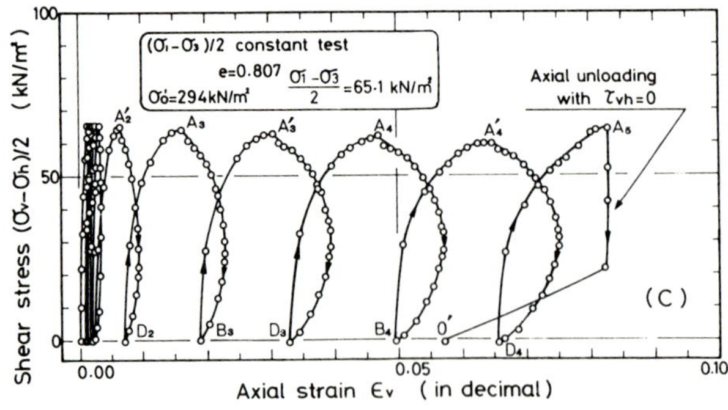

Fig. 9(c). Stress difference versus axial strain plot (Undrained test)

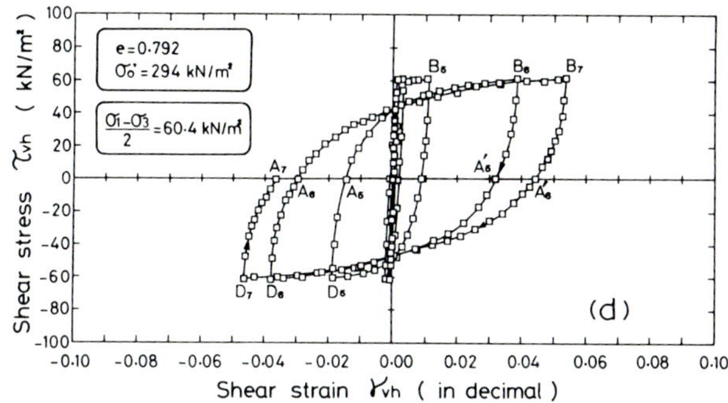

Fig. 9(d). Stress-strain relation in the torsional shear mode (Undrained test)

deformation are exactly the same as were noticed in the previous test result.

In order to obtain a better insight into the mechanism of deformation, a drained test was carried out on a loose sand specimen in which the direction of the principal stress was rotated continuously so that the stress path goes around the half-circle as shown in Fig.3(c). The test specimen was consolidated isotropically under a confining pressure of σ_0' = 147 kN/m^2 to a void ratio of e = 0.792 and then subjected to a vertical stress of σ_v = 153.4 kN/m^2 under drained conditions. In this state of stress, a deviator stress of (σ_1 - σ_3)/2 = 76.7 kN/m^2 was applied to the specimen in the mode of triaxial compression corresponding to point A in the stress space of Fig.3(c). Next, the torsional stress was increased while reducing the vertical stress so as to keep the deviator stress constant all the way through. After reaching a state of purely torsional stress application, the loading was continued in the opposite direction coming down to the stress point D via point A in Fig.3(c). The stress application was continued back and forth cyclically until the specimen deformed to a strain on the order of 1 percent. The results of this test are presented in Fig.10. Fig.10(a) plotting the volumetric strain versus the shear stress shows that the volume contraction occurred continuously during the rotation of the principal stress direction even though the deviator stress was maintained constant. The performance of strain is demonstrated in Fig.10(b) in which half of the torsional strain, $\gamma_{vh}/2$, is plotted versus the strain difference, (ε_v - ε_h)/2. It may be seen that, while the torsional strain changes back and forth around the zero strain, the component of the strain difference has a nature of increasing progressively as the cyclic loading goes on. Fig.10(b) shows that the amplitude of the torsional strain tends to dwindle and the rate of development of the strain difference also becomes smaller as the cyclic loading

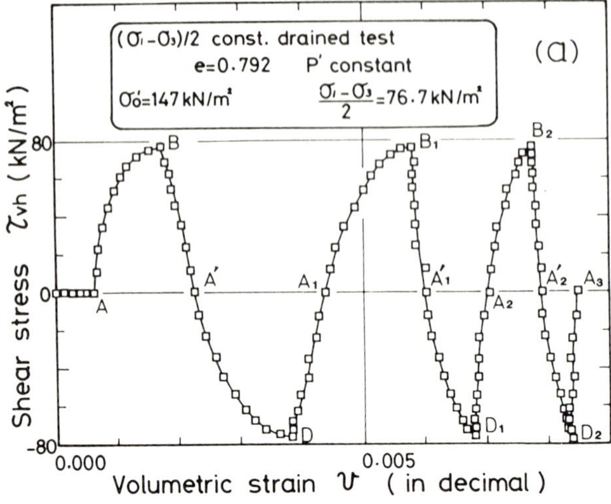

Fig. 10(a). Torsional shear stress versus volumetric strain (Drained test)

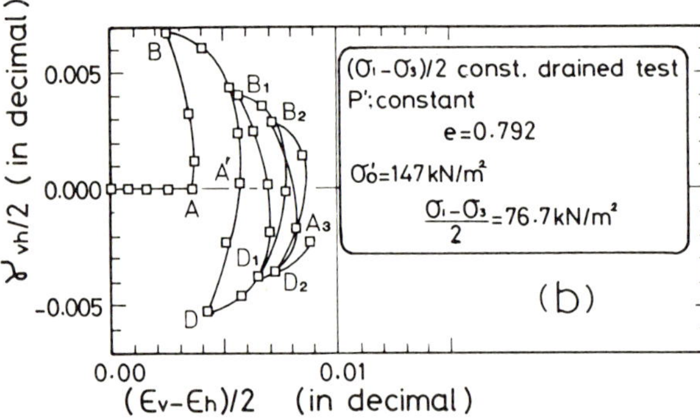

Fig. 10(b). Torsional shear strain versus strain difference (Drained test)

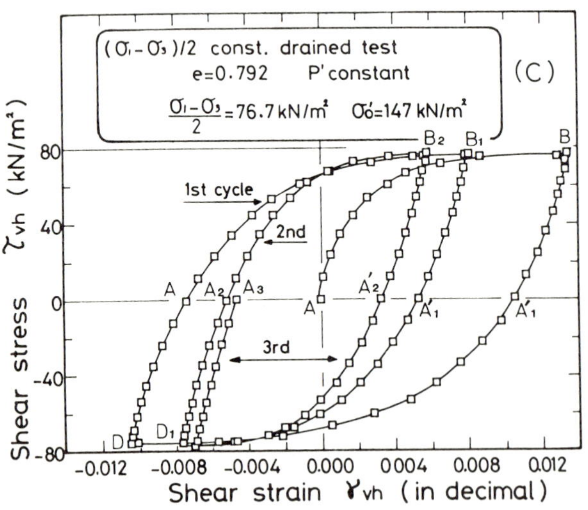

Fig. 10(c). Stress-strain relation in the torsional shear mode (Drained test)

proceeds. The stress-strain behavior in the torsional mode is shown in Fig.10(c), where it can be seen that the amplitude of torsional strain tends to diminish as the cycle proceeds.

PLASTIC DEFORMATION DURING THE ROTATION OF THE PRINCIPAL STRESS AXES

The results of a series of undrained and drained cyclic loading tests on saturated loose sand have shown that the pore water pressure and the volumetric strain can build up during the rotation of the principal stress axes even when the amplitude of the deviator stress is held constant throughout the cyclic loading test. This fact would appear to indicate that the plastic irrecoverable deformation can occur in the sand in any loading phase involving rotations of the principal stress directions.

In order to clarify this aspect of deformation, consider the established relation between the increment of shear stress and increment of shear strain for the deformation of an elastic body. Expressed in a vector form, the increments of the deviator strain, $d(\varepsilon_v - \varepsilon_h)/2$ and shear strain, $d\gamma_{hv}/2$, are related, respectively, with the increments of the deviator stress, $d(\sigma_v - \sigma_h)/2$, and shear stress, $d\tau_{vh}$, as follows,

$$d\begin{pmatrix} \dfrac{\varepsilon_v - \varepsilon_h}{2} \\ \dfrac{\gamma_{hv}}{2} \end{pmatrix} = \frac{1}{2G} \cdot d\begin{pmatrix} \dfrac{\sigma_v - \sigma_h}{2} \\ \tau_{vh} \end{pmatrix} \tag{6}$$

where G denotes the shear modulus. When the two components of shear strain are represented superimposed on the stress space consisting of $(\sigma_v - \sigma_h)/2$ and τ_{vh} as coordinates, then what is meant by Eq.(6) is that a stress increment vector with components, $d(\sigma_v - \sigma_h)/2$ and $d\gamma_{vh}$, is oriented in the same direction as the strain increment vector formed by $d(\sigma_v - \sigma_h)/2$ and $d\gamma_{vh}/2$. Therefore, the deformation which is elastic in nature is characterized by the parallelism of the strain increment vector and stress increment vector. In contrast to this, if the deformation is perfectly plastic, it is well known that the shear strain increment vector is oriented in the same direction as the vector of the existing shear stresses.

From the results of the drained test shown in Fig.10, two components of shear strain increment $d(\varepsilon_v - \varepsilon_h)/2$ and $d\gamma_{hv}/2$, were determined for several steps of cyclic loading during the rotation of the principal stress axes. The shear strain increments thus determined are shown in Fig.11 in a form of vectors laid off at the tip of the existing shear stress vectors. In Fig.11 the strain increment vectors emerging from the points marked by open circles indicate the strain increment obtained in the virgin loading of the first cycle, whereas the

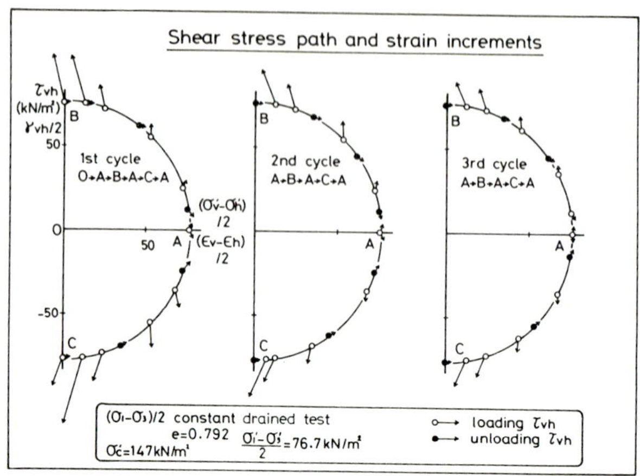

Fig. 11. Strain increment vectors plotted superimposed on the stress space (Drained test)

strain increment vector pertaining to the unloading phase is designated by the vectors with solid circles. It may be seen in Fig.11 that the strain increment vectors are neither parallel to the vector of stress increment nor oriented in the same direction of the current stress vectors. Consequently, it may be mentioned that the deformation of the sand is neither purely elastic nor competely plastic. The deformation contains both elastic and plastic components. A scrutiny of the picture of the vectors in Fig.11 appears to indicate further that the strain increment vectors tend to deviate outwardly from the stress increment vector significantly in the virgin loading phase, but at the beginning of the unloading near the point of stress reversal the direction of the strain increment vectors becomes almost perpendicular to the direction of the current stress vectors. This implies that in the virgin loading phase the plastic component is more dominant, but at the beginning of the stress reversal, the deformation is predominantory of elastic nature. In order to visualize this feature of deformation characteristics more quantitatively, other plots are provided as shown in Fig.12 where the orientations of the stress increment vector, strain increment vector and the current stress vector are plotted versus the torsional shear strain, $\gamma_{vh}/2$. The shear strain, $\gamma_{vh}/2$, in this figure may be viewed as a parameter indicating the progression of the cyclic loading. The figure on the left side shows the changes in the direction of the three vectors during the first cycle, whereas the variations of the vector direction in the third cycle of loading are shown on the right figure. It can be seen more clearly in Fig.12 that during the virgin loading in the first cycle the strain increment vectors are oriented rather closely to the stress vectors, but at the

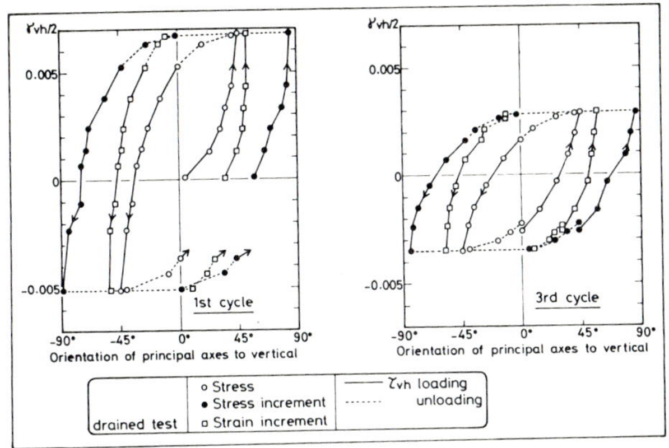

Fig. 12. Changes of direction of the principal axes of stress, stress increment and strain increment vectors during the cyclic loading

beginning of the stress reversal the orientation of the strain increment vector coincides with the stress increment vector.

In the theory of soil plasticity, it has been customarily assumed that the components of the plastic strain increment, $d\varepsilon_{ij}$ can be expressed in terms of the hardening function, h, plastic potential function, g, and yield function, f, as follows,

$$d\varepsilon^p_{ij} = h\frac{\partial g}{\partial \sigma_{ij}}df \qquad (df > 0)$$
$$d\varepsilon^p_{ij} = 0 \qquad (df \leqq 0 \text{ or } f < 0) \tag{7}$$

where $d\varepsilon^p_{ij}$ denotes the plastic strain increment. Eq.(7) indicates that the plastic strain can occur only when stress components are changed so as to produce a positive increase in the yield function, f, starting from the stress lying on the most recent yield surface. Otherwise the deformation is regarded as being elastic. In developing the plasticity theory as above, it has often be assumed that the yield function, f, is a function of the principal stresses. For example, in the two-dimensional plane strain condition, the yield function for cohesionless materials based on the Extended Tresca type failure criterion is written as,

$$f(\sigma_1', \sigma_3') = \frac{\sigma_1' - \sigma_3'}{\sigma_1' + \sigma_2' + \sigma_3'} \tag{8}$$

Other yield functions proposed by many investigators are also expressed as functions of only the principal stresses. When the yield function is specified in terms of the principal stresses alone, it is apparent that there occurs no change in the yield function if stress components are changed so as to maintain the principal stresses constant. In such conditions the theory of plasticity as formulated in Eq.(7) indicates that df = 0 and therefore no plastic deformation should occur. This is apparently in contradiction to what was observed in the cyclic loading tests involving the rotation of the principal stress axes as mentioned in the foregoing section. To be more precise, introduce Eq.(1) into (8), then one obtains

$$f(\sigma_1, \sigma_3) = \frac{\sqrt{(\sigma_v - \sigma_h)^2 + 4\tau^2_{vh}}}{\sigma_v' + 2\sigma_h'} \tag{9}$$

In the drained test as described above, both values of the numerator and denominator of Eq.(9) were maintained constant throughout the cyclic loading test employing the rotation of the principal stress direction. Therefore, there was no change in the value of the yield function as defined by Eq.(9). Hence, the deformation should be elastic all the way around, if the theory of plasticity is to be valid. However, as manifested in Figs.10(a) and 10(b), the test results disclosed a significant amount of permanent plastic deformation taking place particularly in the strain difference component as well as in the volume contraction. From these observation, it may be argued that the plastic deformation can be induced in sand even when the increment of the yield function is zero, i.e., df = 0, if the loading scheme involves the rotation of the principal stress directions. Hence, it may be generally mentioned that any yield function defined as a function of the principal stresses alone is not adequate for describing the plastic stress-strain relation for the general case of cyclic loading involving the rotation of the principal stress directions.

CONCLUSIONS

Cyclic triaxial torsion shear tests were performed in a statical manner on loose saturated specimens of sand, both under drained and undrained conditions, using a triaxial torsion shear test apparatus. Both the stress difference (difference between the axial and horizontal stresses) and torsional shear stress were varied cyclically so that the axis of the principal stresses could be rotated continously, while maintaining the amplitude of the combined shear stress (deviator stress) constant throughout the cyclic loading test. The results of such tests indicated that the deformation of plastic nature as represented by the progressive development of pore water pressure in undrained loading and the volume contraction in drained loading could take place, even when the deviator stress was maintained unchanged, if the rotation of the principal stress axes is executed on the test specimen. On the basis of the test results as above, it was argued that the conventional plastic stress-strain relations formulated as a function of the principal stresses alone may not be valid for the case of loading conditions involving the rotation of the principal stress directions.

ACKNOWLEDGEMENTS

The manufacture of the triaxial torsional test apparatus used in this study was sponsored by the grant-in-aid of the Ministory of Education of Japanese Government. Messrs. Y. Kikuchi and S. Nakazato assisted in conducting the tests in the laboratory. Dr. Kenji Mori kindly reviewed the draft with valuable comments. The authors wish to express their gratitude to the organization and individuals as cited above.

REFERENCES

Arthur, J.R.F., Chua, K.S., Dunstan, T. and Rodoriguez, J.I. (1980), "Principal Stress Rotation : A Missing Parameter, " Proc. ASCE , Vol.106, GT.4, pp.419-433.
Arthur, J.R.F., Bekenstein, S., Germaine, J.T. and Ladd, C.C. (1981), "Stress Path Tests with Controlled Rotation of Principal Stress Directions," Laboratory Shear Strength of Soil, ASTM, STP 740, pp.516-540.
Ishibashi, I. and Sherif, M.A. (1974), "Soil Liquefaction by Torsional Simple Shear Device," Proc. ASCE, Vol.100, GT.8, pp.871-888.
Ishihara, K., Tatsuoka, F., and Yasuda, S. (1975), "Undrained Deformation and Liquefaction of Sand under Cyclic Stresses," Soils and Foundations, Vol.15, No.1, pp.29-44.
Ishihara, K. and Yasuda, S. (1975), "Sand Liquefaction in Hollow Cylinder Torsion under Irregular Excitation," Soils and Foundations, Vol.15, No.1 pp.45-99.
Ishihara, K. and Towhata, I. (1983), "Effects of Rotation of Principal Stress Directions on Cyclic Response of Sand, "Submitted to International Conference on Constitutive Laws for Engineering Materials, University of Arizona, U.S.A.

Mechanics of Granular Materials: New Models and Constitutive Relations, edited by
J.T. Jenkins and M. Satake, 1983
Elsevier Science Publishers B.V., Amsterdam — Printed in The Netherlands

DISCUSSION ON YIELD LOCI FOR SANDS

F. TATSUOKA[1] and F. MOLENKAMP[2]
[1]Institute of Industrial Science, Univ. of Tokyo, Minato-ku, Tokyo 106 (Japan)
[2]Delft Soil Mechanics Laboratory, P.O.Box 69, 2600 AB Delft (The Netherlands)

ABSTRACT

This paper presents the results of triaxial compression tests on sand using various stress paths. These tests were performed to see the effects of preloading in isotropic or triaxial compression on the shear deformation and volume change.

To account for all of the measured phenomena the existing models should be modified. A possible modification involving the implementation of the Granta Gravel type of model into a double hardening type of model is discussed.

INTRODUCTION

In order to predict the deformation of sand using an elasto-plastic model, a yield surface (or yield locus) and a plastic potential have to be defined. For the case of conventional triaxial tests, in which independent principal stresses are axial and radial, basically three different types of yield surfaces (or yield loci) have been proposed for cohesionless soils. The yield locus and the coinciding plastic potential of the Granta Gravel model, which is an elasto-plastic model for cohesionless soils proposed by Roscoe and his colleagues (1963, 1968), have a shape as shown in Fig. 1 (type 1). Using this type of yield locus and plastic potential, both the shear strains which accompany volumetric strains (dilatancy) and the volume strains due to compression can be readily calculated for any monotoneously increasing stress path. On the other hand, it has been shown experimentally by many researchers that the yield locus of the Granta Gravel model differs considerably from many measured yield loci of sands. Poorooshasb,

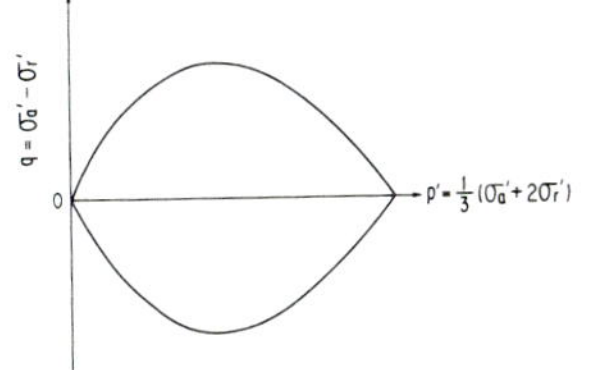

Fig. 1. Yield locus of Granta Gravel model (type 1).

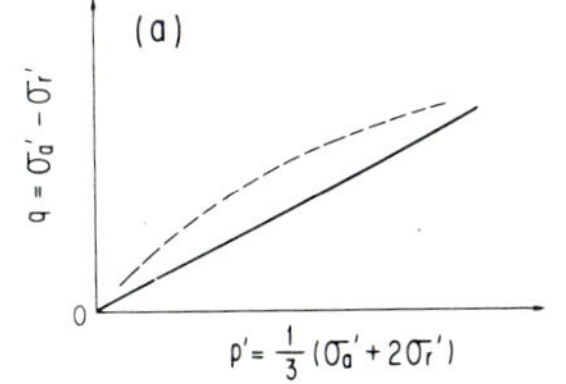

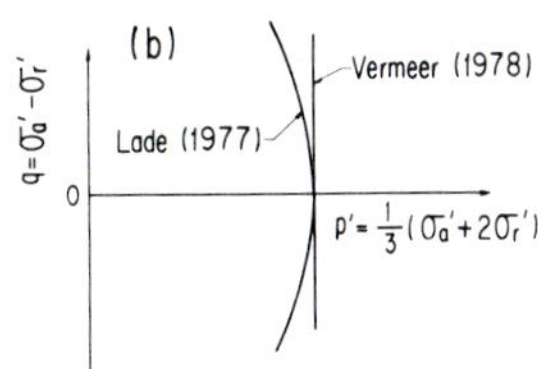

Fig. 2. Yield loci of double hardening models ((a) type 2 and (b) type 3).

et al. (1966), Cole (1967) and Barden, et al. (1969) proposed for a yield locus on the p'-q stress plane a straight line originating from the origin (p'= q = 0) and Stroud (1971) and Tatsuoka and Ishihara (1974) proposed a yield locus of a slightly curved line (Fig. 2(a)). While this kind of yield locus shown in Fig. 2 (a) is known to describe the yielding properties for shear strains and accompanying volumetric strains of pre-sheared sands, it fails to predict the yielding properties for volumetric strains due to compression. For this reason, another kind of yield locus with coinciding plastic potential (type 3) was proposed for the volumetric strains due to compression shown in Fig. 2(b). In the double hardening model (Lade (1977), Vermeer (1978), Molenkamp (1980)), these two different kinds of yield loci are used to describe the yielding properties of shear strains and dilatancy for an extending yield locus of type 2 together with the irreversible volumetric strains due to isotropic compression. In these double hardening models, it is assumed that these two kinds of yielding take place independently; there is no coupling between shear deformations with accompanying dilatancy and deformations due to isotropic compression. In reality there probably is some kind of coupling between these two different kinds of deformation.

The purpose of this paper is to discuss the coupling between shear deformation with accompanying dilatancy and deformations due to isotropic compression using the triaxial tests and employing various stress paths. This paper is a preliminary one because of the limited number of stress paths employed. For better modifications of double hardening model, additional experiments are necessary.

TEST PROGRAM AND RESULTS

The triaxial specimens were 10 cm in height and 5 cm in diameter. Only loose specimens were tested. These were prepared by spooning a de-aired mixture of fresh sand and distilled water into a mold filled with de-aired water. No other compaction efforts were used. The void ratio values of the specimens thus obtained at a confining pressure of 0.5 kgf/cm^2 (49 kN/m^2) were around 0.78 to 0.80. The test material was sand from the Fuji river bed. The grain shape was subangular, the mean diameter D_{50} was 0.41 mm, the coefficient of uniformity U_c was 2.21, the specific gravity G_s was 2.68 and the maximum and minimum void ratio values were 1.08 and 0.53 (see Tatsuoka (1973) for detail).

The stress parameters used here are the effective mean principal stress $p' = (\sigma'_a + 2\sigma'_r)/3$, and the shear stress, $q = \sigma'_a - \sigma'_r$, where σ'_a and σ'_r are the effective axial and radial stresses. The strain parameters used here are the volumetric strain, $v = \varepsilon_a + 2\varepsilon_r$, and the shear strain, $\gamma = \varepsilon_a - \varepsilon_r$, where ε_a and ε_r are the axial and radial strains, taken to be positive in compression.

Effects of deformations under isotropic stress condition on shear deformation and dilatancy

TABLE 1

List of test on pre-sheared and virgin samples.

TEST No.	STRESS PATH	VOID RATIO	SHEAR STRAIN FOR AB, γ_{AB} (%)	Δe	$\Delta\gamma$ (%)
1	ABCDD'	e_A=0.790	4.2	-0.006	-0.2
2	ABCEE'	e_A=0.794	3.9	-0.018	-0.5
3	ABCFF'	e_A=0.795	4.3	-0.034	-0.9
4	ABCGG'	e_A=0.785	4.2	-0.055	-1.6
5	ADD'	e_D=0.784	0.0	—	—
6	AEE'	e_E=0.762	0.0	—	—
7	AFF'	e_F=0.743	0.0	—	—
8	AGG'	e_G=0.711	0.0	—	—

Tests 1 through 4; Pre-sheared as ABC.
Tests 5 through 8; No pre-shearing.

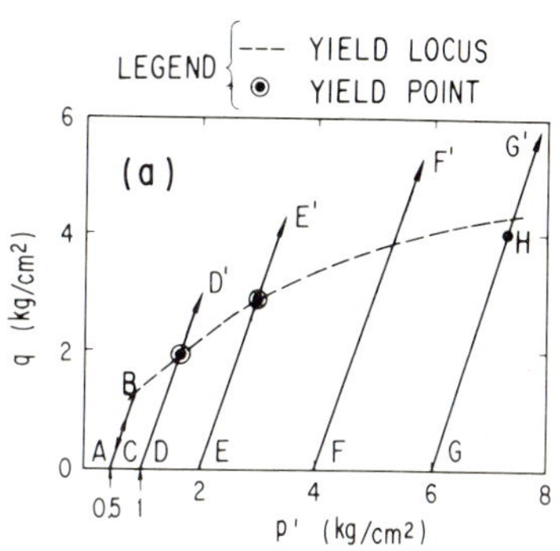

Fig. 3(a). Stress paths for Tests 1 through 8 to evaluate effects of isotropic compression on deformation during shear re-loading.

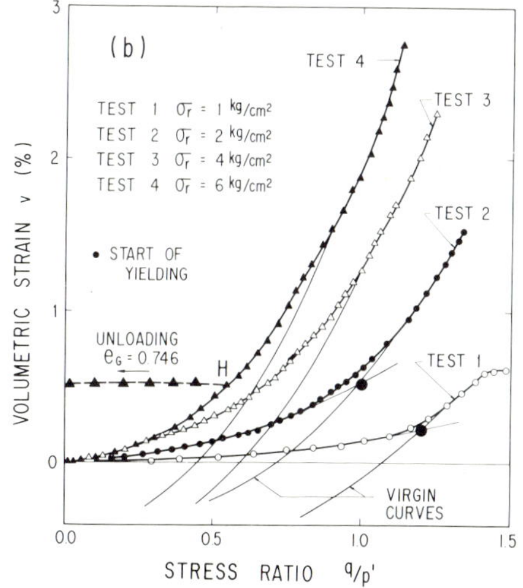

Fig. 3(b). Volumetric strain versus stress ratio relationships of pre-sheared and virgin samples.

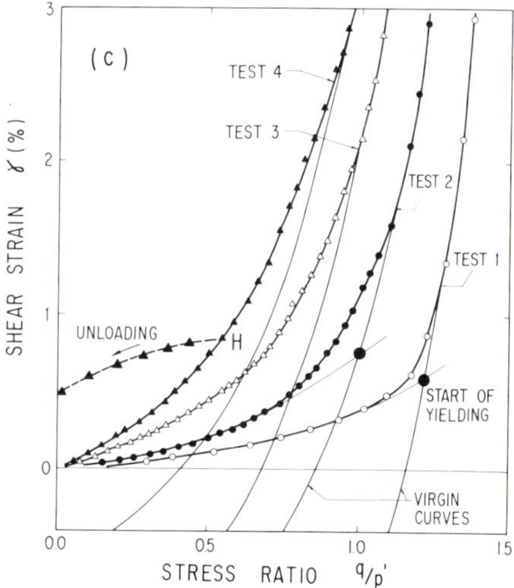

Fig. 3(c). Shear strain versus stress ratio relationships of pre-sheared and virgin samples.

Tests 1 through 8 were performed to clarify the effects of isotropic compression on the deformation during shear reloading (see Table 1). In tests 1 through 4, each sample was loaded and unloaded along the common stress path ABC (σ'_r was 0.5 kgf/cm^2 (49 kN/m^2) and q/p' at B was 1.32). Then each sample was compressed isotropically to the different values of p' and reloaded as DD', EE', FF' and GG'. For references, Tests 5 through 8 with virgin loading were performed. In these the stress path ABC was not included, but the material was isotropically compressed directly to D, E, F or G and then sheared. The values of γ_{AB} listed in Table 1 are shear strains observed at the stress point B. The values of γ was defined as zero at the stress point A. The values of Δe and $\Delta\gamma$ are changes in void ratio and shear strain observed during isotropic compressions between the first and second shear loadings. The stress-strain curves during reloading of pre-sheared samples are compared with virgin curves in Figs. 3(b) and (c) with the value of shear strain at the start of reloading being defined as zero. In these figures, the origins of curves are shifted along the ordinates so that the virgin curves overlap the stress-strain curves of pre-sheared samples for stress ratio values larger than the maximum value for preloading. Tatsuoka and Ishihara (1974) performed a series of triaxial compression tests as Tests 1 through 4 to determine the shape of yield loci of pre-sheared samples of this test sand. The yield locus of type 2 which passes through B shown in Fig. 3(a) was obtained following Tatsuoka and Ishihara (1974).

From the test results shown in Figs. 3(b) and 3(c) the following points can be noticed. (1) The point where yielding starts becomes less clear with the increase in change of p' during isotropic consolidation. In particular, it is rather difficult to determine the point where yielding starts in Tests 3 and 4. This behavior is probably due to the change in the fabric during the isotropic compression; namely, the fabric at the stress point B had a certain resistance against shear deformation and this fabric was changed during the isotropic compression. As seen from Table 1, the shear strain values which were induced during the isotropic compression between the first and second shear loadings were negative. This fact supports the decrease in resistance of the fabric against shear deformation during the isotropic compression. Therefore, it seems natural that the strains in the initial part of the second shear loading involves larger plastic parts in the case where deformations during the isotropic compression are larger too. The unloading curves from the point H on the reloading curves for Test 4 shown in Figs. 3(b) and (c) were obtained by using another specimen. As shown in Fig. 3(a), the stress point H is located slightly below the yield locus of type 2. It is clearly seen from Figs. 3(b) and (c) that there are large plastic strains during reloading for Test 4 even up to the stress point below the yield locus of type 2. (2) For Test 4 the decrease in void ratio during the isotropic compression (CG) was 0.055 which was much larger than the decrease in void ratio during the stress path ABC of 0.021. This fact above means that since a large plastic strain occured during the isotropic compression CG, following the Granta Gravel model, the stress point G is located well out of the instantaneous yield locus which involves the stress point B and located on the present yield locus. Therefore, the deformation properties from G should be of a virgin sample when the yield locus of type 1 is applicable. However, as shown in Figs. 3(b) and (c), the deformation properties along GG' observed in Test 4 were not identical to those observed along GG' in the virgin sample. Therefore, it is apparent that void ratio or plastic volumetric strain is not the state parameter that it is assumed to be in the Granta Gravel model.

In summary, it seems that the fabric which is formed during shear deformation is changed by the plastic deformations during the following isotropic compression and the resistance against shear deformations during shear reloading is reduced to some extent with the increase in the amount of deformation during the isotropic compression.

Five overconsolidated samples were tested under the triaxial compression stress condition (see Table 2). The test results were compared with those of virgin (normally consolidated and unpresheared) samples. Shown in Fig. 4 is a typical relationship between void ratio and effective mean principal stress p' during isotropic compression (AB), isotropic rebound (BC) and triaxial compression with a constant value of σ'_r of 2 kgf/cm^2 (196 kN/m^2) (CD) for Test 16. The value of p' at the state point B, p'_B, is the preconsolidation stress and the value of p' at the state points A and C, p'_A, is the value of σ'_r during the triaxial compression test. In the following figures, the volumetric strain v will be defined as zero at the state point A. The stress-strain relationships during the triaxial compression tests are shown in Figs. 5 and 6. In Figs. 5(b) and 6(b), the origins of overconsolidated specimens are shifted along the γ-axis so that the stress-strain curves of overconsolidated specimens for stress ratio values larger than certain values overlap the virgin stress-strain curves for the normally consolidated specimens. It can be seen that the decrease in volume change (contraction) of overconsolidated samples during the initial part of triaxial compression test is smaller than that of normally consolidated sample. The amount of decrease increases with the increase in overconsolidation ratio. It can also be seen that the shear rigidity of overconsolidated specimens is larger than that of normally consolidated sample.

The points where yielding starts were defined in the shear strain-stress ratio curves, although these points were not as clear as for Tests 1 and 2 (see Fig. 3). It is also to be noted that the stress-strain curves after the start of yielding are very similar to those of normally consolidated specimens. Furthermore, it can be seen from Figs. 5(a) and 6(a) that except for Test 14 (OCR = 13, p'_A = 0.5 kgf/cm^2) the maximum volumetric strains of overconsolidated specimens at a stress

TABLE 2

List of tests on over-and normally consolidated samples.

TEST NO.	p'_A (kgf/cm^2)	p'_B (kgf/cm^2)	OCR $= p'_B/p'_A$	e_A	e_C
11	0.5	0.5	1	0.799	0.799
12		1.0	2	0.740	0.734
13		2.0	4	0.752	0.739
14		6.5	13	0.751	0.738
15	2.0	2.0	1	0.762	0.762 *
16		4.0	2	0.744	0.731
17		6.0	3	0.756	0.732

* same with Test 6.

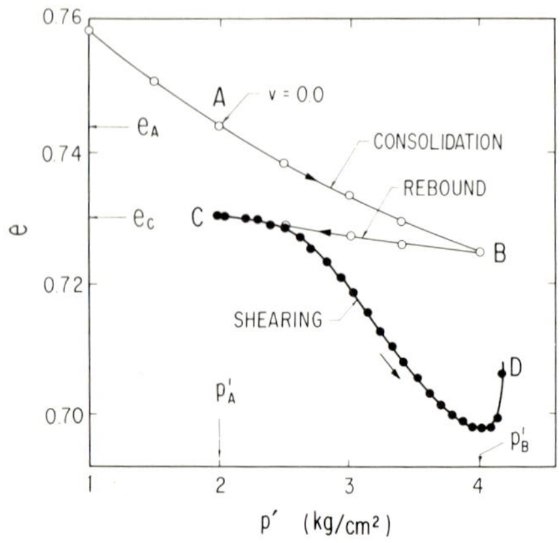

Fig. 4. Void ratio versus effective mean principal stress relationships during isotropic compression, rebound and triaxial compression for Test 16.

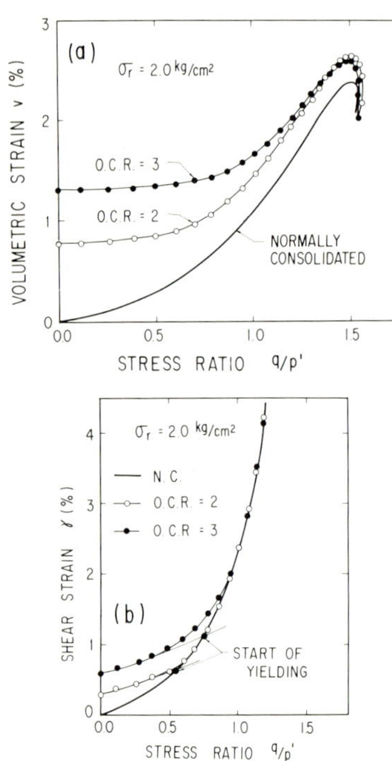

Fig. 5. (above) (a) Volumetric strain and (b) Shear strain versus shear stress at σ'_r =2.0kgf/cm^2 for virgin and overconsolidated samples.

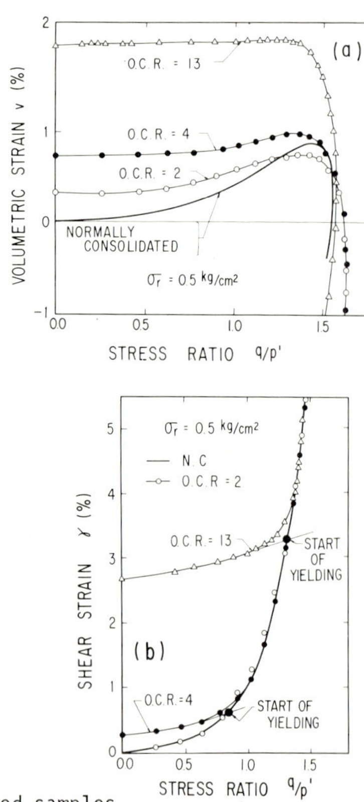

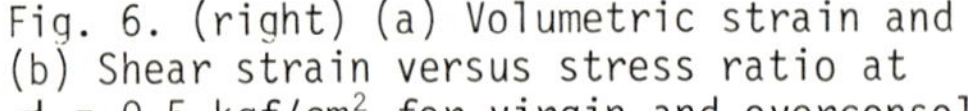

Fig. 6. (right) (a) Volumetric strain and (b) Shear strain versus stress ratio at σ'_r = 0.5 kgf/cm^2 for virgin and overconsolidated samples.

ratio q/p' of around 1.5 are very similar to those of normally consolidated specimens. This means that after yielding starts the density of overconsolidated sample becomes similar to that of normally consolidated sample.

It is apparant that, in order to describe this behavior, the yield locus of type 1 shown in Fig. 1 is more appropriate than the yield locus of type 2. The stress points where yielding began in overconsolidated specimens were plotted in Fig. 7. Also shown in Fig. 7 are the drained failure lines of normally consolidated specimens, the yield loci for pre-sheared samples which pass the yielding stress points A, B, C and D and the fractions of yield locus of overconsolidated specimens AA', CC', DD' as solid lines. The latter were constructed by connecting two stress points by slightly curved lines. It can be seen from Figs. 6(a) and (b) that even for such a large OCR of 13, the stress-strain curves for increasing volume is very similar to those of virgin samples. This may imply that an overconsolidation history can affect neither the deformation properties after the volume of sample starts increasing nor the drained strength. Therefore, it seems that the yield locus which involves the stress points B and B' for overconsolidated samples may consist of such a straight line BB" and a curved line B"B' which should have a shape similar to AA', CC' and DD'. From these observations, it is likely that the yield locus of type 1 can not totally explain the yielding properties of heavily overconsolidated specimens in which the volumetric strain due to overconsolidation (ABC in Fig. 4) is larger than the maximum positive volumetric strain during a triaxial compression test on a normally consolidated sample.

In summary, it seems that the fabric change during overconsolidation provides an increased resistance against shear deformation and volume change below a certain stress ratio where for triaxial compression still contraction occurs in samples. Therefore, it may be inferred that the fabric changes during isotropic compression and during shearing where contraction occurs in a sample have a similar aspect, probably which may be the degree of homogeneity in the sample.

It is interesting to know the effect on the yielding properties of sand due to both overconsolidation and preshearing. For this purpose, the fractions of yield loci which were obtained by using the stress paths ABCDEF (Test 31) and AGCHIJ (Test 32) as shown in Fig. 8 were compared with two kinds of yield loci (types 1 and 2). The stress-strain curves along DEF and HIJ are shown in Figs. 9(a) and 9(b). The yield stress points E and I obtained from these figures are plotted on Fig. 8. It can be seen that the yielding stress points E and I are not located on the yield loci of type 1 for samples overconsolidated to A nor on the yield loci of type 1 which involve the stress points B and G (these are not shown in Fig. 8). It may be seen that the fractions of yield loci BE and GI are similar to the yield locus of type 2, but strictly speaking, they are somewhat different from the yield loci of type 2. It can be considered that when the stress points B

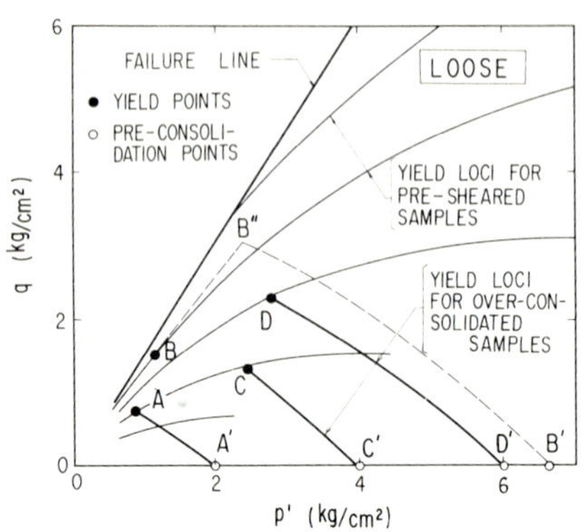

Fig. 7. Comparisons of yield loci between for pre-sheared samples (type 2) and for overconsolidated samples (type 1).

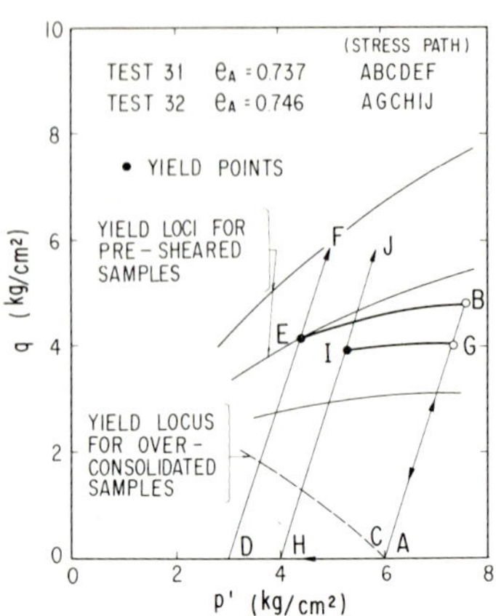

Fig. 8. Yield loci for overconsolidated and pre-sheared samples.

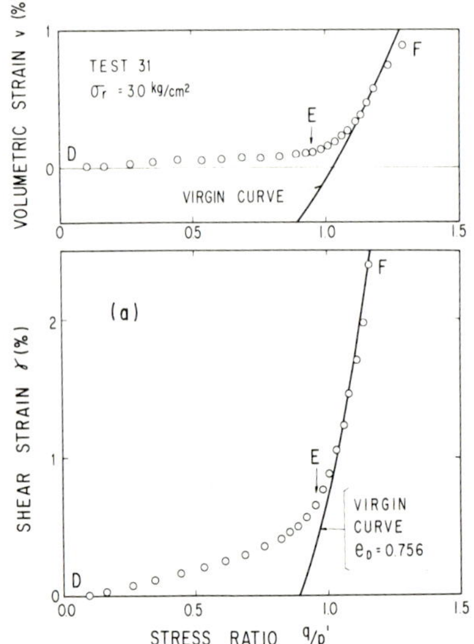

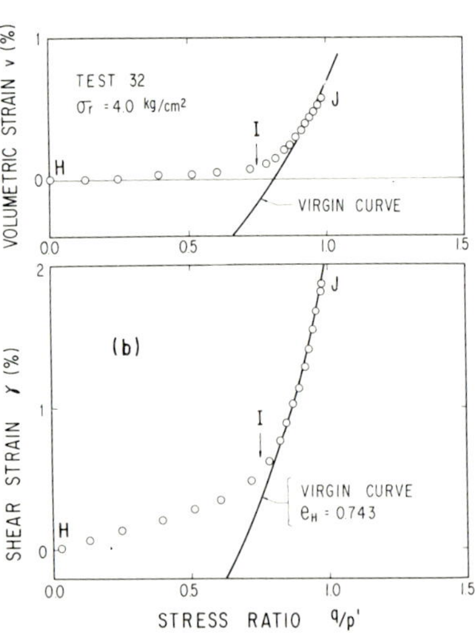

Fig. 9. Volumetric strain and shear strain versus stress ratio relationships for overconsolidated and pre-sheared samples, (a) Test 31 and (b) Test 32.

or G are located closer to the stress point A, the shapes of BE or GI become more similar to the yield locus of type 1 starting from stress points B or G and when the stress points B or G are located more further from the stress point A, the shapes of BE or GI become more similar to the yield locus of type 2 starting from the stress points B or G. This point will be discussed more in detail later.

Effects of shear deformation on deformation under the isotropic compression

To know the characteristics of coupling between the deformations due to triaxial and isotropic compression, a test was performed, the stress path of which is shown in Fig. 10(a). First, isotropic compression and unloading were performed ($1 \rightarrow 2 \rightarrow 3 \rightarrow 4$). The relationship between the void ratio e and the effective mean principal stress p' during this test is shown in Fig. 10(b). The deformation along $3 \rightarrow 4$ can be considered rather elastic. Then the sample was sheared as $4 \rightarrow 5 \rightarrow 6$ (the shear strain value at 5 was 3.1%) and again loaded and unloaded under the isotropical stress condition as $6 \rightarrow 7 \rightarrow 8$. If the yielding property of sand during isotropic loading or unloading is fully described by the yield locus of type 1 or type 3, the behavior along $6 \rightarrow 7 \rightarrow 8$ should be elastic. However, the actual behavior was not elastic but around 40% of the plastic volume change for the virgin sample took place (see Fig. 10(d)). At the same time, it is to be noted that, in addition, a considerable negative plastic shear strain was induced during the isotropic compression after shear as shown in Fig. 10(c). Then, two more cyclic loadings of isotropic compression and shearing were repeated as $8 \rightarrow 9 \rightarrow 10 \rightarrow 11 \rightarrow 12$ and $12 \rightarrow 13 \rightarrow 14 \rightarrow 15 \rightarrow 16$. The shear strain values at the stress points 9 and 13 were 10.5% and 27.5%. It can be seen from these figures that the amounts of plastic volume change and plastic shear deformation during isotropic loading after shearing never converge to zero during such cyclic loadings but increase with the increase in shear deformation during the preceding shearing.

These observations described above clearly show that the yield loci of type 2 and 3 can never be independent, but the fabric formed during the isotropic compression is changed during the following shearing. In particular, this fabric change seems larger when larger positive dilatancy occurs during shearing.

POSSIBLE MODIFICATION OF ISOTROPIC DOUBLE HARDENING MODEL

The stress path dependency of the volume change as measured could in principle be accounted for by modifying the existing isotropic double hardening models. To this end another yield surface has to be introduced which will be similar to the yield surface of type 1 for isotropically overconsolidated stress paths as is discussed next. Consider all possible stress increments (see Fig. 11) in a triaxial test on normally consolidated material starting from a stress point A situated in the region of contraction thus with a mobilized friction angle ϕ_{mob}

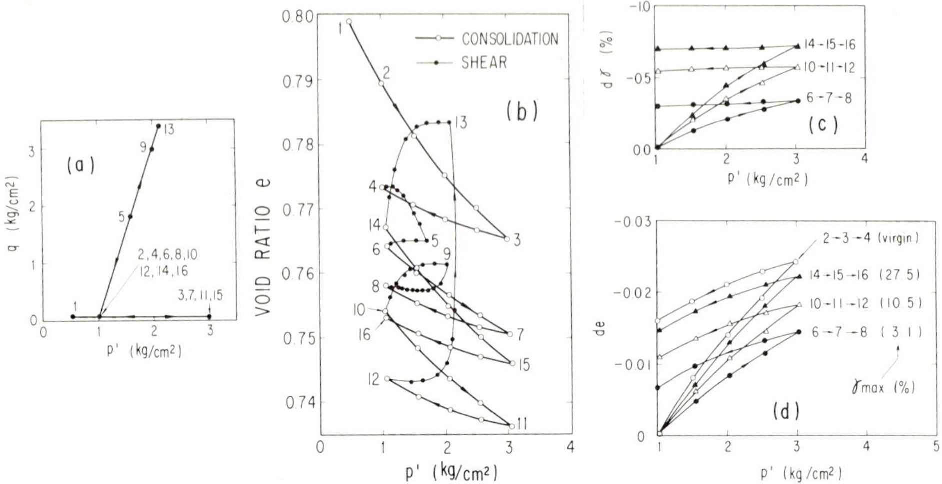

Fig. 10. (a) Stress paths, (b) Void ratio changes, (c) Shear strain changes during isotropic compression and (d) Void ratio changes during isotropic compression for test performed to see effects of pre-shearing on deformation during isotropic compression.

Fig. 11. Four regions in stress space divided by yield loci of types 2 and 3.

Fig. 12. Yield locus of type 4 for overconsolidated and pre-sheared samples.

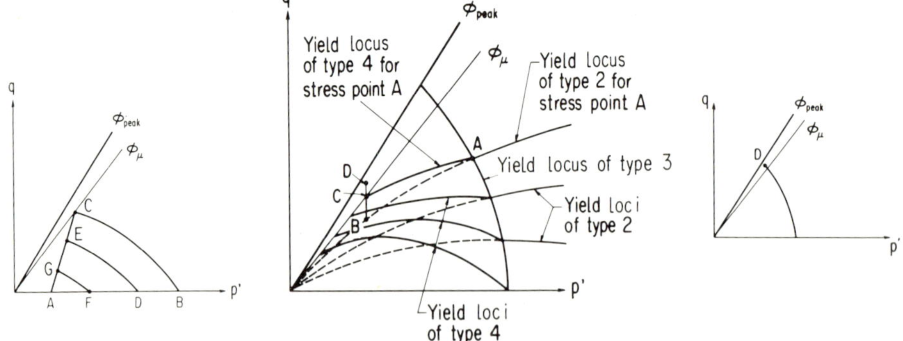

Fig. 13. Yield locus of type 4 for isotropically overconsolidated specimens.

Fig. 14. Possible shapes of yield loci of type 4 for overconsolidated and pre-sheared samples.

Fig. 15. Possible location of yield locus of type 3 moved by dilatancy.

smaller than the inter-particle friction angle ϕ_μ (Rowe, 1962) or $\phi_{mob} < \phi_\mu$. For an isotropic double hardening model two independent yield loci pass through this stress point A, namely the yield loci of types 2 and 3. These yield loci divide this stress space in 4 regions, namely;

region 1 : non-linear elastic region.

region 2 : in this region the plastic model with the yield locus of type 2 is active.

region 3 : in this region the plastic model with the yield locus of type 3 is active.

region 4 : region where both plastic models are active.

The modification needed to improve the simulation of the stress path dependency of the deformation involves a partition of region 2. In region 2 a new yield surface is defined of type 4 (see Fig. 12). An arbitrary stress point C on this yield surface can be reached along the stress path ABC (see Fig. 12), where point B is situated on the same yield surface of type 2 as stress point A while the isotropic stress p of points B and C are equal. Now the deviator stress q of point C might be defined by the condition:

$$\Delta f^4 = -\Delta V^3{}_{BA} + \Delta V^2{}_{BC} + \alpha \times \Delta\gamma^2{}_{BC} = 0 \quad (1)$$

in which:

$\Delta V^3{}_{BA}$: plastic volume change (compression) according to plastic model with yield locus of type 3 for virgin loading from point B to point A.

$\Delta V^2{}_{BC}$: plastic volume change according to plastic model with yield locus of type 2 for a load increment from point B to point C.

α : constant (positive)

$\Delta\gamma^2{}_{BC}$: plastic shear strain increment according to plastic model with yield locus of type 2 for a load increment from point B to point C.

Thus Eq. (1) divides region 2 into two different parts. In the lower part of region 2 below the newly defined yield locus of type 4 thus for

$$\Delta f^4 = -\Delta V^3{}_{BA} + \Delta V^2{}_{BC} + \alpha \times \Delta\gamma^2{}_{BC} < 0 \quad (2)$$

no plastic strains occur, so in this subregion the behavior is still elastic like in region 1. In the upper part of region 2 above the newly defined yield locus of type 4 thus for

$$\Delta f^4 > 0 \quad (3)$$

the same plastic strain increments occur as for the original double hardening models. In fact Δf^4 in Eq. (2) is a measure of the remaining plastic deformation experienced by the material during preloading which a virgin sample still has to experience during further loading along the stress path BC before the virgin sample and the preloaded sample can behave similarly again. With this modification the effects of isotropic overconsolidation and pre-shearing as shown in Figs. 4

through 10 can be accounted for in principle. For stress paths ABAC, ADAE and AFAG as shown in Fig. 13 the shape of the yield locus of type 4 can be described by Eq. (1). In this case the shape of the yield locus of type 4 will be similar to the yield locus of type 1. It should be noted that stress point C is situated on the line of maximum contraction of the normally consolidated sample (ϕ_{mob} = ϕ_μ). As shown in Figs. 6(a) and 6(b) the overconsolidation has no effect in the dilatant region ($\phi_{mob} > \phi_\mu$). To account for this property of sand the yield surface of type 4 should not extend into the dilatant region of this stress space but it should stop at the surface with $\phi_{mob} = \phi_\mu$ at least or it should stay underneath this line and reach the origin of stress space as can be expected for very loose sands which can liquefy in undrained triaxial compression. For higher shear stress levels the yield surface of type 2 should be used only.

However during loading into the dilatant region not only has the overconsolidation little effect but also its effect in the region of contraction itself seems to be reduced and finally even erased. To account for this property for loading into the region with $\phi_{mob} > \phi_\mu$ the yield surface of type 3 could be displaced to lower isotropic stress for instance by using:

$$\Delta f^3 = -\Delta V^3{}_{BA} - \beta \times \Delta V^2{}_{CD} \tag{4}$$

in which

β : constant (positive).

$\Delta V^2{}_{CD}$: plastic volume change due to dilatancy (negative because of dilatancy), point C is supposed to be situated on the line with the maximum contraction (Fig. 14).

Thus during the application of increased shear from point C to point D the dilatant volume change increases and Δf^3 becomes positive, which simulates the erasure of the effects of preloading. Thereafter the yield surface of type 3 will pass through the instantaneous stress point as illustrated in Fig. 15.

With the described modifications of the existing isotropic double hardening models the effects of preloading will be simulated much better. The described modifications seem to bridge the gap between the different kinds of yield locus, which have been mentioned before. For a proper support by experimental evidence more high quality experimental research is needed.

CONCLUSIONS

No existing constitutive model seems able to simulate the stress strain behavior for both virgin loading and loading in shear after preloading by isotropic compression; usually only one or the other is simulated.

With a combination of those different types of models better simulation for more general stress paths seems possible. More detailed high quality experimental research is needed to validate the discussed combination.

ACKNOWLEDGEMENT

The tests were performed as a part of the first author's Doctor of Engineering thesis work at the University of Tokyo under Professors Mogami and Ishihara. Professor Jenkins of Cornell University reviewed the draft. The authors are grateful to them for their kind suggestions.

The authors are also grateful to Miss Michie Torimitsu for typing this paper.

REFERENCES

1 L. Barden, H. Ismail and P. Tong, Plane Deformation of Granular Materials at Low and High Pressures, Géotechnique, 19, 4, 1969, pp. 441-452.
2 E.R.L. Cole, The behavior of Soils in the Simple Shear Apparatus, Ph.D. Thesis, University of Cambridge, 1967.
3 P.V. Lade, Elasto-Plastic Stress-Strain Theory for Cohesionless Soil with Curved Yield Surface, J. Soils Structures, 13, 1977, pp. 1019-1035.
4 F. Molenkamp, Elasto-Plastic Double Hardening Model MONOT, Delft Soil Mechanics Laboratory, Co. 218595, October, 1980.
5 H.B. Poorooshasb, I. Holubec and A.N. Sherbourne, Yielding and Flow of Sand in Triaxial Compression, Part I., Canadian Geotechnical Journal, 3, 4, 1966, pp. 179-190.
6 K.H. Roscoe, A.N. Schofield and C.P. Wroth, On the Yielding of Soils, Géotechnique, 8, 1, 1958, pp. 22-52.
7 P.W. Rowe, The Stress-Dilatancy Relation for Static Equilibrium of an Assembly of Particles in Contact, Proc. Royal Soc. 269, 1962, pp. 500-527.
8 A.N. Schofield and C.P. Wroth, Critical State Soil Mechanics, McGraw-Hill, London, 1968.
9 M.A. Stroud, The Behavior of Sand at Low Stress Levels in the Simple Shear Apparatus, Ph.D. Thesis, University of Cambridge, 1971.
10 F. Tatsuoka, A Fundamental Study on the Deformation of a Sand by Triaxial Tests, Dr. Eng. Thesis, University of Tokyo, 1972 (in Japanese).
11 F. Tatsuoka and K. Ishihara, Yielding of Sand in Triaxial Compression, Soils and Foundations, 14, 2, 1974, pp. 63-76.
12 Vermeer, A Double Hardening Model of Sand, Géotechnique, 28, 4, 1978, pp. 413 -433.

Mechanics of Granular Materials: New Models and Constitutive Relations, edited by
J.T. Jenkins and M. Satake, 1983
Elsevier Science Publishers B.V., Amsterdam — Printed in The Netherlands

A SIMPLE PROCEDURE TO PREDICT PORE-PRESSURE-RISE OF SATURATED GRANULAR SOILS DURING EARTHQUAKES

Isao ISHIBASHI

School of Civil and Environmental Engineering, Cornell University, Ithaca, N.Y.

ABSTRACT

A simple technique is proposed for the prediction of the increase of pore-pressure of saturated granular soils subjected to random earthquake-type cyclic loads. The model involves the determination of two material constants from a single liquefaction test. The predicted rise in pore-pressure is in fairly good agreement with that observed by other researchers.

INTRODUCTION

Prediction of the pore-pressure-rise in saturated granular soils during earthquakes is particularly important in evaluating liquefaction potential and performing dynamic effective stress analysis. Several researchers have proposed analytical, semi-empirical, or empirical models which predict time histories of the pore-pressure-rise in saturated sands due to earthquake-type loads (refs. 1, 4, 5, 6, 7, 9, 10, 12, 15 and 16).

For a given soil, prediction models require the determination of the effective stress paths for undrained loadings, or the volume change in drained cyclic loading tests and the rebounding characteristics of the soil. These models should eventually lead to the same results when the material parameters which appeared in each model are adequately determined.

Based on the accumulated knowledge and observations of the liquefaction phenomenon of saturated sands provided by many researchers during the past decade, Ishibashi et al. have proposed an empirical model for predicting the pore-pressure-rise (refs. 6, 7 and 15). The model is simple and includes a few material parameters, which can be determined from actual cyclic shear testings for given soil conditions.

In this paper, this model is briefly outlined and compared to the earlier analytical model of Martin et al. (ref. 12). A simple but general procedure for determining the material constants that appear in the model from the results of a single cyclic shear test is presented, and comparison between the calculated pore-pressure-rise and actual laboratory test results reported by other researchers in the past are given.

PORE-PRESSURE-RISE MODEL

Ishibashi et al. (refs. 6, 7 and 15) proposed a model which predicts the pore-pressure-rise of saturated granular soils under any type of cyclic load. They assume that the pore-pressure-rise is a product of the stress history H, a function $\overline{N}$ of the number-of-cycles N, and applied stress intensity function I:

$$\Delta U_N^* = H \cdot \overline{N} \quad I \tag{1}$$

where ΔU_N^* is the increment in pore-pressure-rise during the Nth cycle divided by the initial effective confined pressure, σ_o'. For uniform cyclic shear stress the functions H, $\overline{N}$, and I are determined in terms of the stresses applied in the experiment, the number of cycles, and the parameters characterizing the material by

$$H = 1 - U_{N-1}^* \tag{2}$$

$$\overline{N} = \frac{C_1 \ N}{N^{C_2} - C_3} \tag{3}$$

$$I = [\frac{\tau_N}{\sigma'_{N-1}}]^{\alpha} \tag{4}$$

where U_{N-1}^* is the normalized (by σ_o') residual pore-pressure at the end of the (N-1)th cycle, τ_N is the amplitude of the applied cyclic shear stress during the Nth cycle, σ'_{N-1} is the effective confining pressure at the end of the (N-1)th cycle, and C_1, C_2, C_3 and α are the material constants. With these, Eq. 1 becomes

$$\Delta U_N^* = (1 - U_{N-1}^*) \cdot \frac{C_1 N}{N^{C_2} - C_3} \cdot [\frac{\tau_N}{\sigma'_{N-1}}]^{\alpha} \tag{5}$$

For earthquake-type random excitations, the pore-pressure- rise is evaluated separately for the positive shear stress and the negative shear stress regions as shown in fig. 1. The final expressions are summarized below:

$$\Delta U^*_{Np} = (1 - U^*_{N-1}) \cdot \frac{C_1(N_{eq})_p}{(N_{eq})_p^{C_2} - C_3} \cdot [\frac{\tau_{Np}}{\sigma'_{N-1}}]^\alpha \quad (6)$$

$$\Delta U^*_{Nn} = (1 - U^*_{N-1}) \cdot \frac{C_1(N_{eq})_n}{(N_{eq})_n^{C_2} - C_3} \cdot [\frac{\tau_{Nn}}{\sigma'_{N-1}}]^\alpha \quad (7)$$

$$(N_{eq})_p = \sum_{i=1}^{N} [\frac{\tau_{ip}}{\tau_{Np}}]^\beta \quad (8) \qquad (N_{eq})_n = \sum_{i=1}^{N} [\frac{\tau_{in}}{\tau_{Nn}}]^\beta \quad (9)$$

$$\Delta U^*_N = \frac{1}{2} (\Delta U^*_{Np} + \Delta U^*_{Nn}) \quad (10) \qquad U^*_N = U^*_{N-1} + \Delta U^*_N \quad (11)$$

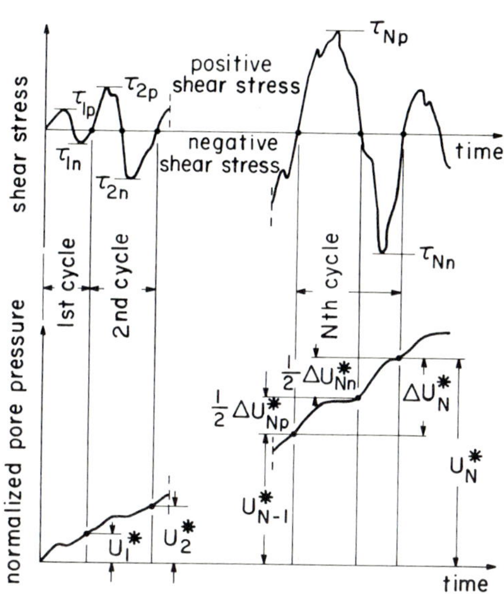

Fig. 1. Illustration of normalized pore-pressure-rise with earthquake shear stress cycles (ref. 7).

where $(N_{eq})_p$ and $(N_{eq})_n$ are, respectively, the equivalent numbers of positive and negative shear stress cycles, and the subscriptions p and n stand for positive and negative shear stress regions as seen in Fig. 1, respectively. Eqs. 6 through 11 do indeed adequately predict pore-pressure-rise under earthquake-type loadings as demonstrated in Fig. 2.

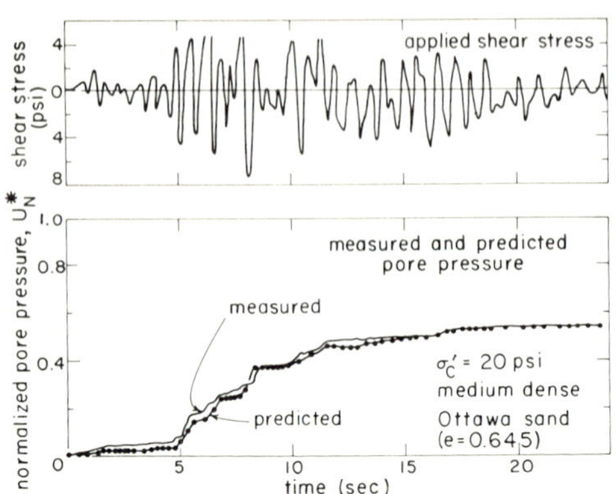

Fig. 2. Comparison between predicted and measured pore-pressure-rise under earthquake-type loading (ref. 14).

The constants C_2 and C_3 were found to have unique values regardless of soil type and density; i.e., $C_2 = 2.0$ and $C_3 = 0.5$ (ref. 7). Liu et al. (ref. 11) investigated the variations of the C_2 value relative to the initial horizontal to vertical stress ratio, K_o ($= \sigma_{ho}/\sigma_{vo}$) in a cyclic triaxial device and found that the C_2 value changes randomly from 2.1, 1.89, 2.0 to 1.66 (with an average value of 1.91) for $K_o = 1.0$, 0.67, 0.5 and 0.4, respectively. This, coupled with the fact that the change in the value of C_2 is not sensitive to pore-pressure-rise (ref. 2), indicates that the use of 2.0 for the value of C_2 is reasonable regardless of soil type, density, K_o value, and the type of cyclic shear equipment.

The value of α varies with soil density, and the constant C_1 changes with soil type and density. It should be noted that the value of the constant β in Eqs. 8 and 9 was considered to be the same as that of α in the previous literature (refs. 7 and 15), but that it was found to be nearly constant, i.e., $\beta = 2.0$ for all cases in this investigation.

In summary, the material constants appeared in the pore-pressure-rise prediction equations (Eqs. 6 through 11) are

C_1 = a function of soil type, density, and test conditions

α = a function of density, and test conditions (12)

$C_2 = 2.0$, $C_3 = 0.5$, and $\beta = 2.0$

COMPARISONS WITH VOLUMETRIC STRAIN MODEL

The volumetric strain model by Martin et al. (ref. 12) is the most comprehensive model to predict pore-pressure-rise. Ignoring the compressibility of water, the model couples the volume change characteristics in drained cyclic loading tests and the pore-water pressure development in the corresponding undrained tests:

$$\Delta u = \Delta\varepsilon_{vd} \cdot \bar{E}_r \tag{13}$$

where Δu is the increment of residual pore-water pressure, $\Delta\varepsilon_{vd}$ is the volumetric strain increment occuring during a drained load cycle having the same shear strain amplitude and initial effective stress conditions, and $\bar{E}_r$ is the tangent modulus of the one-dimensional rebound or unloading curve of the soil at a point corresponding to the initial vertical effective stress. The values of $\Delta\varepsilon_{vd}$ and $\bar{E}_r$ are to be determined experimentally. Martin et al. (ref. 12) used following empirical relations:

$$\Delta\varepsilon_{vd} = a_1(\gamma - a_2\,\varepsilon_{vd}) \frac{a_3\,\varepsilon_{vd}^2}{\gamma + a_4\,\varepsilon_{vd}} \tag{14}$$

and

$$\bar{E}_r = \frac{(\sigma'_v)^{1-m}}{m\,k_2(\sigma'_{vo})^{n-m}} \tag{15}$$

where γ is the cyclic shear strain, ε_{vd} is the total accumulated volumetric strain, and σ'_{vo} and σ'_v are the initial effective and effective vertical stresses, respectively. The experimental constants a_1, a_2, a_3,a_4, m, n and k_2 are to be determined from laboratory testings for a given soil and test conditions.

Another empirical formula for the volumetric strain introduced by Oh-oka (ref. 13) is

$$\Delta\varepsilon_{vd} = \frac{k}{(N+b_1)(N+b_2)} [\frac{\tau_d}{\sigma'_v}]^{\delta} \tag{16}$$

where τ_d is the applied dynamic shear stress and b_1, b_2, k and δ are material constants. Eq. 16 is equivalent to Eq. 14 and is used in following analysis for ease of comparison. By substituting Eqs. 15 and 16 into Eq. 13, Δu is obtained as

$$\Delta u = [\frac{\tau_d}{\sigma_v'}]^{\delta} \frac{k}{(N+b_1)(N+b_2)} \frac{(\sigma_v')^{1-m}}{m\, k_2 (\sigma_{vo}')^{n-m}}$$

$$= \sigma_v' \frac{F}{(N+b_1)(N+b_2)} \cdot \frac{(\tau_d)^{\delta}}{(\sigma_v')^{\delta+m} (\sigma_{vo}')^{n-m}} \qquad (17)$$

where F is a new material constant and is equal to k/mk_2. By dividing Δu in Eq. 17 by σ_{vo}' and upon using $\sigma_v'/\sigma_{vo}' = 1-U^*$, Eq. 17 becomes

$$\Delta U_N^* = \frac{\Delta u}{\sigma_{vo}'} = \frac{\sigma_v'}{\sigma_{vo}'} \frac{F}{(N+b_1)(N+b_2)} \cdot \frac{(\tau_d)^{\delta}}{(\sigma_v')^{\delta+m} (\sigma_{vo}')^{n-m}}$$

$$\doteqdot (1-U^*) \cdot \frac{F}{(N+b_1)(N+b_2)} \cdot [\frac{\tau_d}{\sigma_v'}]^{\delta} \qquad (18)$$

According to Martin et al. and Oh-oka's experimental results, typical values of m, n and δ are 0.43, 0.62 and 5.0, respectively. In the last form of Eq. 18, therefore, m was small enough to be neglected in comparison with the δ value and n-m was also neglected due to its near zero value. It is interesting to see that Eq. 18 is essentially the same as Eq. 5.

DETERMINATION OF CONSTANTS C_1 AND α

In the proposed equations (Eqs. 6 - 11) for the pore-pressure-rise prediction, the constants C_1 and α are the only undermined parameters. To determine these values, it is proposed to run at least one constant stress cyclic shear test under given conditions (a soil type, density, and test type). From the pore-pressure measurement in the above test, N simultaneous equations can be obtained from Eq. 5:

$$\Delta U_i^* = (1-U_{i-1}^*) \frac{C_1[i]}{[i]^2-0.5} [\frac{\tau_i}{\sigma_{i-1}'}]^{\alpha} \quad , \quad i = 1, \ldots, N \qquad (19)$$

where C_1 and α are only unknown values and N is the maximum number of stress cycles. First, by taking the ratio of any two ΔU_i^* and ΔU_j^* values (i>j), C_1 is eliminated and the α_{ij} value is determined as follows:

$$\alpha_{ij} = \frac{\log\left(\frac{\Delta U_i^*}{\Delta U_j^*} \cdot \frac{1-U_{j-1}^*}{1-U_{i-1}^*} \cdot \frac{[j]}{[i]} \cdot \frac{[i]^2-0.5}{[j]^2-0.5}\right)}{\log\left(\frac{\tau_i}{\tau_j} \cdot \frac{1-U_{j-1}^*}{1-U_{i-1}^*}\right)} \tag{20}$$

i,j = 1, ..., N (i>j)

since $\alpha_{ij} = \alpha_{ji}$, the total number of independent α values is $(N^2-N)/2$ and the average α value is calculated from

$$\alpha = \frac{2}{N^2-N} \sum_{i,j=1}^{N} \alpha_{ij} \qquad (i>j) \tag{21}$$

Next, to solve for the C_1 value, the ΔU_i^* (i=1,N) in Eq. 19 are all added together:

$$\sum_{i=1}^{N} \Delta U_i^* = U_N^* - U_o^* = U_N^* = C_1 \sum_{i=1}^{N} (1-U_{i-1}^*) \cdot \frac{[i]}{[i]^2-0.5} \left[\frac{\tau_i}{\sigma'_{i-1}}\right]^{\alpha} \tag{22}$$

Therefore, the average C_1 value is

$$C_1 = \frac{U_N^*}{\sum_{i=1}^{N} (1-U_{i-1}^*) \frac{[i]}{[i]^2-0.5} \cdot \left[\frac{\tau_i}{\sigma'_{i-1}}\right]^{\alpha}} \tag{23}$$

Note that the values, α and C_1 obtained from Eqs. 21 and 23, respectively, are the average values of those based on the data from one cyclic shear test by assuming that those two values are constant through the entire pore-pressure-rise process. In the following section, the validity of the proposed model and the technique for the determination of α and C_1 will be evaluated by using previously reported data by other investigators.

APPLICATION EXAMPLES

Cyclic Triaxial Test Result

Fig. 3 is a cyclic triaxial test data of a saturated medium dense Niigata sand under a nearly uniform shear stress reported by Ishihara and Yasuda (ref. 8). Eqs. 21 and 23 were utilized on the above data. The average values of α = 4.158 and C_1 = 65.57 were obtained by using data up to $U^{\star}$ $(=\Delta u/\sigma_0')$ = 0.859 or N = 8. With C_2 = 2.0, C_3 = 0.5, β = 2.0 and the above C_1 and α values, Eqs. 6 through 11 were used to calculate the pore-pressure-rise of the above same soil under an irregular shear stress application shown in Fig. 4 (a). Fig. 4 (b) compares the computed pore-pressure-rise prediction and the experimentally obtained results by Ishihara and Yasuda. They agree fairly well with each other.

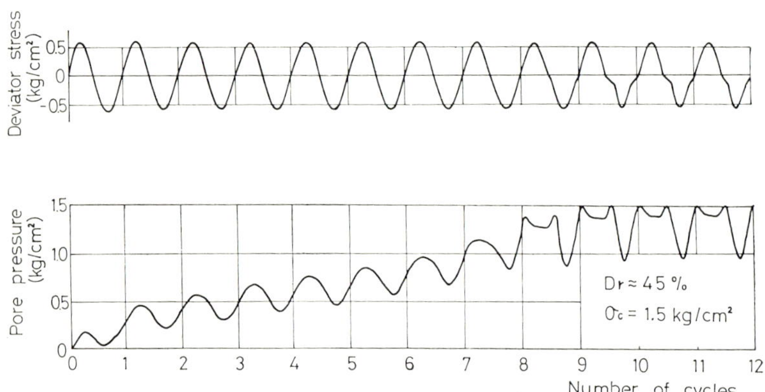

Fig. 3. Pore-pressure-rise in uniform cyclic triaxial test (ref. 8).

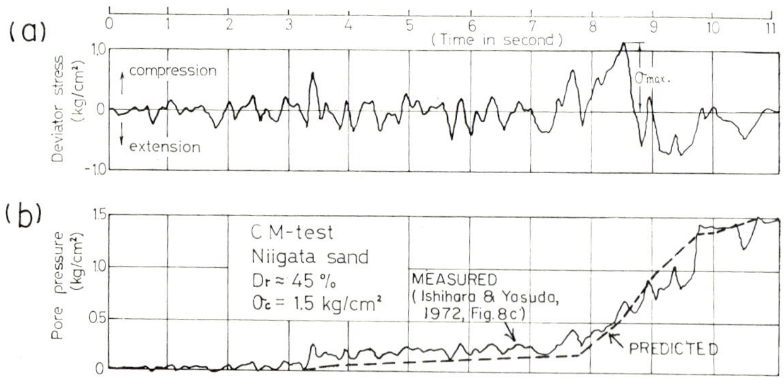

Fig. 4. Measured and predicted pore-pressure-rise in cyclic triaxial experiment.

Shaking Table Test Result

Fig. 5 shows comparisons between the predicted pore-pressure-rise based on the same procedure outlined earlier and measured values from shaking table liquefaction test results reported by DeAlba et al. (ref. 3). The liquefaction test data under a nearly uniform shear stress up to $U^* = 0.85$ were used to calculate the average α and C_1 values. $\alpha = 3.169$ and $C_1 = 5.460$ were obtained for the Dr = 90% sample and $\alpha = 3.487$ and $C_1 = 15.10$ were for the Dr = 82% sample. The predicted and measured values show reasonably good agreement.

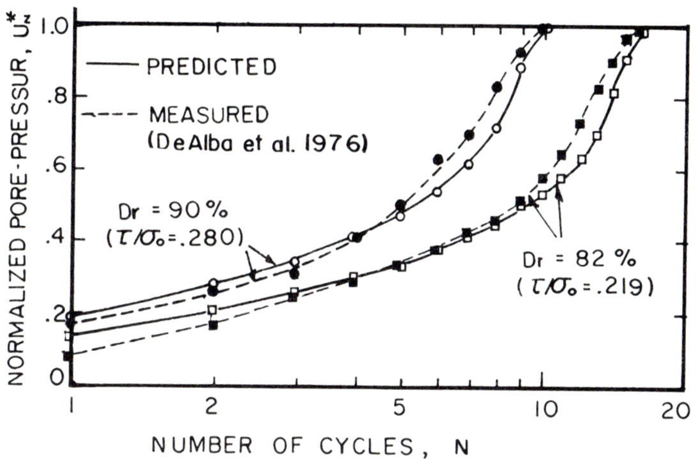

Fig. 5. Measured and predicted pore-pressure-rise in shaking table experiment.

CONCLUSIONS

A simple procedure to predict pore-pressure-rise of saturated granular soils during irregular cyclic loadings was proposed in Eqs. 6 through 11. These equations involve five material constants, three of which are already known and the rest of which, α and C_1, should be determined for a given individual case depending on the soil type, density, K_0-value, test type, etc. Eqs. 21 and 23 were proposed to calculate the C_1 and α values from a single cyclic shear test.

Previously reported data by other researchers were reanalyzed to test the validity of the newly proposed pore-pressure-rise prediction technique. The predicted and observed results show reasonably good agreement as indicated in Figs. 4 and 5.

This paper also showed that the fundamental model of Martin et al. and the proposed experimental model are essentially the same. The author believes that a simple empirical model such as the one proposed in this paper has a certain merit over more complicated analytical models, which typically require far more information about the mechanical behavior of the soil than is easily available.

REFERENCES

1 Z. P. Bazant and R. J. Krizek, Endochronic Constitute Law for Liquefaction of Sand, Proc. ASCE, Vol. 102, No. EM2, 1976, pp. 225-238.

2 W. L. Cheng, Soil Parameters Affecting Pore Pressure Increments in Saturated Sands During Cyclic Loading, M.S. thesis, Department of Civil Engineering, University of Washington, Seattle, 1978.

3 P. DeAlba, H. B. Seed and C. K. Chan, Sand Liquefaction in Large-Scale Simple Shear Tests, Proc. ASCE, Vol. 102, No. GT9, 1976, pp. 909-927.

4 W. D. L. Finn and S. K. Bhatia, Prediction of Seismic Porewater Pressures, Proc. 10th Int. Conf. on Soil Mechanics and Foundation Engineering, Vol. 3, Stockholm, 1981, pp. 201-206.

5 J. Ghaboussi and S. U. Dikmen, Liquefaction Analysis of Horizontally Layered Sands, Proc. ASCE, Vol. 104, No. GT3, 1978, pp. 341-356.

6 I. Ishibashi, M. A. Sherif and C. Tsuchiya, Pore-Pressure Rise Mechanism and Soil Liquefaction, Soils and Foundations, Vol. 17, No. 2, 1977, pp. 17-27.

7 I. Ishibashi, M. A. Sherif and W. L. Cheng, The Effects of Soil Parameters on Pore-Pressure-Rise and Liquefaction Prediction, Soils and Foundations, Vol. 22, No. 1, 1982, pp. 39-48.

8 K. Ishihara and S. Yasuda, Sand Liquefaction Due to Irregular Excitation, Soils and Foundations, Vol. 12, No. 4. 1972, pp. 65-77.

9 K. Ishihara, F. Tatsuoka and S. Yasuda, Undrained Deformation and Liquefaction of Sand During Cyclic Stresses, Soils and Foundations, Vol. 15, No. 1, 1975, pp. 29-44.

10 C. P. Liou, V. L. Streeter and F. E. Richart, Jr., Numerical Model for Liquefaction, Proc. ASCE, Vol. 103, No. GT6, 1977, pp. 589-606.

11 Y. Liu, Y. Tong and X. Qi, Residual Pore-Pressure of Saturated Sand Under Cyclic Loadings (in Chinese), Institute of Engineering Mechanics Report No. 80-084, Academia Sinica, Harbin, China, 1980, 17 pp.

12 G. R. Martin, W. D. L. Finn and H. B. Seed, Fundmentals of Liquefaction Under Cyclic Loading, Proc. ASCE, Vol. 101, No. GT5, 1975, pp. 423-438.

13 H. Oh-oka, Drained and Undrained Stress-Strain Behavior of Sands Subjected to Cyclic Shear Stress Under Nearly Plane Strain Condition, Soils and Foundations, Vol. 16, No. 3, 1976, pp. 19-31.

14 M. A.Sherif, I. Ishibashi and C. Tsuchiya, Pore-Pressure Prediction During Earthquake Loadings, Soils and Foundations, Vol. 18, No. 4, 1978, pp. 19-30.

15 O. C. Zienkiewicz, C. T. Chang and E. Hinton, Non-Linear Seismic Response and Liquefaction, Int. J. for Numerical Analytical Methods in Geomechanics, Vol. 2, 1978, pp. 381-404.

Mechanics of Granular Materials: New Models and Constitutive Relations, edited by
J.T. Jenkins and M. Satake, 1983
Elsevier Science Publishers B.V., Amsterdam — Printed in The Netherlands

A STRESS-STRAIN MODEL FOR GRANULAR MATERIALS CONSIDERING THE MECHANISM OF FABRIC CHANGE

H. MATSUOKA

Department of Civil Engineering, Nagoya Institute of Technology, Gokiso, Showa-ku, Nagoya (Japan)

ABSTRACT

A stress-strain model for granular materials is proposed by consideration of the mechanism of fabric change under shear. A new relationship between strains and fabric change of granular materials is derived from paying attention to "two particles" across the potential sliding plane and considering the mechanism of disappearance and generation of interparticle contacts with the third particle. Combining the above mentioned relationship between strains and fabric change with that between stresses and fabric change which has been proposed by Matsuoka (1974a), a stress-strain relationship is formulated under two-dimensional stress conditions. The concept of "Spatial Mobilized Plane (SMP)" is then introduced to extend the stress-strain relationship to that in the three-dimensional state. For example, the stress-strain behaviours under cyclic loading and the liquefaction phenomena are analyzed by the proposed model.

INTRODUCTION

It is one of the most interesting and difficult problems in the mechanics of granular materials to derive their stress-strain relationship from the mechanism of fabric change in granular materials. Direct shear tests and simple shear tests have been performed using a stack of aluminium rods (ϕ 3, 5 and 9mm) or photoelastic rods (ϕ 6.2 and 10mm) as a two-dimensional model of granular materials as shown in Figs. 1 and 2. Based on the observation of these shear tests, the relationships among the shear-normal stress ratio (τ/σ_N), the shear

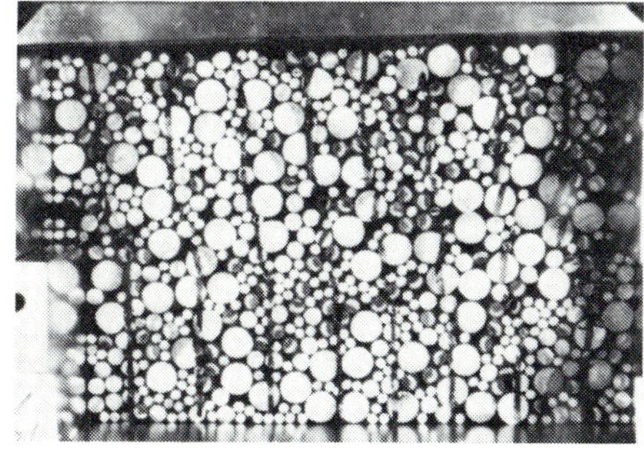
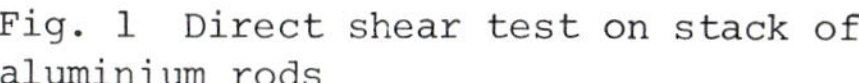

Fig. 1 Direct shear test on stack of aluminium rods

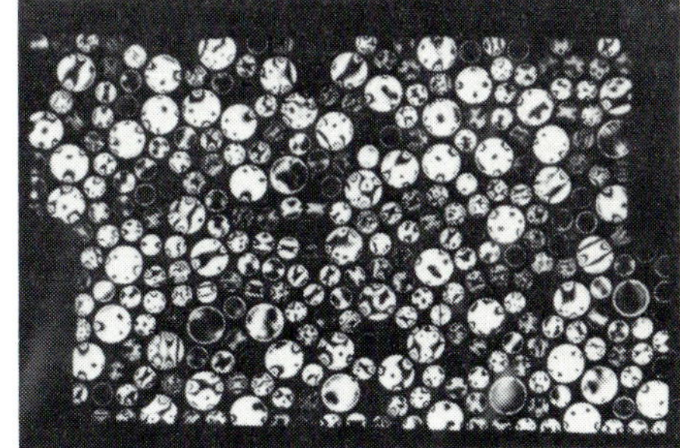

Fig. 2 Direct shear test on stack of photoelastic rods

strain (γ) and the normal strain (ε_N) on the potential sliding plane are derived by the medium of the angles of interparticle contacts (θ) along the potential sliding plane. Introducing "Spatial Mobilized Plane (SMP)" as such a plane on which particles are most mobilized on the average under three-dimensional stress conditions, the above mentioned stress-strain relationship is extended to that in the three-dimensional state. The stress-strain behaviours under cyclic loading is also analyzed by the proposed model.

RELATIONSHIP BETWEEN SHEAR-NORMAL STRESS RATIO (τ/σ_N) ON POTENTIAL SLIDING PLANE AND AVERAGE ANGLE OF INTERPARTICLE CONTACTS ($\bar{\theta}$)

Paying attention to the interparticle contacts along the potential sliding plane as shown in Fig. 3 and denoting the angle of interparticle contact (abbreviated "contact angle") by θ_i, the interparticle force by f_i and the average value of the mobilized interparticle friction angle by ϕ_μ, the following equation can be obtained from the equilibrium of interparticle forces on the potential sliding plane.

$$\frac{\tau}{\sigma_N} = \frac{\sum_{i=1}^{n} f_i \cdot \sin(\theta_i + \phi_u)}{\sum_{i=1}^{n} f_i \cdot \cos(\theta_i + \phi_u)} \tag{1}$$

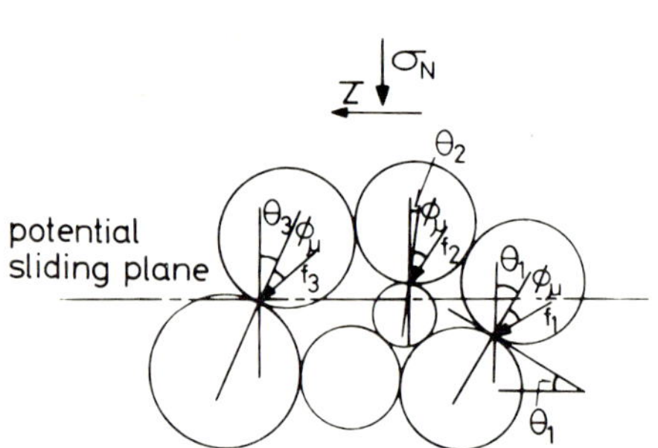

Fig. 3 Equilibrium of interparticle forces on the potential sliding plane

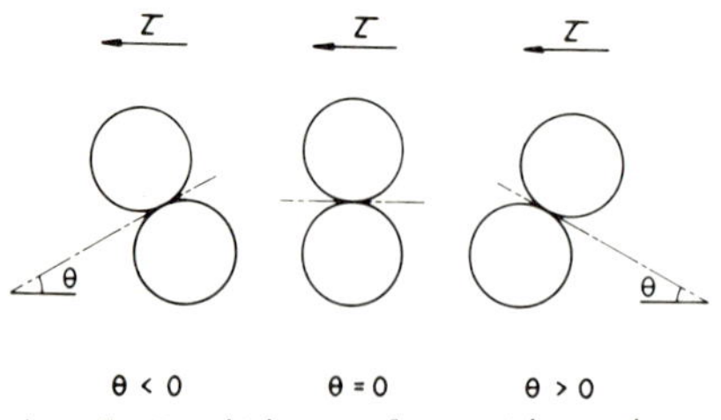

Fig. 5 Positive and negative signs for angle of interparticle contact θ

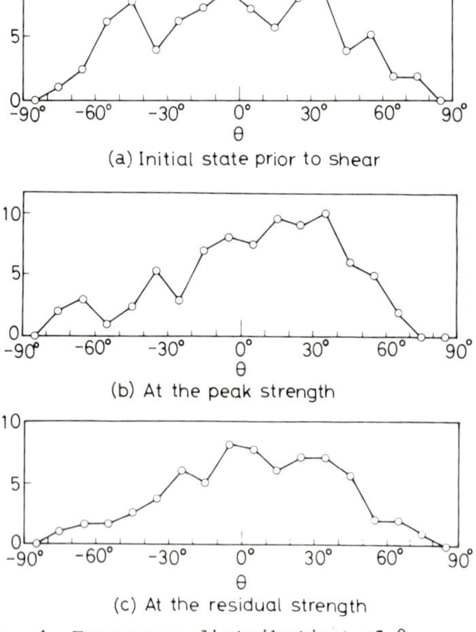

Fig. 4 Frequency distribution of θ during shear test on stack of aluminium rods

where n means the number of the interparticle contacts along the potential sliding plane. It is assumed on the basis of the results of direct shear tests on a stack of photoelastic rods (see Fig. 2) that the interparticle forces f_i are the same at all of the interparticle contacts to first approximation. Then, if the frequency distribution of contact angles θ is denoted by $N(\theta)$, Eq. 1 can be transformed as follows:

$$\frac{\tau}{\sigma_N} = \frac{\int_{-\pi/2}^{\pi/2} N(\theta)\cdot\sin(\theta+\phi_\mu)\cdot d\theta}{\int_{-\pi/2}^{\pi/2} N(\theta)\cdot\cos(\theta+\phi_\mu)\cdot d\theta} \qquad (2)$$

Figure 4 shows the variation of the frequency distribution $N(\theta)$ during a direct shear test on a stack of aluminium rods. Figures 4 (a), (b) and (c) correspond to the frequency distributions at the initial state prior to shear, at the peak strength and at the residual strength respectively. At the initial state, $N(\theta)$ tends to show a symmetrical distribution with respect to $\theta=0°$. The frequency is usually low at the both ends of θ (at $\theta \fallingdotseq \pm 90°$) because there is little probability that the center line of two particles concerned with the equilibrium of interparticle forces intersects the potential sliding plane (see Fig. 3). At the peak strength, $N(\theta)$ has the distribution one-sided to the positive zone of θ where the angle θ is effective to resist the applied shear stress (see Fig. 5). At the residual strength, $N(\theta)$ returns to a symmetrical distribution again.

Now, we consider the reason why $N(\theta)$ is one-sided to the positive zone of θ, though the particles on the potential sliding plane slide to the direction of the applied shear stress, namely, the negative direction of θ. Figure 6 expresses the frequency distribution of θ obtained by a simple shear test on a stack of aluminium rods, in which the disappeared and generated contact angles during shear (shear strain γ=5 to 15%) are shown. It is seen from this figure that the bias of the frequency distribution of θ toward its positive side is caused by disappearance of the negative angles and generation of the positive angles of θ. The main mechanism of these disappearance and generation of θ may

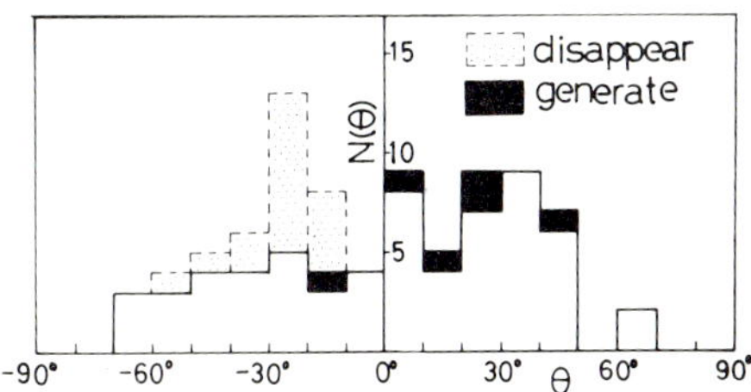

Fig. 6 Frequency distribution of θ showing disappeared and generated angles during shear

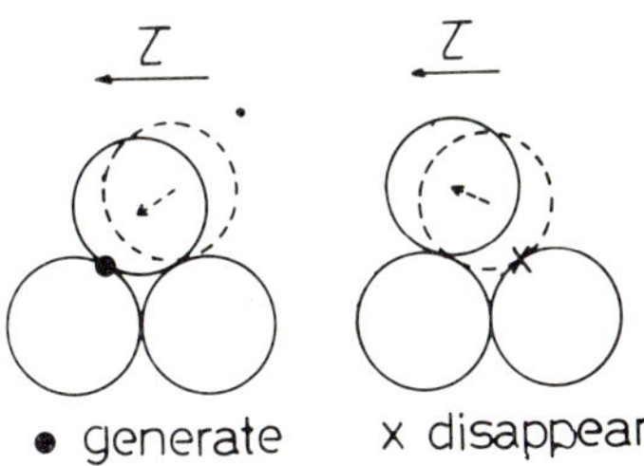

Fig. 7 Disappeared and generated mechanism of interparticle contacts

be visualized by Fig. 7.

If the frequency distribution of θ is approximated by a triangular distribution as shown in Fig. 8, $N(\theta)$ can be expressed by the following equations.

$$\left.\begin{aligned} N(\theta) &= h\cdot\frac{\theta+\pi/2}{\theta_p+\pi/2} \qquad (-\pi/2\leqq\theta\leqq\theta_p) \\ N(\theta) &= h\cdot\frac{\theta-\pi/2}{\theta_p-\pi/2} \qquad (\theta_p\leqq\theta\leqq\pi/2) \end{aligned}\right\} \tag{3}$$

The following equation is obtained by substituting Eq. 3 into Eq. 2.

$$\frac{\tau}{\sigma_N} = \frac{\pi\cdot\sin(3\bar{\theta}+\phi_\mu)-6\bar{\theta}\cdot\cos\phi_\mu}{\pi\cdot\cos(3\bar{\theta}+\phi_\mu)+6\bar{\theta}\cdot\sin\phi_\mu} \tag{4}$$

where $\bar{\theta}$ is the average value of θ ($\bar{\theta}=\int N(\theta)\cdot\theta\cdot d\theta/\int N(\theta)\cdot d\theta=\theta_p/3$). Equation 4 can be approximated within the practical range of $\bar{\theta}$ by the following formula (Matsuoka, 1974a).

$$\frac{\tau}{\sigma_N} \fallingdotseq \lambda\bar{\theta} + \mu \tag{5}$$

where λ and μ are parameters ($\mu=\tan\phi_\mu$).

RELATIONSHIP BETWEEN SHEAR STRAIN (γ) AND NORMAL STRAIN (ε_N) ON POTENTIAL SLIDING PLANE, AND AVERAGE ANGLE OF INTERPARTICLE CONTACTS ($\bar{\theta}$)

Considering these disappeared, generated and "steady" (interparticle contact is kept during shear) behaviours of the contact angles, the sliding displacements and the sliding strains due to shear are formulated. When the particles on the potential sliding plane slide averagely to the amount of $\bar{d}\cdot|\Delta\theta|$ to the shearing direction, the horizontal and vertical sliding displacement increments at a contact point having a certain value of θ can be expressed as follows (see Fig. 9):

$$\Delta\dot{x} = \bar{d}\cdot|\Delta\theta|\cdot N(\theta)\cdot\cos\theta \tag{6}$$

$$\Delta\dot{y} = -\bar{d}\cdot|\Delta\theta|\cdot N(\theta)\cdot\sin\theta \tag{7}$$

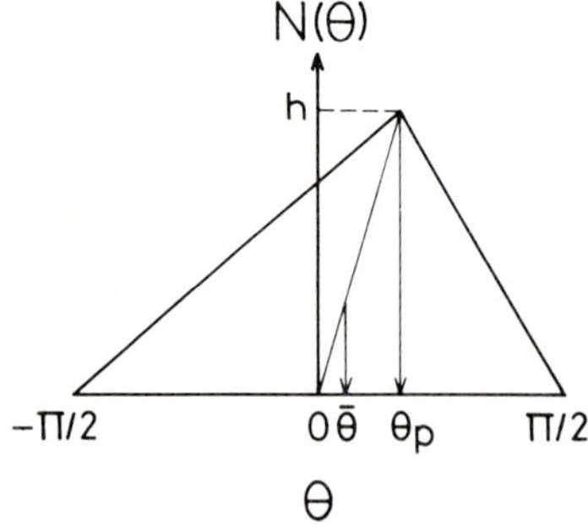

Fig. 8 Triangular frequency distribution of θ

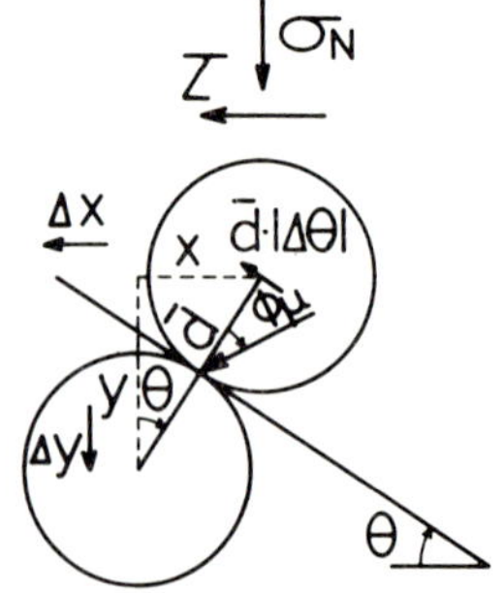

Fig. 9 Unit element in granular materials—"two particles model"

where $\bar{d}$ is the average distance between centers of two adjacent particles. Substituting Eq. 3 into Eqs. 6 and 7, and integrating them in the range of $-\pi/2 \leq \theta \leq \pi/2$, the sliding displacement increments for a certain frequency distribution of $\bar{\theta}$ are obtained as follows:

$$\Delta x = \int_{-\pi/2}^{\pi/2} \bar{d} \cdot |\Delta\theta| \cdot N(\theta) \cdot \cos\theta \cdot d\theta$$

$$= \bar{d} \cdot h \cdot \frac{\pi \cdot \cos 3\bar{\theta}}{(\pi/2+3\bar{\theta})(\pi/2-3\bar{\theta})} \cdot |\Delta\theta| \tag{8}$$

$$\Delta y = -\int_{-\pi/2}^{\pi/2} \bar{d} \cdot |\Delta\theta| \cdot N(\theta) \cdot \sin\theta \cdot d\theta$$

$$= \bar{d} \cdot h \cdot \frac{6\bar{\theta} - \pi \cdot \sin 3\bar{\theta}}{(\pi/2+3\bar{\theta})(\pi/2-3\bar{\theta})} \cdot |\Delta\theta| \tag{9}$$

Equations 8 and 9 must be integrated with respect to $\bar{\theta}$ in order to obtain the sliding displacements, because θ_p $(=3\bar{\theta})$ in Fig. 8 is varied with shear as described in the previous section. For this integration, it is necessary to get the relationship between $|\Delta\theta|$ and $\Delta\bar{\theta}$.

If any change in the average contact angle $\Delta\bar{\theta}$ is produced by the disappeared, generated and "steady" behaviours of the contact angles, the following equation is obtained (Matsuoka and Takeda, 1980).

$$\Delta\bar{\theta} \cdot h = -\int_{-\pi/2}^{0} N_d(\theta) \cdot \theta \cdot d\theta + \int_{-\pi/2}^{0} N_g(\theta) \cdot (-\theta) \cdot d\theta - |\Delta\theta| \cdot \left(n - \int_{-\pi/2}^{0} N_d(\theta) \cdot d\theta\right) \tag{10}$$

where n is the total number of the contact angles $(n=\pi h/2)$. $N_d(\theta)$ and $N_g(\theta)$ are the frequencies of the disappeared and generated contact angles respectively. The left-hand side of Eq. 10 corresponds to the total change in θ. The first term on the right-hand side expresses the total increase in θ caused by disappearance of the negative contact angles. The second term expresses the total increase in θ caused by generation of the positive contact angles. The range of integration in this term is chosen $-\pi/2$ to 0 from the following assumption. The main mechanism of the generation of θ is assumed that a particle having one negative contact angle generates a new contact at the symmetrically positive angle with respect to $\theta=0°$ through movement of the particle to the shearing direction. Finally, the third term means the total decrease in θ caused by sliding of the "steady" contacts to the negative direction.

It is seen from the results of direct shear tests on a stack of photoelstic rods that the main mechanism of disappearance and generation of contact angles can be illustrated by Fig. 10. At the initial state prior to shear, the interparticle contact points are stable under the normal stress σ_N on the potential sliding plane. If the shear stress τ is applied, the interparticle force at the positive contact angle becomes large and that at the negative contact angle becomes small generally. These phenomena may be considered a kind of generation and disappearance of contact angles. Therefore, the number of the generated and disappeared contact angles is estimated to be quite large at the beginning of shear. As it is seldom that the generated (or disappeared) contact angles are disappeared (or generated) during the same shearing process, the number of the generated and disappeared contact angles is considered to decrease with shear. Figure 11 shows the relationship between $\bar{\theta}$ and P, which is a ratio defined by the following equation (Geka and Matsuoka, 1981).

$$P = \frac{1}{|\Delta\theta|} \cdot \frac{N_g(\theta)+N_d(\theta)}{N(\theta)} \tag{11}$$

P means the ratio of the number of the generated and disappeared contact angles ($N_g(\theta)+N_d(\theta)$) to the number of all contact angles ($N(\theta)$) per $|\Delta\theta|=1$ radian. The plots in this figure indicate the observed values from the direct shear tests on a stack of photoelastic rods. It is seen from this figure that P decreases extremely with shear (with increase in $\bar{\theta}$). Then, the relationship between P and $\bar{\theta}$ is assumed to be expressed by the following exponential form.

$$P = a \cdot \exp(-b \cdot \bar{\theta}) \tag{12}$$

where a and b are parameters to express the degree of fabric change with shear. The plots in Fig. 11 can be approximated by the parameters of a=100 and b=3. From Eqs. 11 and 12, the following relationship can be obtained.

$$N_g(\theta)+N_d(\theta) = a \cdot \exp(-b \cdot \bar{\theta}) \cdot N(\theta) \cdot |\Delta\theta| \tag{13}$$

By substituting Eq. 13 into Eq. 10, the following relationship among $|\Delta\theta|$, $\bar{\theta}$ and $\Delta\bar{\theta}$ is obtained.

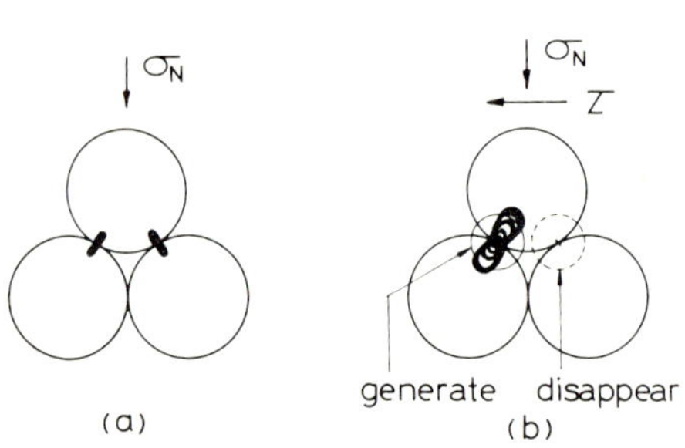

Fig. 10 Disappearance and generation of interparticle contacts observed by photoelastic rods

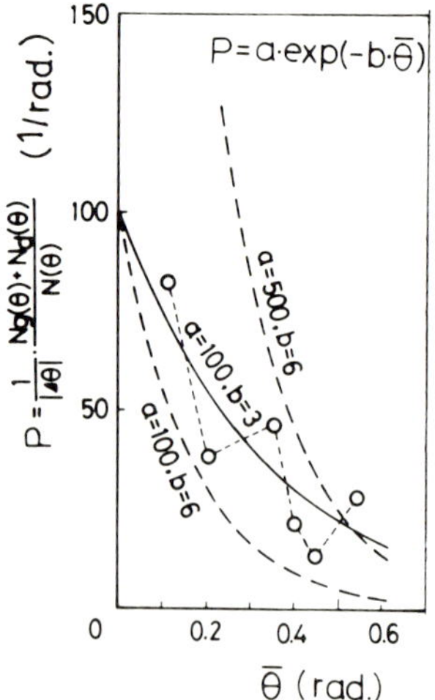

Fig. 11 Variation of disappeared and generated interparticle contacts with shear

$$|\Delta\theta| = \frac{\pi/2+3\bar{\theta}}{(\pi^2/24)\,a\cdot\exp(-b\bar{\theta})-(\pi/2+3\bar{\theta})}\cdot\Delta\bar{\theta} \qquad (\bar{\theta}\geqq 0)$$

$$|\Delta\theta| = \frac{\pi/2-3\bar{\theta}}{a\cdot\exp(-b\bar{\theta})\cdot(3\bar{\theta}^2-(\pi/2)\cdot\bar{\theta}+(\pi^2/24))-(\pi/2-3\bar{\theta})}\cdot\Delta\bar{\theta} \qquad (\bar{\theta}<0) \tag{14}$$

The sliding displacement increments per one contact point can be obtained by substituting Eq. 14 into Eqs. 8 and 9, and dividing them by the total number of the contact angles ($\pi h/2$). If these displacement increments are divided by the average width of one shear layer D, which corresponds to the average normal component of the distances between two adjacent sliding particles to the potential sliding plane, they may be converted to the strain increments. Thus, the shear strain increment $\Delta\gamma$ and the normal strain increment $\Delta\varepsilon_N$ on the potential sliding plane can be expressed as follows:

$$\Delta\gamma = \frac{\pi/2+3\bar{\theta}}{(\pi^2/24)\cdot a\cdot\exp(-b\bar{\theta})-(\pi/2+3\bar{\theta})}\cdot\Delta\bar{\theta} \qquad (\bar{\theta}\geqq 0)$$

$$\Delta\gamma = \frac{\pi/2-3\bar{\theta}}{a\cdot\exp(-b\bar{\theta})\cdot(3\bar{\theta}^2-(\pi/2)\cdot\bar{\theta}+(\pi^2/24))-(\pi/2-3\bar{\theta})}\cdot\Delta\bar{\theta} \qquad (\bar{\theta}<0) \tag{15}$$

$$\Delta\varepsilon_N = \frac{6\bar{\theta}-\pi\cdot\sin 3\bar{\theta}}{\pi\cdot\cos 3\bar{\theta}}\cdot\Delta\gamma \tag{16}$$

where

$$D = \frac{\int_{-\pi/2}^{\pi/2} N(\theta)\cdot\bar{d}\cdot\cos\theta\cdot d\theta}{\int_{-\pi/2}^{\pi/2} N(\theta)\cdot d\theta} = \frac{2\bar{d}\cdot\cos 3\bar{\theta}}{(\pi/2+3\bar{\theta})(\pi/2-3\bar{\theta})} \tag{17}$$

From Eq. 4 or 5 and Eqs. 15 and 16, a stress-strain relationship for granular materials under shear can be obtained by the medium of the average contact angle $\bar{\theta}$ which is considered to be an index to express the fabric of granular materials.

Figures 12 (a), (b) and (c) show the calculated relationships among τ/σ_N, γ and ε_N ($\tan\phi_\mu$=0.27). The reason why τ/σ_N is not equal to 0 at γ=0 in Fig. 12 (a) is due to the assumption that the interparticle friction angle ϕ_μ is fully mobilized from γ=0. In Fig. 12 (a) $\bar{\theta}_0$ denotes the average contact angle at the

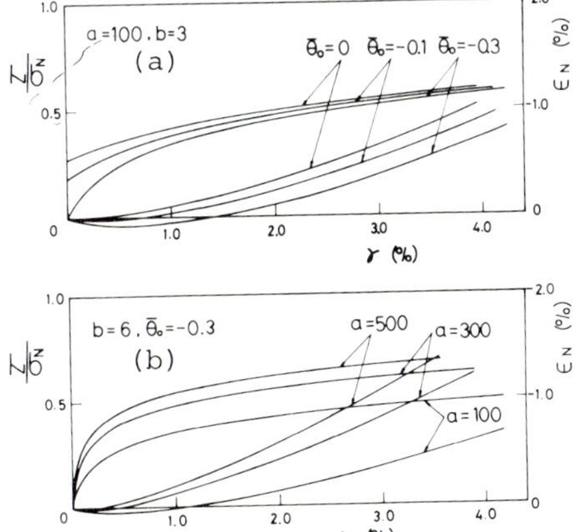

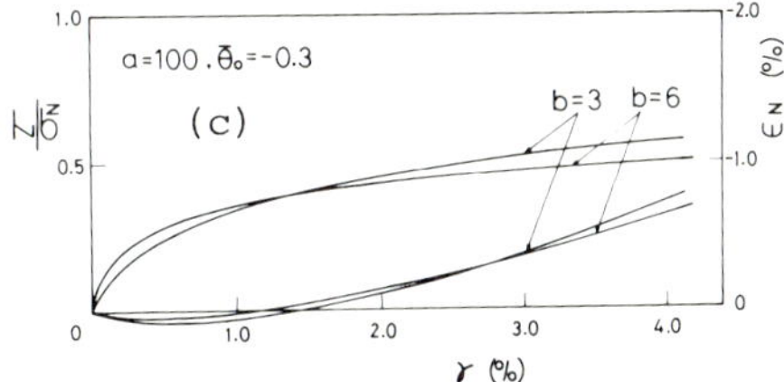

Fig. 12 Calculated relationships among τ/σ_N, γ and ε_N ($\tan\phi_\mu$=0.27)

start of shear. It can be seen from Figs. 12 (b) and (c) that the parameters a and b have close connection with strain hardening behaviour. The strain hardening is considered to be originally caused by generation and disappearance of the contact angles (see Fig. 11). If $\bar{\theta}$ at failure is denoted by $\bar{\theta}_f$(>0), the following relationship can be obtained from the condition that the denominator in Eq. 14 becomes zero at failure ($\Delta\gamma \rightarrow \infty$).

$$(\pi^2/24)\cdot a\cdot \exp(-b\cdot\bar{\theta}_f)-(\pi/2+3\bar{\theta}_f) = 0 \tag{18}$$

Therefore, the parameters in the proposed stress-strain relationship are "ϕ_μ" to denote the interparticle friction angle, "$\bar{\theta}_0$ and $\bar{\theta}_f$" to express the fabric at the start of shear and at failure, and "a" (or b) to express the degree of fabric change during shear.

BASIC STRESS-STRAIN RELATIONSHIP ON POTENTIAL SLIDING PLANE

The author has already proposed the following shear-normal stress ratio (τ/σ_N) vs. normal-shear strain increment ratio ($-d\varepsilon_N/d\gamma$) relationship and shear-normal stress ratio (τ/σ_N) vs. normal-shear strain ratio ($-\varepsilon_N/\gamma$) relationship on the "mobilized plane" (Matsuoka, 1974a, 1974b). The "mobilized plane" is such a plane on which the shear-normal stress ratio (τ/σ_N) is maximized and is considered to correspond to the potential sliding plane approximately.

$$\frac{\tau}{\sigma_N} = \lambda\cdot\left(-\frac{d\varepsilon_N}{d\gamma}\right)+\mu \tag{19}$$

$$\frac{\tau}{\sigma_N} = \lambda\cdot\left(-\frac{\varepsilon_N}{\gamma}\right)+\mu' \tag{20}$$

where λ, μ and μ' are soil parameters ($\mu=\tan\phi_\mu$). Combining Eqs. 19 and 20 and solving the differential equation, the following equations are obtained.

$$\frac{\tau}{\sigma_N} = (\mu'-\mu)\cdot\ln\frac{\gamma}{\gamma_0} + \mu \qquad \left(\gamma=\gamma_0\cdot\exp\left(\frac{\tau/\sigma_N-\mu}{\mu'-\mu}\right)\right) \tag{21}$$

$$\varepsilon_N = \frac{\mu-\mu'}{\lambda}\cdot\gamma\cdot\left\{\ln\frac{\gamma}{\gamma_0}-1\right\} \tag{22}$$

where γ_0 denotes γ at the maximum compression point of ε_N and is a parameter to express the fabric of granular materials (Matsuoka, 1974b).

Figure 13 shows the τ/σ_N vs. ($-d\varepsilon_N/d\gamma$) relationship calculated from Eqs. 5, 15 and 16. It is seen from this figure that Eq. 19 holds good (λ=0.9 and μ=0.27). If the average contact angle at failure $\bar{\theta}_f$ is given, the sets of the values of a and b are determined from Eq. 18. The relationships among τ/σ_N, γ and ε_N, and between τ/σ_N and ($-\varepsilon_N/\gamma$) are shown in Figs. 14 and 15 which are calculated from Eqs. 5, 15 and 16, using the four sets of the values of a and b determined for the case of $\bar{\theta}_f$=0.60 ($(\tau/\sigma_N)_f$=0.81). It can be seen from these two figures that the four different stress-strain curves in Fig. 14 are arranged within a narrow zone by the τ/σ_N vs. ($-\varepsilon_N/\gamma$) relationships in Fig. 15. The four τ/σ_N vs. γ curves in Fig. 14 coincide at a large value of γ because they have the same shear strength ($\bar{\theta}_f$=0.60). The schematic figure of the τ/σ_N vs. ($-d\varepsilon_N/d\gamma$) relationship and the τ/σ_N vs. ($-\varepsilon_N/\gamma$) relationships are shown in Fig. 16. It is seen from this figure that the τ/σ_N vs. ($-\varepsilon_N/\gamma$) relationships become convex curves between ($-\varepsilon_N/\gamma$)=$\bar{\theta}_0$ and $\bar{\theta}_f$. Equation 20 shown by the broken lines in Fig. 16 can be explained to be an approximate linear relationship for the main parts of these convex curves. As these convex curves are considered to be similar figures approximately, the following equation is obtained (Matsuoka et al., 1982).

$$\mu' - \mu = \kappa \cdot (\bar{\theta}_f - \bar{\theta}_0) \tag{23}$$

where the parameter κ is nearly constant ($\kappa \fallingdotseq 0.2$) as can be seen from Fig. 15. Therefore, μ' is considered a parameter to reflect the fabric parameters $\bar{\theta}_0$ and $\bar{\theta}_f$.

Substituting Eqs. 5 and 23 into Eq. 21, and denoting $\bar{\theta}$ and γ at failure by $\bar{\theta}_f$ and γ_f respectively, the following equation is obtained.

$$\gamma_0 = \gamma_f \cdot \exp\left(-\frac{\lambda}{\kappa}\right) \cdot \exp\left(\frac{\lambda \cdot (-\bar{\theta}_0)}{\kappa \cdot (\bar{\theta}_f - \bar{\theta}_0)}\right) \tag{24}$$

If it is assumed that γ_f is the same no matter what the initial fabric ($\bar{\theta}_0$) is, the following relationship for γ_0 is derived from Eq. 24.

$$\gamma_0 = \gamma_{0\,(\bar{\theta}_0=0)} \cdot \exp\left(\frac{\lambda \cdot (-\bar{\theta}_0)}{\kappa \cdot (\bar{\theta}_f - \bar{\theta}_0)}\right) \tag{25}$$

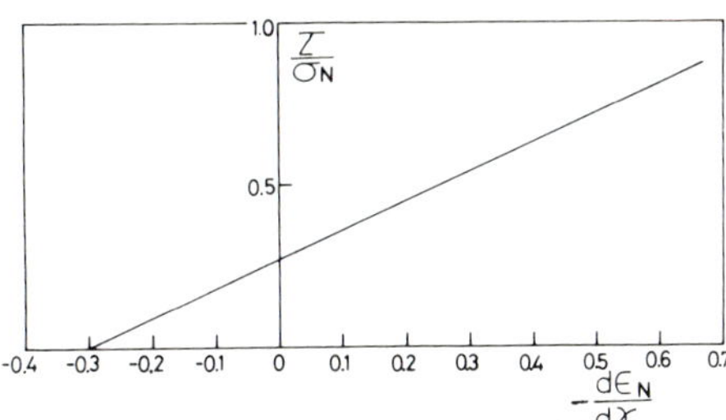

Fig. 13 Calculated relationships between τ/σ_N and $-d\varepsilon_N/d\gamma$ (λ=0.9, μ=0.27)

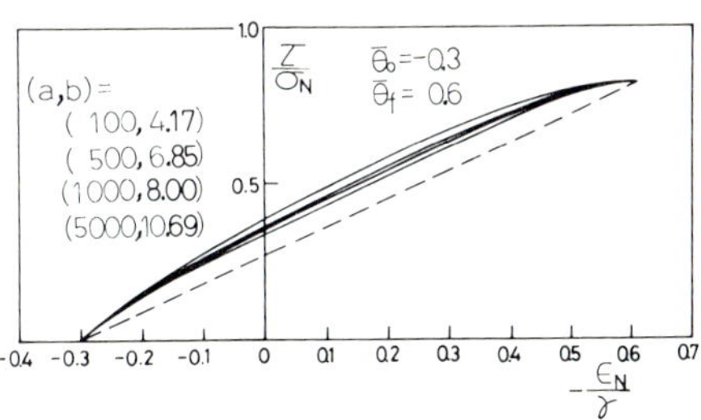

Fig. 15 Calculated relationships between τ/σ_N and $-\varepsilon_N/\gamma$ (λ=0.9, μ=0.27)

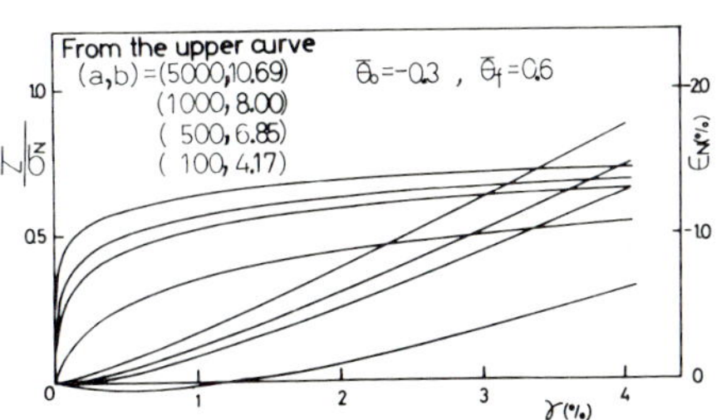

Fig. 14 Calculated relationships among τ/σ_N, γ and ε_N (λ=0.9, μ=0.27)

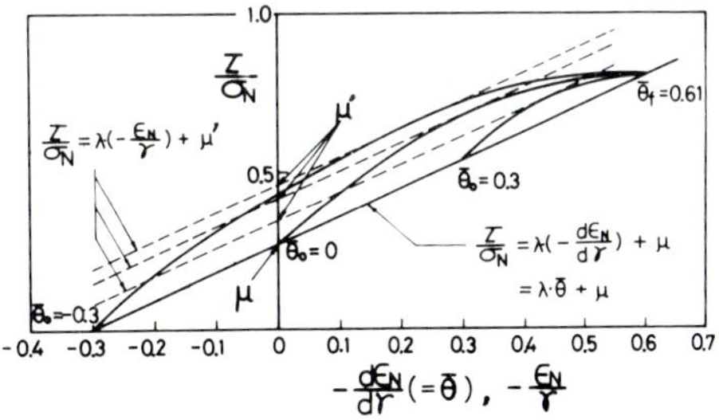

Fig. 16 Schematic relationships between τ/σ_N and $-d\varepsilon_N/d\gamma$, and between τ/σ_N and $-\varepsilon_N/\gamma$

where $\gamma_{0(\bar{\theta}_0=0)}$ means γ_0 at $\bar{\theta}_0=0$. It is seen from Eq. 25 that γ_0 is also a parameter to reflect the fabric of granular materials.

As an example of shear stress history, the stress-strain behaviour under repetitional shear stress is analyzed here considering the mechanism of fabric change as mentioned before. Figure 17 shows the conceptional relationship between a stress-strain behaviour under cyclic loading and the variation of the triangular frequency distribution of contact angles. As the particles are rearranged so as to resist the shear stress τ_{1st} by the 1st loading, the frequency distribution is one-sided to the right hand side of θ. If the particles are then subject to τ_{2nd} in the reverse direction, they slide down easily as shown in Fig. 18, and thus the stress-strain curve under the 2nd loading becomes gentle comparing with that under the 1st loading. For example, $\bar{\theta}_0$ under the 2nd loading is considered to be $-\bar{\theta}_{1st}$ where $\bar{\theta}_{1st}$ is $\bar{\theta}$ at the end of the 1st loading (see Fig. 17). Generally $\bar{\theta}_0$ under cyclic loading is given from Eq. 5 by the following equation.

$$\bar{\theta}_0 = -(\tau/\sigma_N-\mu)/\lambda \tag{26}$$

where τ/σ_N in Eq. 26 represents τ/σ_N at the end of the preceding loading.

Usually $\bar{\theta}_0$ under the 1st loading is considered to be nearly equal to zero ($\bar{\theta}_0 \fallingdotseq 0$), so if μ' under the 1st loading is denoted by μ'_{1st}, the coefficient κ is given from Eq. 23 as follows:

$$\kappa = (\mu'_{1st}-\mu)/\bar{\theta}_f \tag{27}$$

If γ_0 under the 1st loading is denoted by $\gamma_{0_{1st}}$, the parameter γ_0 is similarly expressed from Eq. 25 by the following equation.

$$\gamma_0 = \gamma_{0_{1st}} \cdot \exp\left(\frac{\lambda\cdot(-\bar{\theta}_0)}{\kappa\cdot(\bar{\theta}_f-\bar{\theta}_0)}\right) \tag{28}$$

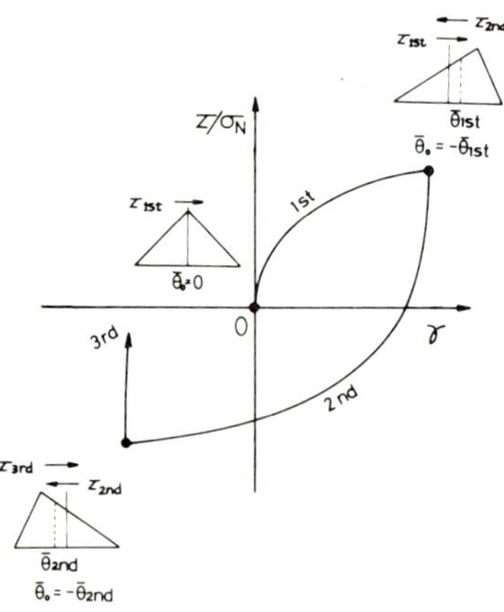

Fig. 17 Conceptional relationship between cyclic stress-strain behaviour and variation in frequency distribution of θ

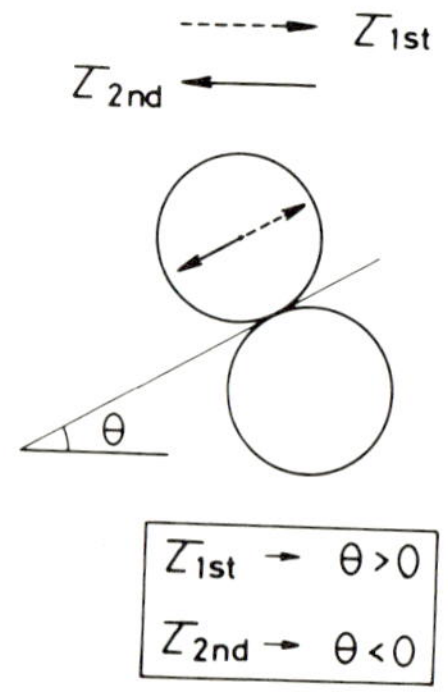

Fig. 18 Positive and negative signs of θ under cyclic shear stress

The parameters λ and μ are considered to be constant and $\bar{\theta}_f$ can be assumed to be constant under cyclic loading. Therefore, the parameters μ' and γ_0 under cyclic loading can be estimated from Eqs. 23, 26, 27 and 28 using parameters such as λ, μ, μ'_{1st}, $\gamma_{0\,1st}$ and $\bar{\theta}_f$ which are obtained from the 1st loading test (the conventional monotonous loading test).

EXTENSION TO THREE-DIMENSIONAL STATE BY "SPATIAL MOBILIZED PLANE (SMP)"

In order to extend the two-dimensional stress-strain relationship derived from the microscopic consideration in the preceding sections to the three-dimensional one, three two-dimensional sliding planes of soil particles (called "mobilized planes") are supposed between the respective two principal stress directions. The stress states on the three mobilized planes correspond to the three points at which three straight lines from the origin are tangent to three Mohr's stress circles as shown in Fig. 19. When three different principal stresses σ_1, σ_2 and σ_3 are applied to the directions of I, II and III axes, three two-dimensional mobilized planes between the respective two principal stress directions are expressed as shown in Fig. 20. The above mentioned two-dimensional stress-strain relationship has been applicable to each of the three mobilized planes, which have been named "Compounded Mobilized Planes" (Matsuoka, 1974b). Paying further attention to a plane ABC with these three mobilized planes as the three sides, it has been defined as "Spatial Mobilized Plane (SMP)" (Matsuoka and Nakai, 1974, 1978). The SMP is considered to be a plane on which soil particles are most mobilized on the average in the three-dimensional stresses. The direction cosines (a_1, a_2, a_3) of the normal of the SMP are expressed as follows:

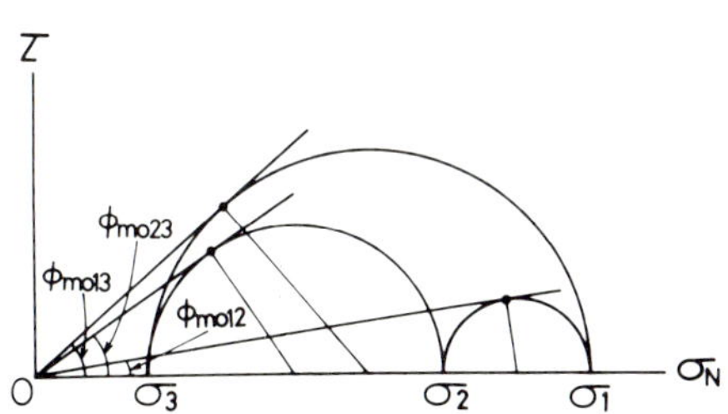

Fig. 19 Three mobilized planes where shear-normal stress ratio is maximized under respective two principal stresses

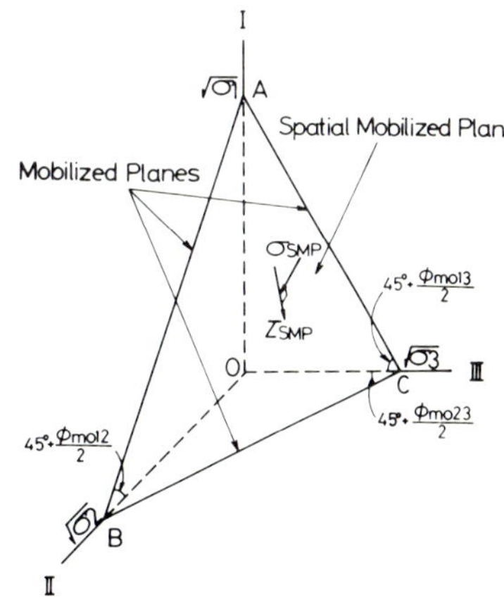

Fig. 20 Three two-dimensional mobilized planes (AB, BC and AC) and spatial mobilized plane (ABC)

$$a_i = \sqrt{J_3/(\sigma_i J_2)} \qquad (i=1,2,3) \tag{29}$$

where J_1, J_2 and J_3 are the first, second and third effective stress invariants which are expressed by the following equations.

$$J_1 = \sigma_1+\sigma_2+\sigma_3\ , \quad J_2 = \sigma_1\sigma_2+\sigma_2\sigma_3+\sigma_3\sigma_1\ , \quad J_3 = \sigma_1\sigma_2\sigma_3 \tag{30}$$

The normal stress σ_{SMP}, the shear stress τ_{SMP} and the shear-normal stress ratio τ_{SMP}/σ_{SMP} on the SMP can be expressed as follows:

$$\sigma_{SMP} = \sigma_1 \cdot a_1^2+\sigma_2 \cdot a_2^2+\sigma_3 \cdot a_3^2 = 3J_3/J_2 \tag{31}$$

$$\tau_{SMP} = \sqrt{(\sigma_1-\sigma_2)^2 \cdot a_1^2 \cdot a_2^2 + (\sigma_2-\sigma_3)^2 \cdot a_2^2 \cdot a_3^2 + (\sigma_3-\sigma_1)^2 \cdot a_3^2 \cdot a_1^2}$$

$$= \sqrt{J_1J_2J_3-9J_3^2}\ /J_2 \tag{32}$$

$$\tau_{SMP}/\sigma_{SMP} = \sqrt{(J_1J_2-9J_3)/(9J_3)} \tag{33}$$

Under axially symmetric stress conditions ($\sigma_1>\sigma_2=\sigma_3$ or $\sigma_1=\sigma_2>\sigma_3$), Eq. 33 is expressed as follows:

$$\tau_{SMP}/\sigma_{SMP} = \sqrt{2}\ (\sqrt{\sigma_1/\sigma_3} - \sqrt{\sigma_3/\sigma_1}\ )/3 \tag{34}$$

As the average sliding direction of soil particles is considered to coincide with the direction of the principal strain increment vector ($d\varepsilon_1$, $d\varepsilon_2$, $d\varepsilon_3$), the normal and parallel components of the principal strain increment vector to the SMP are marked and denoted by $d\varepsilon^*_{SMP}$ and $d\gamma^*_{SMP}$ respectively (Nakai and Matsuoka, 1980).

$$d\varepsilon^*_{SMP} = d\varepsilon_1 \cdot a_1+d\varepsilon_2 \cdot a_2+d\varepsilon_3 \cdot a_3 \tag{35}$$

$$d\gamma^*_{SMP} = \sqrt{(d\varepsilon_1 \cdot a_2-d\varepsilon_2 \cdot a_1)^2+(d\varepsilon_2 \cdot a_3-d\varepsilon_3 \cdot a_2)^2+(d\varepsilon_3 \cdot a_1-d\varepsilon_1 \cdot a_3)^2} \tag{36}$$

If Eqs. 19 and 20 are applicable to the SMP, the following equations are obtained.

$$\frac{\tau_{SMP}}{\sigma_{SMP}} = \lambda^* \cdot \left(-\frac{d\varepsilon^*_{SMP}}{d\gamma^*_{SMP}}\right)+\mu^* \tag{37}$$

$$\frac{\tau_{SMP}}{\sigma_{SMP}} = \lambda^* \cdot \left(-\frac{\varepsilon^*_{SMP}}{\gamma^*_{SMP}}\right)+\mu'^* \tag{38}$$

Combining Eqs. 37 and 38, and solving the differential equation, the following equations are derived.

$$d\gamma^*_{SMP} = \frac{\gamma_0^*}{\mu'^*-\mu^*} \cdot \exp\left(\frac{X - \mu^*}{\mu'^*-\mu^*}\right) \cdot dX \tag{39}$$

$$d\varepsilon^*_{SMP} = \frac{\mu^*-X}{\lambda^*} \cdot d\gamma^*_{SMP} \tag{40}$$

where $X \equiv \tau_{SMP}/\sigma_{SMP}$, and ($\lambda^*$, μ^*, μ'^* and $\gamma_0{}^*$) are soil parameters which have the same characteristics as (λ, μ, μ' and γ_0)in Eqs. 19-22. Under cyclic triaxial compression ($\sigma_1>\sigma_2=\sigma_3$) and triaxial extension ($\sigma_1=\sigma_2>\sigma_3$) tests, the direction of τ_{SMP} is reversed in the same section of the SMP as shown in Fig. 21, so the stress-strain relationships obtained from the two-dimensional analysis are considered to hold good also on the SMP.

ANALYSIS OF STRESS-STRAIN BEHAVIOURS UNDER CYCLIC LOADING

Figure 22 shows the results of triaxial compression and triaxial extension tests on the Toyoura sand (the average grain size D_{50}=0.2mm, the uniformity coefficient U_c=1.3, the specific gravity G_s=2.65, the maximum void ratio e_{max}=0.95, the minimum void ratio e_{min}=0.58 and the initial void ratio prior to shear $e_0 \fallingdotseq 0.68$). It can be seen from this figure that the shear strength is nearly constant ($\bar{\theta}_f \fallingdotseq$const.) and γ^*_{SMP} from the start of shear up to failure is also nearly constant ($(\gamma^*_{SMP})_f \fallingdotseq$const.) regardless of shear stress history, which are assumptions for this analysis as mentioned before. Figure 23 shows the τ_{SMP}/σ_{SMP} vs. $-d\varepsilon^*_{SMP}/d\gamma^*_{SMP}$ relationship arranged from repetitional triaxial compression and extension test results on the Toyoura sand. In this test the

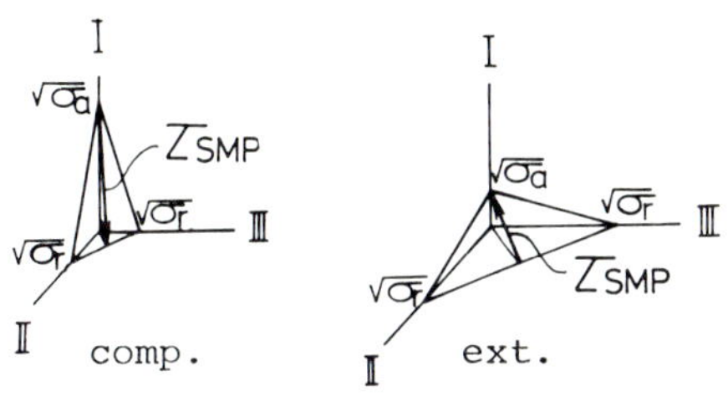

Fig. 21 Direction of τ_{SMP} under triaxial compression and triaxial extension conditions

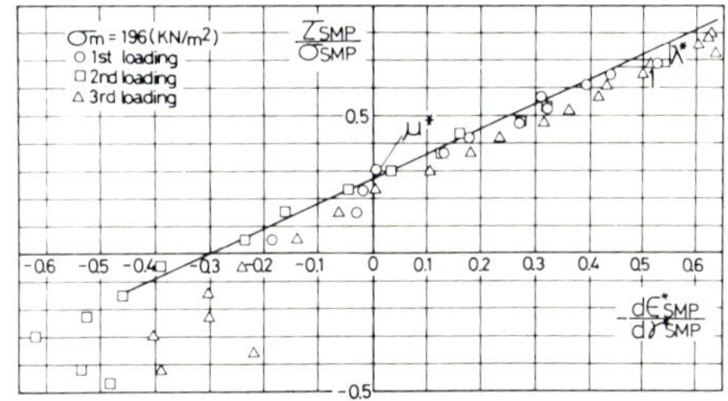

Fig. 23 τ_{SMP}/σ_{SMP} vs. $-d\varepsilon^*_{SMP}/d\gamma^*_{SMP}$ relationships obtained from cyclic triaxial compression and extension test on Toyoura sand

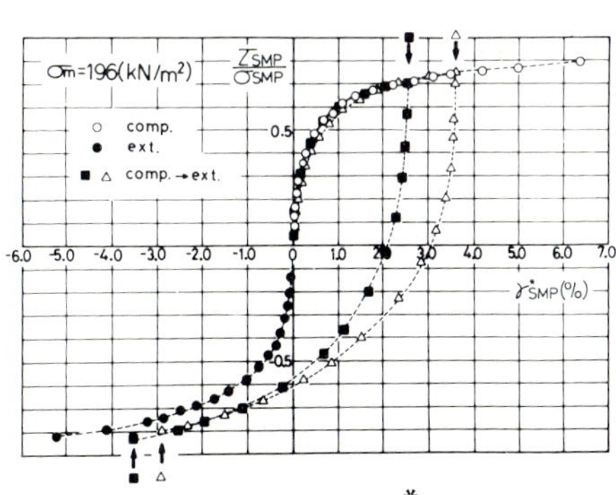

Fig. 22 τ_{SMP}/σ_{SMP} vs. γ^*_{SMP} relationships obtained from cyclic triaxial compression and extension tests on Toyoura sand

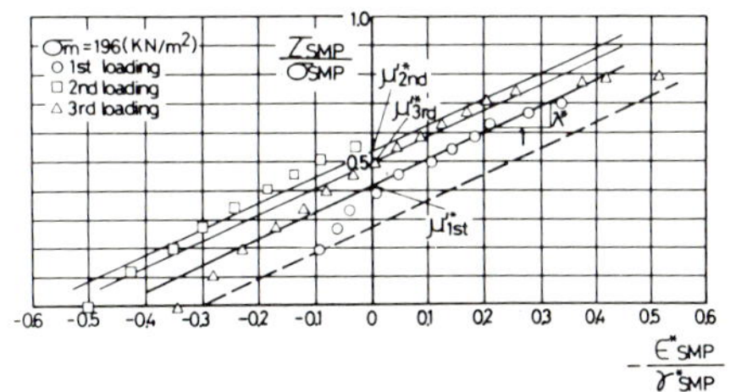

Fig. 24 τ_{SMP}/σ_{SMP} vs. $-\varepsilon^*_{SMP}/\gamma^*_{SMP}$ relationships obtained from cyclic triaxial compression and extension test on Toyoura sand

principal stress ratio is applied up to 4.0 at the compression side for the 1st loading, and then it is applied up to 3.0 at the extension side for the 2nd loading. Finally it is applied up to failure at the compression side for the 3rd loading. It is seen from this figure that λ^* and μ^* are nearly constant under cyclic loading (λ^*=0.9 and μ^*=0.27). As the principal stress ratio at failure R_f is 4.7, $\bar{\theta}_f$=0.60 is obtained from Eqs. 5 and 34. Figure 24 shows the τ_{SMP}/σ_{SMP} vs. $-\varepsilon^*_{SMP}/\gamma^*_{SMP}$ relationship arranged from the same test results as in Fig. 23. From this figure μ'^*_{1st} is seen to be 0.41, and so κ=0.23 is calculated from Eq. 27 for $\bar{\theta}_f$=0.60. The values of μ'^* for the 2nd and 3rd loadings are thus calculated from Eqs. 23, 26 and 34, and the upper two solid lines in Fig. 24 can be drawn for λ^*=0.9. Figure 25 represents the comparison between the repetitional triaxial compression and extension test results (the maximum principal ratio R_{max}=4.0) on the Toyoura sand (Kanaya and Suzuki, 1979), and the calculated values. In Fig. 25 the measured stress-strain curve under the 3rd loading is more gentle than that under the 1st loading, and the calculated curve also shows such tendency of the measured values.

Combining Eqs. 23 and 28 to estimate the parameters μ' and γ_0 with the constitutive equation for soils based on the SMP (Nakai and Matsuoka, 1981), the analyses of the undrained repetitional shear behaviours under triaxial compression and extension stress conditions are performed (Matsuoka et al., 1982). The undrained condition corresponds to the condition that the volumetric strain increment is always zero for saturated samples. The stress-strain parameters used are those for the relatively dense Toyoura sand; λ^*=0.9, μ^*=0.27, μ'^*_{1st}= 0.41, $\gamma_0{}^*_{1st}$=0.10%, $C_c/(1+e_0)$=0.928%, $C_s/(1+e_0)$=0.578%, the principal stress ratio at failure R_f=4.7 (the internal friction angle ϕ_d=40°).

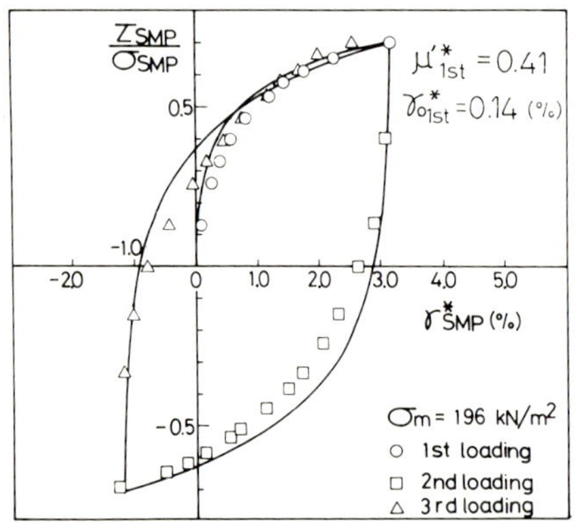

Fig. 25 Comparison between cyclic triaxial compression and extension test results and calculated values

C_c is the compression index, C_s the swell index and e_0 the initial void ratio. C_c is adopted when the effective mean principal stress p (=(σ_1+σ_2+σ_3)/3) is increased and C_s is adopted when p is decreased. The initial effective mean principal stress p_0 is 98 kN/m^2.

Figure 26 shows the analytical result of the effective stress path for the maximum principal stress difference q= 39.2 kN/m^2 in undrained repetitional triaxial compression and extension test (q=σ_a-σ_r, σ_a: axial principal stress, σ_r: radial principal stress). In this figure η_f, η_m and η_μ represent the values of q/p at failure, at the "phase transformation" (Ishihara et al., 1975) and at $\tau_{SMP}/\sigma_{SMP}=\mu^*$ corresponding to the coefficient of interparticle friction respectively. The value of η_m can be determined from the condition of the effective mean principal stress increment dp=0 under undrained conditions (the volumetric strain increment $d\varepsilon_v=0$ under drained conditions). The analytical result in Fig. 26 explains well such experimental tendency as the effective stress path is returned to the positive direction of p after q/p exceeds the η_m-line. It is also seen from Fig. 26 that the increase in the excess pore water pressure is larger at the extension side (at q<0) than at the compression side (at q>0). This is due to the reason that even if q is constant, τ_{SMP}/σ_{SMP} is higher at the extension side than at the compression side. In other words, the yield criterion based on the SMP (τ_{SMP}/σ_{SMP}=const.) is different from the extended Mises criterion (q/p=const.). From the same reason, η_f and η_μ have different values at the compression and extension sides. Figure 27 shows the calculated principal stress difference q vs.

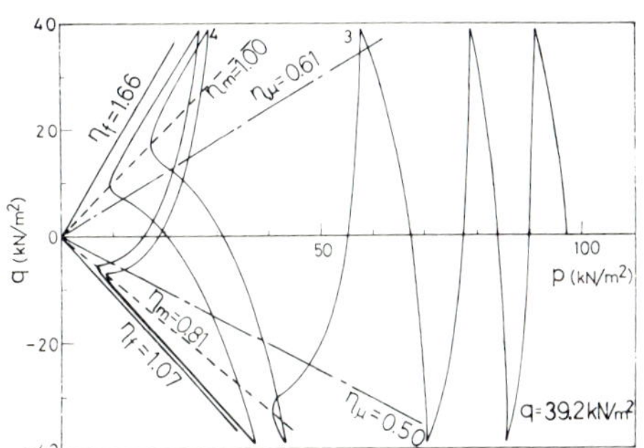

Fig.26 Calculated effective stress path in undrained cyclic triaxial compression and extension test (q=39.2kN/m^2)

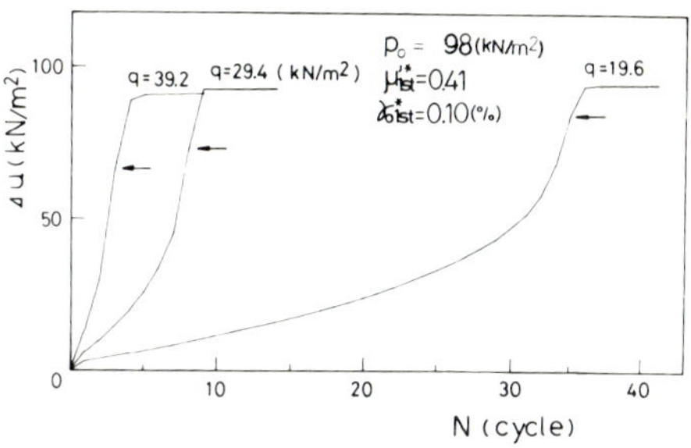

Fig.28 Calculated cumulative excess pore water pressure vs. number of cycles relationship

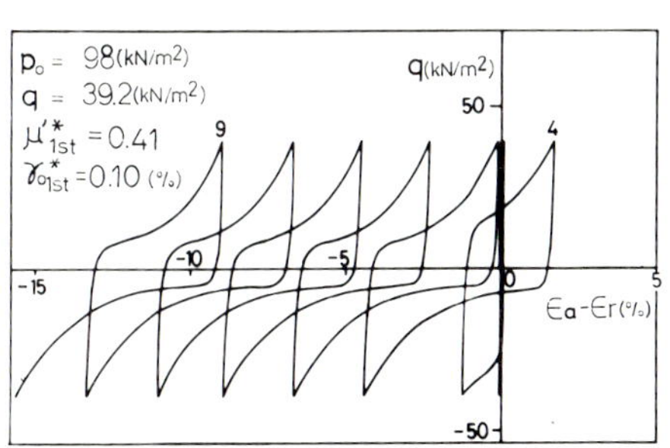

Fig. 27 Calculated stress-strain relationship in undrained cyclic triaxial compression and extension test (q=39.2kN/m^2)

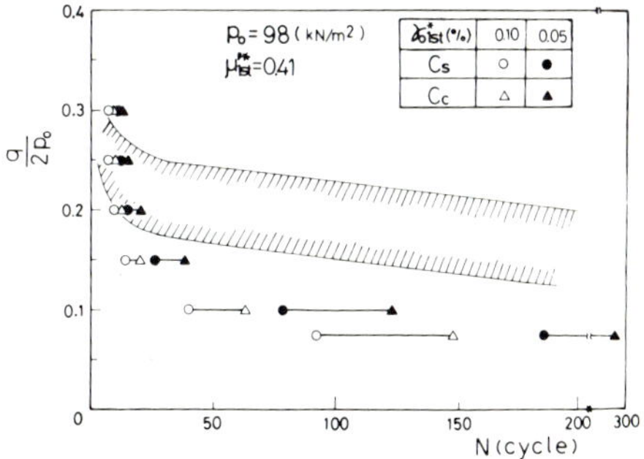

Fig.29 Relationship between cyclic stress ratio and number of cycles to produce initial liquefaction

principal strain difference ($\varepsilon_a-\varepsilon_r$) relationship ($\varepsilon_a$: axial principal strain, ε_r: radial principal strain). After the effective stress path exceeds the η_m-line, the principal strain difference increases suddenly to its negative direction. Figure 28 shows the calculated cumulative excess pore water pressure Δu vs. number of cycles N relationship. The values of q in this figure mean the maximum principal stress differences under cyclic loading. The arrow marks in Fig. 28 correspond to the points that the effective stress paths exceed the η_μ-line. As Δu reaches the peak value in one cycle after the arrow mark, the η_μ-line may be considered to be an index to predict the threshold of liquefaction phenomena. Figure 29 represents the relationship between the cyclic stress ratio $q/(2p_0)$ and the number of cycles to produce the initial liquefaction. The initial liquefaction is here defined as $|\varepsilon_a-\varepsilon_r|\geqq 15\%$. In this figure the analytical results are shown by four kinds of plots corresponding to the four cases of $\gamma_0{}^*_{1st}=0.10\%$ and $\gamma_0{}^*_{1st}=0.05\%$ using C_s or C_c. The shadowed portion represents the range of experimental results on the Toyoura sand (Ishihara, 1976). The analytical values do not agree with the experimental values at a large number of cycles. However, considering that the wave numbers by real earthquakes are no more than 20, Fig. 29 may be considered to be good enough for the practical use. It is interesting to have analyzed the problem of liquefaction by the parameters obtained from drained monotonous triaxial tests without performing undrained vibrational triaxial tests.

CONCLUSIONS

The main results are summarized as follows:

1) A new relationship between strains and fabric change of granular materials has been derived from paying attention to "two particles" across the potential sliding plane and considering the mechanism of disappearance and generation of interparticle contacts with the third particle. This relationship is considered to be one of the expressions for strains of granular materials.

2) Combining the above relationship between strains and fabric change with that between stresses and fabric change proposed by Matsuoka (1974a), a stress-strain relationship has been formulated under two-dimensional stress conditions.

3) Supposing three two-dimensional sliding planes of soil particles (called "mobilized planes") between the respective two principal stress directions, and introducing the new concept of "Spatial Mobilized Plane (SMP)" which is a plane with these three mobilized planes as the three sides, the stress-strain relationship has been extended to that in the three-dimensional state.

4) The stress-strain behaviours under cyclic loading and the liquefaction phenomena have been analyzed by the proposed model and the analytical results have been compared with the results of cyclic triaxial compression and extension tests on sand. The parameters used for these analyses can be obtained only from

conventional monotonous loading tests.

ACKNOWLEDGEMENT

The author wishes to thank Prof. T. Yamauchi and Dr. T. Nakai at Nagoya Institute of Technology, and Prof. F. Tatsuoka at University of Tokyo for their helpful advice. The author is also deeply indebted to Messrs. K. Takeda and H. Geka for their experimental support and discussion.

REFERENCES

1 Geka, H. and Matsuoka, H. (1981): "A stress-strain model for granular materials derived from microscopic shear mechanism", Proc., 36th Annual Meeting of JSCE, pp. 13-15 (in Japanese).
2 Ishihara, K. (1976): "Fundamentals of soil dynamics", Kajima-shuppankai (in Japanese).
3 Ishihara, K., Tatsuoka. F. and Yasuda, S. (1975): "Undrained deformation and liquifaction of sand under cyclic stresses", Soils and Foundations, Vol.15, No.1, pp. 29-44.
4 Kanaya, K. and Suzuki, M. (1979): "Deformation characteristics of a sand under axi-symmetric stress condition", Graduation Thesis at Nagoya Institute of Technology (in Japanese).
5 Matsuoka, H. (1974a): "A microscopic study on shear mechanism of granular materials", Soils and Foundations, Vol.14, No.1, pp. 29-43.
6 Matsuoka, H. (1974b): "Stress-strain relationships of sand based on the mobilized plane", Soils and Foundations, Vol.14, No.2, pp. 47-61.
7 Matsuoka, H., Geka, H. and Yasui, O. (1982): "Cyclic shear behavior of granular materials considering mechanism of fabric change", Proc., 17th Japan National Conf. SMFE, pp. 1673-1676 (in Japanese).
8 Matsuoka, H. and Nakai, T. (1974): "Stress-deformation and strength characteristics of soil under three different principal stresses", Proc., JSCE, No.232, pp. 59-70.
9 Matsuoka, H. and Nakai, T. (1978): "A generalized frictional law for soil shear deformation", Proc., U.S.-Japan Seminar on Continuum Mechanics and Statistical Approaches in the Mechanics of Granular Materials, Sendai, pp. 138-154.
10 Matsuoka, H. and Takeda, K. (1980): "A stress-strain relationship for granular materials derived from microscopic shear mechanism", Soils and Foundations, Vol.20, No.3, pp. 45-58.
11 Nakai, T. and Matsuoka, H. (1980): "A unified law for soil shear behavior under three-dimensional stress condition", Proc., JSCE, No.303, pp. 65-77 (in Japanese).
12 Nakai, T. and Matsuoka, H. (1981): "A unified law for soil deformation behavior under various stress paths", Proc., JSCE, No. 306, pp. 23-34 (in Japanese).

Mechanics of Granular Materials: New Models and Constitutive Relations, edited by
J.T. Jenkins and M. Satake, 1983
Elsevier Science Publishers B.V., Amsterdam — Printed in The Netherlands

A GENERALIZED RELATIONSHIP BETWEEN THE STRESS AND THE DILATANCY IN GRANULAR MATERIALS

Y. KISHINO

Civil Engineering Dept., Tohoku University, Sendai (Japan)

ABSTRACT

The theoretical relationship between stress and dilatancy is obtained through the consideration of the mechanism of dissipation of energy in the cohesionless granular materials. The proposed relationship shows good agreements with experimental data of different states of stress, so that it may be said that this theory is the generalization of Roscoe's theory which neglects the difference of stress states in the formulation of the relationship between stress and dilatancy. The proposed relationship is shown to be related with the yield locus which is adjustable to the Mohr-Coulomb criterion for the triaxial stress states. The appendix is the discussion of the inhomogeneity of deformation field which is important for the statistical specification of dissipative energy.

INTRODUCTION

Roscoe and his co-workers developed the so-called Granta Gravel Model (ref.1) for the prediction of plastic behavior of cohesionless soils. In their model, the relationship between stress and dilatancy was successfully derived through an assumption about the frictional dissipation of energy. However, it has been pointed out by many experimental researchers that the frictional constant in their model is not uniquely determined for different stress states. This inconsistency seems to arise from the specification of the dissipative energy in their model. On the other hand, Rowe (ref.2) presented the stress-dilatancy relation from the consideration of the sliding condition of spherical grains in two dimensional arrangements. Even though the resulting relation has been verified by experimental data of several stress states (ref.3), it can not be appropriately expressed in the tensor form.

In this paper, the generalized relationship between stress and dilatancy is derived from tensorial and statistical consideration of the dissipation of energy in cohesionless granular materials. Throughout this paper, grains are assumed to be rigid and the elastic deformation of the granular material is neglected.

DISSIPATION OF ENERGY IN GRANULAR MATERIAL

RATE OF WORK

In spite of inhomogeneous characteristics of stress and deformation in the granular material, the rate of work per unit volume may be expressed in terms of the stress $\underset{\sim}{\sigma}$ and the strain rate $\underset{\sim}{d}$:

$$\dot{W} = \underset{\sim}{\sigma} \cdot\cdot \underset{\sim}{d} = \frac{1}{3} \operatorname{tr} \underset{\sim}{\sigma} \operatorname{tr} \underset{\sim}{d} + \underset{\sim}{\sigma}' \cdot\cdot \underset{\sim}{d}', \qquad (1)$$

where primed quantities denote the deviatoric parts of tensors. If we assume that the principal axes of stress and strain rate coinside, then we get the equation,

$$\dot{W} = p\,\dot{\varepsilon}_v + \{(3+\mu\nu)/\sqrt{(3+\mu^2)(3+\nu^2)}\}\; q\,\dot{\gamma}\;, \qquad (2)$$

where

$$p = \frac{1}{3} \operatorname{tr} \underset{\sim}{\sigma}\;,\quad \dot{\varepsilon}_v = \operatorname{tr} \underset{\sim}{d}\;,\quad q = \frac{\sqrt{6}}{2} \sqrt{\underset{\sim}{\sigma}' \cdot\cdot \underset{\sim}{\sigma}'},\quad \dot{\gamma} = \frac{\sqrt{6}}{3} \sqrt{\underset{\sim}{d}' \cdot\cdot \underset{\sim}{d}'},$$

$$\mu = (2\sigma_2 - \sigma_3 - \sigma_1)/(\sigma_3 - \sigma_1)\;,\quad \nu = (2d_2 - d_3 - d_1)/(d_3 - d_1)$$

and σ_i (i=1,2,3) and d_i (i=1,2,3) are the principal values of $\underset{\sim}{\sigma}$ and $\underset{\sim}{d}$, respectively. In the following, the relationships,

$$\sigma_1 \geq \sigma_2 \geq \sigma_3\;, \qquad (3)$$

are assumed without the loss of generality.

Because the elastic deformation is neglected, all of the work is dissipated by frictional sliding between the grains. Therefore, it is necessary to specify a form for the rate of dissipation in granular material. The simplest specification has been given by Roscoe, et al. in their Granta Gravel Model (ref.1). The rate of dissipative energy in their model is in the form,

$$D = M\,p\,\dot{\gamma}\;, \qquad (4)$$

where M is a frictional constant. Because the frictional dissipation at each contact point is proportional to the product of contact force and relative slide, and because the latter quantities seem to be statistically proportional to the mean stress and the octahedral shear strain, the form of the above expression is valid as the first approximation. However, it should be noted that the product of the averages of distributed quantities is not equal to the average of the product of their values.

In the following sections we take into account the effect of spatial distribution of contact forces and relative slides in order to obtain more precise specification of dissipative energy.

DISSIPATIVE ENERGY OF MEAN QUANTITIES

In Fig. 1, typical grains A and B are in contact with each other at the contact point α. In terms of the normal component of contact force F_α, the rate of relative slide $\dot{S}_\alpha$ and the frictional coefficient k_α, the rate of dissipative energy per unit volume is

$$D = \frac{1}{V} \sum_\alpha k_\alpha F_\alpha \dot{S}_\alpha , \quad (5)$$

where V is the volume of the region under consideration.

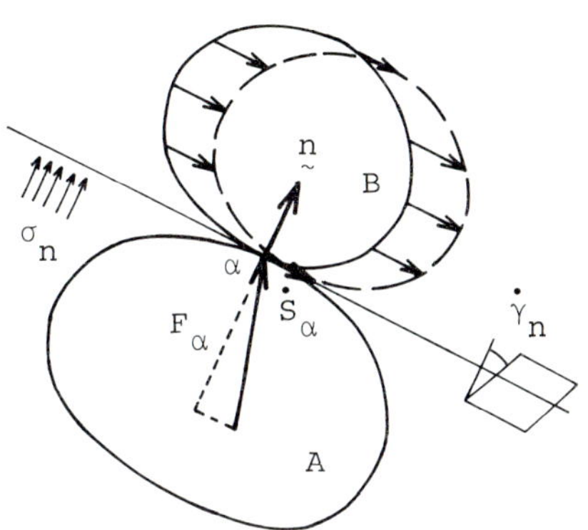

Fig. 1. Relative sliding between two grains

We assume that the average value of F_α is proportional to the normal component of stress σ_n acting on the contact plane with normal vector $\underset{\sim}{n}$, and that the average value of $\dot{S}_\alpha$ is proportional to the rate of shear strain $\dot{\gamma}_n$ in the direction of that contact plane. In general, the frictional coefficient for σ_n and $\dot{\gamma}_n$ should be given in the form of a density function $f(\underset{\sim}{n})$, so that the rate of dissipative energy of the mean quantities is expressed as

$$\bar{D} = \oint_A f(\underset{\sim}{n}) \; \sigma_n \; \dot{\gamma}_n \; da , \quad (6)$$

where A is the surface of the sphere with unit radius.

If we assume the isotropic function for $f(\underset{\sim}{n})$ and the coaxiality of principal directions of stress and strain rate, (6) becomes

$$\bar{D} = \frac{K}{2\pi} \sum_{i=1}^{3} \sigma_i \oint_A n_i^2 \sqrt{\sum_{j=1}^{3} (n_j d_j)^2 - (\sum_{j=1}^{3} n_j^2 d_j)^2} \; da , \quad (7)$$

where K is a constant. It is noted that $\bar{D}$ is always positive, because σ_i is always compressive in granular materials.

To calculate the integral of the right hand side of (7), we write $\bar{D}$ in the form,

$$\bar{D} = \frac{K}{4} \left(p\, I_0 + \sum_{i=1}^{3} \sigma_i' I_i \right) |d_1 - d_3| \;, \tag{8}$$

where

$$I_0 = \frac{1}{\pi} \oint_A \Gamma \, da \;, \tag{9}$$

$$I_i = \frac{1}{\pi} \oint_A n_i^2 \, \Gamma \, da \qquad (i=1,2,3) \tag{10}$$

and

$$\Gamma = \sqrt{(1+\nu)^2 n_1^2 + (1-\nu)^2 n_3^2 - \{(1+\nu) n_1^2 - (1-\nu) n_3^2\}^2} \;. \tag{11}$$

Clearly,

$$I_0 = I_1 + I_2 + I_3 \;. \tag{12}$$

Using the polynomial expression of $\Gamma = \Gamma(\nu)$, we obtain the following approximate expressions of (9) and (10):

$$\begin{aligned} I_0 &= \frac{1}{3}(7+\nu^2) \;; \\ I_1 &= \frac{1}{120}(101+19\nu+11\nu^2-3\nu^3) \;; \\ I_2 &= \frac{1}{20}(13+3\nu^2) \;; \\ I_3 &= \frac{1}{120}(101-19\nu+11\nu^2+3\nu^3) \;; \end{aligned} \tag{13}$$

which give the exact integrated values for Γ and $\partial\Gamma/\partial\nu$ at $\nu = \pm 1$. Substitution of (13) into (8) gives

$$\bar{D} = K \left\{ \frac{7+\nu^2}{4\sqrt{3+\nu^2}}\, p + \frac{23\mu+57\nu-7\mu\nu^2-9\nu^3}{240\sqrt{(3+\mu^2)(3+\nu^2)}}\, q \right\} \dot{\gamma} \;. \tag{14}$$

DISSIPATIVE ENERGY OF RESIDUAL QUANTITIES

In the specification of the dissipative energy in granular materials, the inhomogeneity of mechanical quantities (ref.4) should be taken into account. Let $\bar{F}_\alpha$ and $\dot{\bar{S}}_\alpha$ be the mean quantities related to σ_n and $\dot{\gamma}_n$, and ΔF_α and $\Delta \dot{S}_\alpha$ be the residual quantities, so that F_α and $\dot{S}_\alpha$ are given by the following equations:

$$F_\alpha = \bar{F}_\alpha + \Delta F_\alpha \; ; \tag{15}$$

$$\dot{S}_\alpha = \bar{\dot{S}}_\alpha + \Delta \dot{S}_\alpha \; . \tag{16}$$

If we assume that the frictional coefficient k_α has no correlation with F_α and $\dot{S}_\alpha$, the substitution of (15) and (16) into (5) gives

$$D = \bar{D} + \Delta D \; , \tag{17}$$

where

$$\bar{D} = \frac{\bar{k}}{V} \sum_\alpha \bar{F}_\alpha \, \bar{\dot{S}}_\alpha \; , \tag{18}$$

$$\Delta D = \frac{\bar{k}}{V} \sum_\alpha \Delta F_\alpha \, \Delta \dot{S}_\alpha \tag{19}$$

and $\bar{k}$ is the mean value of k_α.

As may be observed in (19), the contribution of ΔD to the macroscopical dissipative energy is depend on the degree of statistical correlation between ΔF_α and $\Delta \dot{S}_\alpha$. The contact forces in granular materials, even if they are acting on the contact planes facing the same direction, are not homogeneous, and it is quite natural to assume that the smaller relative sliding takes place for the larger contact force and vice versa. Therefore, it may be concluded that the random variables ΔF_α and $\Delta \dot{S}_\alpha$ have negative correlation. There has been no definite statistical predictions about the above variables, but it seems from the test of granular model (see Appendix) that the standard deviation of ΔF_α and $\Delta \dot{S}_\alpha$ are proportional to p and $\dot{\gamma}$ respectively.

Thus we assume that the dissipative energy of residual quantities is given in the form,

$$\Delta D = - \lambda \, p \, \dot{\gamma} \; , \tag{20}$$

where λ is a constant. From (14), (17) and (20), a necessary condition for the positiveness of the dissipative energy is given by

$$\lambda \leqq K \; . \tag{21}$$

RELATIONSHIP BETWEEN STRESS AND DILATANCY

GENERAL STRESS STATE

If all of the work is dissipated with the internal friction, the equilibrium condition of energy is

$$\dot{W} = \overline{D} + \Delta D \,. \tag{22}$$

The substitution of (2), (14) and (20) into the above equation gives the generalized relationship between stress and dilatancy,

$$\frac{q}{p} = \frac{240\sqrt{(3+\mu^2)(3+\nu^2)}}{240(3+\mu\nu)-K(23\mu+57\nu-7\mu\nu^2-9\nu^3)}\left\{\frac{K(7+\nu^2)}{4\sqrt{3+\nu^2}} - \lambda - \frac{\dot{\varepsilon}_V}{\dot{\gamma}}\right\} \,. \tag{23}$$

The above equation is a general one, but for the determination of unique relationship between stress and dilatancy the quantities μ and ν must be related with each other. The corresponding relation in the classical plasticity theory is the Mises-Reuss flow law ($\mu = \nu$). In the following we discuss how to relate these quantities for granular materials.

The gradient of the curve relating dilatancy and shear strain,

$$g = \dot{\varepsilon}_V/\dot{\gamma} \,, \tag{24}$$

may be regarded as an index which characterizes the state of the strain hardening. We assume that the value of g for a certain plastic state is constant in the π plane of principal stress space.

Considering the loading process in which p and μ are preserved, the plastic state specified by a certain value of g would be realized by the minimum value of q. Thus, at least for a monotonic loading process, the relationship between μ and ν may be obtained by the condition,

$$\frac{\partial}{\partial\nu}\left(\frac{q}{p}\right) = 0 \,. \tag{25}$$

From (23) and (25), we obtain

$$\mu = \frac{3K\sqrt{3+\nu^2}\{480\nu+K(133-82\nu^2-3\nu^4)\}-4(\lambda+g)\{720\nu+K(171-81\nu^2-18\nu^4)\}}{48K\sqrt{3+\nu^2}(35-5\nu^2+3K\nu)-4(\lambda+g)\{720+K(65\nu+7\nu^3)\}} \,. \tag{26}$$

According to this equation, the quantities μ and ν are uniquely related for the practical values of K, λ and g. Then, the yield locus in the π plane is obtained by the substitution of (26) into (23). In Fig. 2, an example of yielding locus in the π plane is shown.

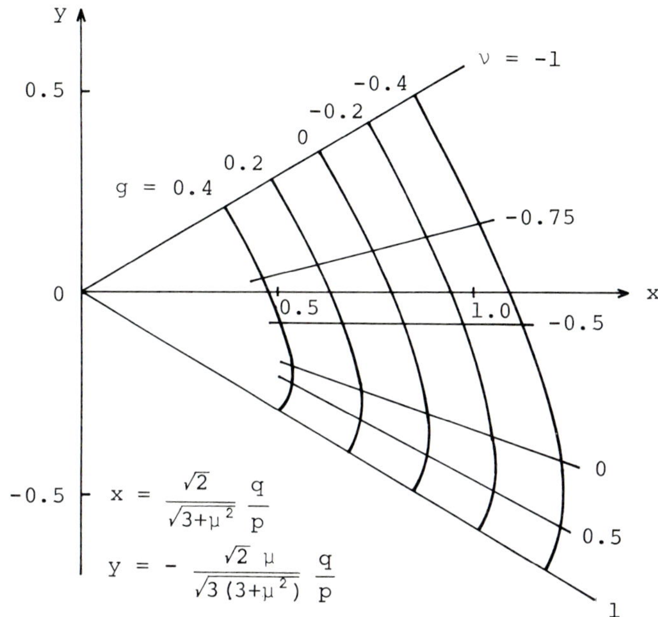

Fig. 2. Yield locus in the π plane ($K=2.5$, $\lambda=1.5$)

TRIAXIAL STRESS STATE

For the special cases of (26), $\mu = \pm 1$ corresponds to $\nu = \pm 1$. Thus, it is concluded for the triaxial stress states that the strain is also triaxial. The relationship between stress and dilatancy for these states is

$$\frac{q}{p} = \frac{1}{1 \mp \frac{K}{15}} (K - \lambda - g) \quad (\mu = \pm 1) . \tag{27}$$

The theoretical relationship is compared with the experimental data of Matsuoka and Nakai (ref.5) in Fig. 3. The quantities attached the subscript "oct" are representing the tensor components of the octahedral plane. From this comparison as well as others, the theoretical relationship presented in this paper shows fairly good agreements with experiments.

The ratio of the values of q for the compressive and extensile stress coincide with that of the Mohr-Coulomb criterion, if

$$K = 5 \sin \phi \ , \quad \lambda = 3 \sin \phi \tag{28}$$

where ϕ is the angle of internal friction.

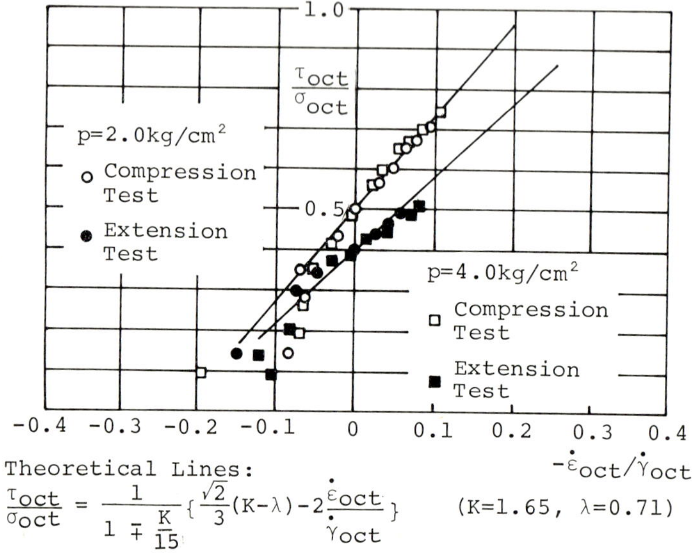

Fig. 3. Comparison with the tests on Toyoura Sand (ref. 5)

CONCLUDING REMARKS

In this paper, the generalized relationship between stress and dilatancy is derived through the consideration of the dissipation of energy in the granular materials. The dissipative energy defined here has the form, $D(\underset{\sim}{\sigma}, \underset{\sim}{d}')$. This function may not necessarily be expressed as $D(\underset{\sim}{d}')$ as is defined by Drucker and Prager (ref.6). The latter expression of the dissipative energy appears in the viscous material model, which seems to be different from the assemblage of cohesionless grains. The definition of dissipative energy, as well as the fundamental idea about the energy balance presented in this paper, will be usefull for the constitution of generalized theory of granular mechanics.

(APPENDIX) MEASUREMENT OF INHOMOGENEITY OF DEFORMATION FIELD IN TWO-DIMENSIONAL GRANULAR MODEL

The tensor which expresses correlation between components of the residual displacement vectors has been defined by Kishino (ref.4). For the discrete system of assemblage of rigid grains, the correlation tensor is calculated by

$$\underset{\sim}{T} = \sum_{R} \Delta\underset{\sim}{u}\, \Delta\underset{\sim}{u} , \tag{A1}$$

where $\Delta\underset{\sim}{u}$ is the residual displacement vector of grain and R is the region for the statistical calculation. The tensor $\underset{\sim}{T}$ was calculated for the series of pictures obtained by the simple shear test of two-dimensional granular model in which about 400 rods with three different diameters (8, 10, 12 mm) are randomly packed. The area in which the statistical calculation was performed was the inner part of circle with diameter of 160 mm. Fig. A1 shows the relationship between the strain of shear frame and the shearing force applied at a corner of the frame. The numbers attached in Fig. A1 represent the stages at which the pictures were taken.

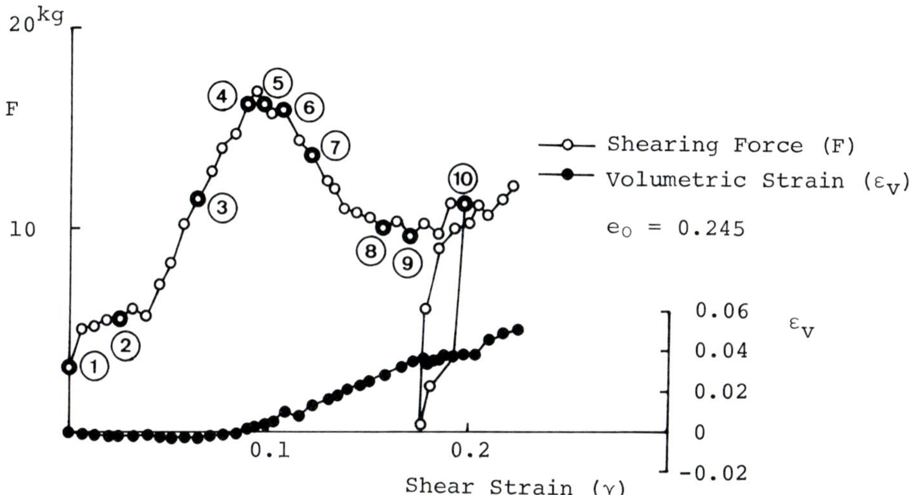

Fig. A1. Relationship between shearing force and strain

The coordinates of grain centers at each stage were obtained with the use of a photo-analysing system. The displacement vectors of grain centers and the tensor $\underset{\sim}{T}$ were calculated from the values of coordinates of grain centers at successive stages. The Mohr's circle of $\underset{\sim}{T}$ were drawn and it was observed that they are located between two envelope curves. Thus, it may be said that the degree of inhomogeneous deformation (the diameter of Mohr's circle) is a variable dependent on the mean inhomogeneity of deformation (tr $\underset{\sim}{T}$).

Fig. A2 shows the relationship between shear strain and parameter of inhomogeneity of deformation ($\sqrt{tr\ \tilde{T}}$). If we take into account the initial unstableness of the model test, it can be assumed that the standard deviation of residual displacement is proportional to the shear strain. From this conclusion, the standard deviation of the residual sliding between particles may be considered to be proportional to the shear strain.

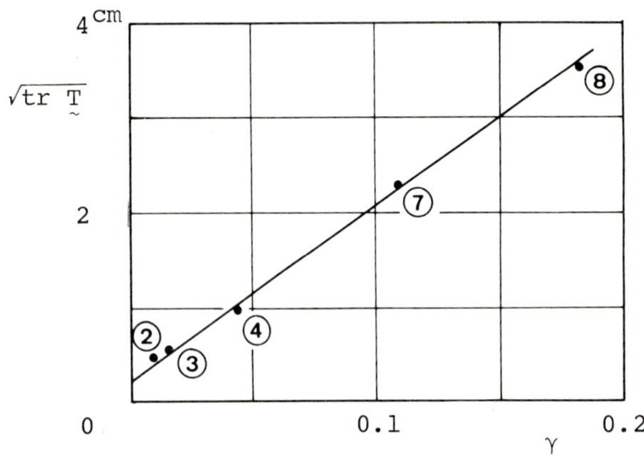

Fig. A2. Relationship between shear strain and parameter of inhomogeneity of deformation

REFERENCES

1 A. Schofield and P. Wroth, Critical State Soil Mechanics, McGraw-Hill, New York, 1968, 89 pp.
2 P.W. Rowe, The Stress-Dilatancy Relation for Static Equilibrium of an Assembly of Particles in Contacts, Proc. Roy. Soc. A., 269 (1962) 500-527.
3 L. Barden and A.J. Khayatt, Incremental Strain Rate Ratios and Strength of Sand in Triaxial Test, Géotechnique, 16, 4 (1966) 338-357.
4 Y. Kishino, Statistical Consideration on Deformation Characteristics of Granular Materials, Proc. of the U.S.-Japan Seminar on Continuum Mechanical and Statistical Approaches in the Mechanics of Granular Materials, Gakujutsu Bunken Fukyu-Kai, Tokyo, 1978, 114-122.
5 H. Matsuoka and T. Nakai, A Generalized Frictional Law for Soil Shear Deformation, Proc. of the U.S.-Japan Seminar on Continuum Mechanical and Statistical Approaches in the Mechanics of Granular Materials, Gakujutsu Bunken Fukyu-Kai, Tokyo, 1978, 138-153.
6 D.C. Drucker and W. Prager, Soil Mechanics and Plastic Analysis or Limit Design, Q. Appl. Math. 10 (1952) 157-165.

Mechanics of Granular Materials: New Models and Constitutive Relations, edited by
J.T. Jenkins and M. Satake, 1983
Elsevier Science Publishers B.V., Amsterdam — Printed in The Netherlands

CONSTITUTIVE EQUATIONS OF GRANULAR MATERIALS

K.HASHIGUCHI

Agricultural Engineering Dept., Kyushu University, Fukuoka 812 (Japan)

ABSTRACT

Various constitutive models for the description of anisotropic hardening have been proposed in the past. Among them the two-surface theory would be one of the most reasonable theories at present and has the simplest form. It has been, however, confined to the special materials, *viz*, metals or granular media. In this paper the two-surface theory is generalized so as to be applicable to most of elastoplastic materials. Further, based on the generalized theory, special constitutive equations of granular media are formulated, and they are compared with some experimental data.

INTRODUCTION

Previously Mróz (ref.1) proposed the model of a *field of hardening moduli* to describe the anisotropic hardening of metals by introducing multiple yield surfaces in a transitional state from an elastic to a distinct-yield state. Thereafter Mróz, Norris and Zienkiewicz (ref.2) formulated constitutive equations of soils according to the similar concept. On the other hand, Krieg (ref.3) and Dafalias and Popov (ref.4) showed that the anisotropic hardening characteristics of metals described by the Mróz model of a field of a hardening moduli can be expressed simply by introducing only two surfaces, and this simplified theory is called a *two-surface theory*. Also Mróz, Norris and Zienkiewicz (ref.5) formulated constitutive equations of soils along the concept of two-surface theory.

Two-surface theory would be one of the most reasonable and the simplest models to describe the anisotropic hardening. In the past this theory has been, however, formulated in special forms confined to metals or to soils.

In this paper the two-surface theory is to be formulated in general forms unconfined to any special material. Further, special constitutive equations of granular media are formulated according to the generalized theory. Finally they are compared with some experimental data.

BASIC CONSTITUTIVE EQUATIONS

A typical stress-strain curve of elastoplastic materials is shematically illustrated in Fig.1. Assume that the surface which represents stresses in the *distinct-yield state* shown by the envelope curve of reloading curves in this figure is described by the following equation (see Fig. 2).

$$f(\boldsymbol{\sigma} - \hat{\boldsymbol{\alpha}}) - F(K) = 0 \tag{1}$$

or

$$f(\hat{\boldsymbol{\sigma}}) - F(K) = 0, \tag{2}$$

where

$$\hat{\boldsymbol{\sigma}} \equiv \boldsymbol{\sigma} - \hat{\boldsymbol{\alpha}}. \tag{3}$$

The second-order tensor $\boldsymbol{\sigma}$ is a stress, and the scalar K and the second-order tensor $\hat{\boldsymbol{\alpha}}$ are parameters to describe, respectively, the extention or contraction and the translation of the surface. Let this surface be called a *distinct-yield surface*.

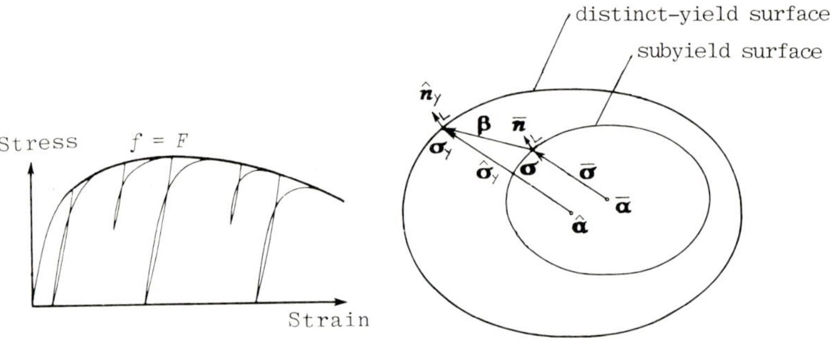

Fig.1 The distinct-yield state ($f = F$) illustrated as the envelope curve of reloading curves

Fig.2 The distinct-yield and the subyield surfaces

Now, we assume that the distinct-yield surface retains a similarity in a stress space. Therefore, the function f is to be a homogeneous function of its arguments. Then, let the degree of f be denoted by n.

Further, we introduce the secondary surface which is similar to the distinct-yield surface and translates within the distinct-yield surface (see Fig.2). And postulate that when a current stress exists on this surface, an elastoplastic deformation can occur but when within it, only an elastic one can. Hence, let the surface be represented by

$$f(\boldsymbol{\sigma} - \bar{\boldsymbol{\alpha}}) - r^{n}F(K) = 0 \tag{4}$$

or

$$f(\bar{\boldsymbol{\sigma}}) - r^{n}F(K) = 0, \tag{5}$$

where we set

$$\bar{\boldsymbol{\sigma}} \equiv \boldsymbol{\sigma} - \bar{\boldsymbol{\alpha}}. \tag{6}$$

r ($0 \leqq r \leqq 1$) is a material constant and the second-order tensor $\bar{\boldsymbol{\alpha}}$ is a parameter to describe a translation of the surface. Let this surface be called a *subyield surface*.

In what follows, the parameters K, $\hat{\boldsymbol{\alpha}}$ and $\bar{\boldsymbol{\alpha}}$ are defined.

Let $\dot{K}$ where a superposed dot designates a material time derivative be a function of stress, plastic strain and plastic strain rate $\dot{\boldsymbol{\varepsilon}}^{\mathrm{P}}$ in degree one, which satisfies the condition $\dot{K} = 0$ when $\dot{\boldsymbol{\varepsilon}}^{\mathrm{P}} = \mathbf{0}$.

Now, suppose that the current stress is on the subyield surface. Further, let the conjugate point on the distinct-yield surface having the same outer normal direction as that of the subyield surface at $\boldsymbol{\sigma}$ be designated by $\boldsymbol{\sigma}_{\mathrm{y}}$.

Then, let $\dot{\hat{\boldsymbol{\alpha}}}$ be given by

$$\dot{\hat{\boldsymbol{\alpha}}} = A\,\mathrm{tr}(\dot{\boldsymbol{\varepsilon}}^{\mathrm{P}}\mathbf{1})\mathbf{1} + B\,\mathrm{tr}(\dot{\boldsymbol{\varepsilon}}^{\mathrm{P}}\frac{\hat{\boldsymbol{\sigma}}_{\mathrm{y}}}{|\hat{\boldsymbol{\sigma}}_{\mathrm{y}}|})\frac{\hat{\boldsymbol{\sigma}}_{\mathrm{y}}}{|\hat{\boldsymbol{\sigma}}_{\mathrm{y}}|}$$

$$= A\dot{\varepsilon}_{\mathrm{v}}^{\mathrm{P}}\mathbf{1} + B\,\mathrm{tr}(\dot{\boldsymbol{\varepsilon}}^{\mathrm{P}}\frac{\hat{\boldsymbol{\sigma}}_{\mathrm{y}}}{|\hat{\boldsymbol{\sigma}}_{\mathrm{y}}|})\frac{\hat{\boldsymbol{\sigma}}_{\mathrm{y}}}{|\hat{\boldsymbol{\sigma}}_{\mathrm{y}}|} \tag{7}$$

or

$$\dot{\hat{\boldsymbol{\alpha}}} = A\dot{\varepsilon}_{\mathrm{v}}^{\mathrm{P}}\mathbf{1} + B\,\mathrm{tr}(\dot{\boldsymbol{\varepsilon}}^{\mathrm{P}}\frac{\bar{\boldsymbol{\sigma}}}{|\bar{\boldsymbol{\sigma}}|})\frac{\bar{\boldsymbol{\sigma}}}{|\bar{\boldsymbol{\sigma}}|} \tag{8}$$

by the relation

$$\hat{\boldsymbol{\sigma}}_{\mathrm{y}} = \frac{1}{r}\bar{\boldsymbol{\sigma}} \tag{9}$$

in accordance with the assumption of the similarity of the distinct-yield and the subyield surfaces. In the these equations

$$\dot{\boldsymbol{\varepsilon}}_{\mathrm{v}}^{\mathrm{P}} \equiv \mathrm{tr}(\dot{\boldsymbol{\varepsilon}}^{\mathrm{P}}),$$

$$\hat{\boldsymbol{\sigma}}_{\mathrm{y}} \equiv \boldsymbol{\sigma}_{\mathrm{y}} - \hat{\boldsymbol{\alpha}}. \tag{10}$$

A and B are scalar functions of K and $\hat{\boldsymbol{\alpha}}$, and the notation $|\;\;|$ is used to represent the magnitude. In (7) or (8) the first and the second term would be significant for granular media and metals, respectively.

Since the subyield surface must not intersect the distinct-yield surface but can come into contact with it, the direction of the relative motion of the conjugate points $\boldsymbol{\sigma}_y$ and $\boldsymbol{\sigma}$ must coincide with that of the vector $\boldsymbol{\sigma}_y - \boldsymbol{\sigma}$ at least when $\boldsymbol{\sigma}$ is near to $\boldsymbol{\sigma}_y$. Then, let the following equation be assumed as is done in the Mróz model (ref.1)

$$\dot{\boldsymbol{\beta}} = \boldsymbol{\beta}\dot{\mu}, \tag{11}$$

where $\dot{\mu}$ is a scalar parameter formulated later and we set

$$\boldsymbol{\beta} \equiv \boldsymbol{\sigma}_y - \boldsymbol{\sigma} \tag{12}$$

which can be written as

$$\boldsymbol{\beta} = (\frac{1}{r} - 1)\bar{\boldsymbol{\sigma}} + \hat{\boldsymbol{\alpha}} - \bar{\boldsymbol{\alpha}} \tag{13}$$

by (9).

Substituting (13) into (11) using (10), we have

$$\dot{\bar{\boldsymbol{\alpha}}} = (1 - r)\dot{\bar{\boldsymbol{\sigma}}} + r\dot{\hat{\boldsymbol{\alpha}}} - r\boldsymbol{\beta}\dot{\mu}. \tag{14}$$

Differentiation of (5) with substitution of (14) yields

$$\dot{\mu} = \frac{r^{n-1}\dot{F} - \mathrm{tr}\{\frac{\partial f}{\partial \bar{\boldsymbol{\sigma}}}(\dot{\bar{\boldsymbol{\sigma}}} - \dot{\hat{\boldsymbol{\alpha}}})\}}{\mathrm{tr}(\frac{\partial f}{\partial \bar{\boldsymbol{\sigma}}}\boldsymbol{\beta})}. \tag{15}$$

Further, substituing (15) into (14), we obtain

$$\dot{\bar{\boldsymbol{\alpha}}} = (1 - r)\dot{\bar{\boldsymbol{\sigma}}} + r\dot{\hat{\boldsymbol{\alpha}}} - r\boldsymbol{\beta}\frac{r^{n-1}\dot{F} - \mathrm{tr}\{\frac{\partial f}{\partial \bar{\boldsymbol{\sigma}}}(\dot{\bar{\boldsymbol{\sigma}}} - \dot{\hat{\boldsymbol{\alpha}}})\}}{\mathrm{tr}(\frac{\partial f}{\partial \bar{\boldsymbol{\sigma}}}\boldsymbol{\beta})} \tag{16}$$

Finally, the plastic strain rate $\dot{\boldsymbol{\varepsilon}}^P$ is formulated in the following.

In the distinct-yield state ($\boldsymbol{\sigma} = \boldsymbol{\sigma}_y$), it can be obtained by differentiating (2) and substituting (7) that

$$\mathrm{tr}(\frac{\partial f}{\partial \hat{\boldsymbol{\sigma}}_y}\dot{\boldsymbol{\sigma}}_y) - \mathrm{tr}[\frac{\partial f}{\partial \hat{\boldsymbol{\sigma}}_y}\{A\,\dot{\varepsilon}_v^P\mathbf{1} + B\,\mathrm{tr}(\dot{\boldsymbol{\varepsilon}}^P\frac{\hat{\boldsymbol{\sigma}}_y}{|\hat{\boldsymbol{\sigma}}_y|})\frac{\hat{\boldsymbol{\sigma}}_y}{|\hat{\boldsymbol{\sigma}}_y|}\}] = F'\dot{K}, \tag{17}$$

where

$$F' \equiv \frac{dF}{dK}. \tag{18}$$

Adopting the associated flow rule

$$\dot{\boldsymbol{\varepsilon}}^{\mathrm{P}} = \dot{\lambda}\hat{\boldsymbol{n}}_{\mathrm{y}} \qquad (\dot{\lambda} > 0), \tag{19}$$

where

$$\hat{\boldsymbol{n}}_{\mathrm{y}} \equiv \frac{\dfrac{\partial f}{\partial \hat{\boldsymbol{\sigma}}_{\mathrm{y}}}}{\left|\dfrac{\partial f}{\partial \hat{\boldsymbol{\sigma}}_{\mathrm{y}}}\right|}, \tag{20}$$

and $\dot{\lambda}$ is proportionality factor, and noting the relation

$$\frac{\partial f}{\partial \hat{\boldsymbol{\sigma}}_{\mathrm{y}}} = \frac{nf}{\mathrm{tr}(\hat{\boldsymbol{n}}_{\mathrm{y}}\hat{\boldsymbol{\sigma}}_{\mathrm{y}})}\hat{\boldsymbol{n}}_{\mathrm{y}}, \tag{21}$$

$\dot{\boldsymbol{\varepsilon}}^{\mathrm{P}}$ is given from (17) as follows:

$$\dot{\boldsymbol{\varepsilon}}^{\mathrm{P}} = \frac{\mathrm{tr}(\hat{\boldsymbol{n}}_{\mathrm{y}}\dot{\boldsymbol{\sigma}}_{\mathrm{y}})}{\frac{1}{n}\mathrm{tr}(\hat{\boldsymbol{n}}_{\mathrm{y}}\hat{\boldsymbol{\sigma}}_{\mathrm{y}})\frac{F'}{F}\hat{\kappa}_{\mathrm{y}} + A(\mathrm{tr}\hat{\boldsymbol{n}}_{\mathrm{y}})^2 + B\{\mathrm{tr}(\hat{\boldsymbol{n}}_{\mathrm{y}}\frac{\hat{\boldsymbol{\sigma}}_{\mathrm{y}}}{|\hat{\boldsymbol{\sigma}}_{\mathrm{y}}|})\}^2}\hat{\boldsymbol{n}}_{\mathrm{y}}, \tag{22}$$

where $\hat{\kappa}_{\mathrm{y}}$ is a scalar function of stress, plastic strain and $\hat{\boldsymbol{n}}_{\mathrm{y}}$ in degree one given by

$$\hat{\kappa}_{\mathrm{y}} = \dot{K}/\dot{\lambda}, \tag{23}$$

that is, $\hat{\kappa}_{\mathrm{y}}$ is given by replacing the $\dot{\boldsymbol{\varepsilon}}^{\mathrm{P}}$ by $\hat{\boldsymbol{n}}_{\mathrm{y}}$ is the function $\dot{K}$.

Now, we extend (22) to the subyield state $\boldsymbol{\sigma} \neq \boldsymbol{\sigma}_{\mathrm{y}}$ as follows:

$$\dot{\boldsymbol{\varepsilon}}^{\mathrm{P}} = H\frac{\mathrm{tr}(\hat{\boldsymbol{n}}_{\mathrm{y}}\dot{\boldsymbol{\sigma}})}{\frac{1}{n}\mathrm{tr}(\hat{\boldsymbol{n}}_{\mathrm{y}}\hat{\boldsymbol{\sigma}}_{\mathrm{y}})\frac{F'}{F}(\hat{\kappa}_{\mathrm{y}} - h) + A(\mathrm{tr}\hat{\boldsymbol{n}}_{\mathrm{y}})^2 + B\{\mathrm{tr}(\hat{\boldsymbol{n}}_{\mathrm{y}}\frac{\hat{\boldsymbol{\sigma}}_{\mathrm{y}}}{|\hat{\boldsymbol{\sigma}}_{\mathrm{y}}|})\}^2}\hat{\boldsymbol{n}}_{\mathrm{y}}, \tag{24}$$

where H and h are monotonically decreasing and increasing scalar functions respectively of the scalar parameter

$$b \equiv \mathrm{tr}(\frac{\boldsymbol{\beta}}{F^{1/n}}\hat{\boldsymbol{n}}_{\mathrm{y}}) \tag{25}$$

satisfying the conditions

$$\begin{aligned} &H < 1 \text{ and } h > 0 \text{ when } b > 0, \\ &H = 1 \text{ and } h = 0 \text{ when } b = 0, \\ &H > 1 \text{ and } h > 0 \text{ when } b < 0. \end{aligned} \tag{26}$$

(24) means that the plastic strain rate produced in the subyield state relates to the parameter b, i.e., the projection of the vector $\boldsymbol{\beta}/F^{1/n}$ to the outer-normal direction of the subyield surface at $\boldsymbol{\sigma}$. While the function H represents the ratio of theplastic strain rate produced in a subyield state to that in a distinct-yield state, the function is indispensable in constitutive equations of granular media which exhibits a softening behavior as will be described in a later section.

The homogeneity of the function and the relation (9) lead to the equality

$$\hat{\boldsymbol{n}}_y = \bar{\boldsymbol{n}}, \tag{27}$$

where $\bar{\boldsymbol{n}}$ is the unit outer-normal vector of the subyield surface, i.e.,

$$\bar{\boldsymbol{n}} = \frac{\dfrac{\partial f}{\partial \bar{\boldsymbol{\sigma}}}}{\left|\dfrac{\partial f}{\partial \bar{\boldsymbol{\sigma}}}\right|}, \tag{28}$$

Substituting (9) and (27) into (16), (24) and (25), we have

$$\dot{\boldsymbol{\varepsilon}}^P = H\frac{\mathrm{tr}(\bar{\boldsymbol{n}}\dot{\boldsymbol{\sigma}})}{\dfrac{1}{rn}\mathrm{tr}(\bar{\boldsymbol{n}}\bar{\boldsymbol{\sigma}})\dfrac{F'}{F}(\bar{\kappa} - h) + A(\mathrm{tr}\bar{\boldsymbol{n}})^2 + B\{\mathrm{tr}(\bar{\boldsymbol{n}}\dfrac{\bar{\boldsymbol{\sigma}}}{|\bar{\boldsymbol{\sigma}}|})\}^2}\bar{\boldsymbol{n}}, \tag{29}$$

$$\dot{\boldsymbol{\alpha}} = (1 - r)\dot{\boldsymbol{\sigma}} + r\dot{\hat{\boldsymbol{\alpha}}} + \frac{\mathrm{tr}\{\bar{\boldsymbol{n}}(r\dot{\boldsymbol{\sigma}} - r\dot{\hat{\boldsymbol{\alpha}}} - \dfrac{1}{n}\dfrac{\dot{F}}{F}\bar{\boldsymbol{\sigma}})\}}{\mathrm{tr}(\bar{\boldsymbol{n}}\boldsymbol{\beta})}\boldsymbol{\beta}, \tag{30}$$

$$b = \frac{1}{F^{1/n}}\mathrm{tr}(\bar{\boldsymbol{n}}\boldsymbol{\beta}), \tag{31}$$

where $\bar{\kappa}$ stands for a function given by replacing the argument $\hat{\boldsymbol{n}}_y$ by $\bar{\boldsymbol{n}}$ in the function $\hat{\kappa}_y$.

CONSTITUTIVE EQUATIONS OF GRANULAR MEDIA

Based on the constitutive equations formulated in the preceding section, special class of constitutive equations for granular media are derived in this section.

First, assume that the loading function f, the hardening parameter K and the hardening function F as follows :

$$f(\bar{\boldsymbol{\sigma}}) = \sqrt{\bar{P}^2 + \left(\frac{|\bar{\boldsymbol{\sigma}}^*|}{m}\right)^2} \equiv \bar{\Pi}, \tag{32}$$

$$K = \varepsilon_v^p, \tag{33}$$

$$F = F_0 \exp\left(-\frac{\varepsilon_v^p}{\rho}\right), \tag{34}$$

where $\bar{P} \equiv \frac{1}{3}\mathrm{tr}(\bar{\boldsymbol{\sigma}})$, m and ρ are material constants, and F_0 is an initial value of F.

Further, the kinematic hardening parameters A and B are selected as follows:

$$A = \frac{m}{\rho M}F, \quad B = 0, \tag{35}$$

where M ($\geqq m$) is a material constant.

The distinct-yield surface corresponding to (32)-(35) is represented by an ellipsoid in (P, $|\boldsymbol{\sigma}^*|$) space as shown in Fig.3 where P and $\boldsymbol{\sigma}^*$ are the mean stress and the deviatoric stress respectively.

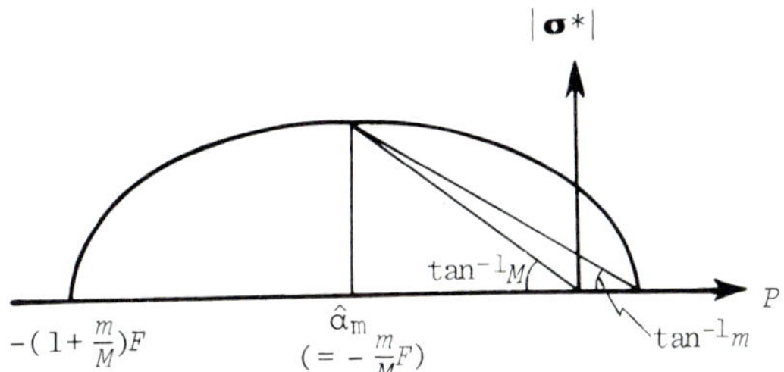

Fig.3 Assumed distinct-yiels surface

Now, from (32) it follows that

$$n = 1, \quad \bar{\Pi} = |\bar{P}|X, \quad \bar{\boldsymbol{n}} = \frac{\frac{1}{3}\mathbf{1} + \bar{\boldsymbol{\chi}}}{Y}\frac{\bar{P}}{|\bar{P}|}, \quad \bar{K} = \frac{1}{Y}\frac{\bar{P}}{|\bar{P}|}, \tag{36}$$

where

$$\bar{\boldsymbol{\chi}} = \frac{1}{m^2}\frac{\bar{\boldsymbol{\sigma}}^*}{\bar{P}}, \quad X = \sqrt{1 + (m|\bar{\boldsymbol{\chi}}|)^2}, \; Y = \sqrt{\frac{1}{3} + |\bar{\boldsymbol{\chi}}|^2} \tag{37}$$

Substitution of (32)-(36) into (29)-(31) leads to a series of constitutive equations of granular media as follows :

$$\dot{\boldsymbol{\varepsilon}}^p = H\frac{\dot{\bar{P}} + \mathrm{tr}(\bar{\boldsymbol{\chi}}\dot{\boldsymbol{\sigma}})}{-\frac{F}{\rho}\left\{\frac{\bar{\Pi}}{|\bar{P}|}\left(\frac{\bar{P}}{|\bar{P}|} - hY\right) - \frac{m}{M}\right\}}\left(\frac{1}{3}\mathbf{1} + \bar{\boldsymbol{\chi}}\right), \tag{38}$$

$$\dot{\lambda} = H\frac{\dot{\bar{P}} + \mathrm{tr}(\bar{\boldsymbol{\chi}}\dot{\boldsymbol{\sigma}})}{-\frac{F}{\rho}\left\{\frac{\bar{\Pi}}{|\bar{P}|}\left(\frac{\bar{P}}{|\bar{P}|} - hY\right) - \frac{m}{M}\right\}}Y\frac{\bar{P}}{|\bar{P}|}, \tag{39}$$

$$\dot{\hat{\boldsymbol{\alpha}}} = \frac{m}{\rho M}F\dot{\varepsilon}_v^p\mathbf{1} = -\frac{m}{M}\dot{F}\mathbf{1}, \quad \hat{\boldsymbol{\alpha}} = -\frac{m}{M}F\mathbf{1}, \tag{40}$$

$$\dot{\bar{\boldsymbol{\alpha}}} = (1 - r)\dot{\bar{\boldsymbol{\sigma}}} + r\dot{\hat{\boldsymbol{\alpha}}} + \frac{\mathrm{tr}\{(\frac{1}{3}\mathbf{1} + \bar{\boldsymbol{\chi}})(r\dot{\bar{\boldsymbol{\sigma}}} - r\dot{\hat{\boldsymbol{\alpha}}} - \frac{\dot{F}}{F}\bar{\boldsymbol{\sigma}})\}}{\mathrm{tr}\{(\frac{1}{3}\mathbf{1} + \bar{\boldsymbol{\chi}})\boldsymbol{\beta}\}}\boldsymbol{\beta}, \tag{41}$$

$$\boldsymbol{\beta} = (\frac{1}{r} - 1)\bar{\boldsymbol{\sigma}} + \hat{\boldsymbol{\alpha}} - \bar{\boldsymbol{\alpha}}, \tag{42}$$

$$b = \frac{1}{FY}\mathrm{tr}\{(\frac{1}{3}\mathbf{1} + \boldsymbol{\chi})\boldsymbol{\beta}\}\frac{\bar{P}}{|\bar{P}|}, \tag{43}$$

Now, we compare with the experimental data on an isotropic loading of Weald clay measured by Henkel (ref.6). The experimental curves for the relation of the void ratio e to P are shown in Fig.4 which involves three reloading curves from -206.82 (-30), -413.64 (-60), and -827.28 KPa (-120 psi) but not a reloading curve from them. These curves are altered to the relation of P to a volumetric strain ε_v shown by the solid line in Fig.5, selecting the point 0 (P = -206.82 KPa (-30 psi), e = 0.636) in Fig.4 as an initial state ($\varepsilon_v = 0$). On the other hand, theoretical curves calculated by (38)-(43) are depicted by the broken lines, while they are suppulemented the the elastic volumetric strain ε_v^e given by the following equation (ref.7).

$$\varepsilon_v^e = -\gamma \ln(\frac{P}{P_0}), \tag{44}$$

where γ is a material constant and P_0 is an initial value of P, which is selected as P_0 = -206.82 KPa (-30 psi) in the present calculation as was dscribed in the foregoing. Material constants and functions in (38)-(43) and (44) are selected as follows:

$$r = 0.3,\ m = M = 0.72,\ \rho = 0.045,\ \gamma = 0.015,$$
$$H = \exp(-b),\ h = \ln(1 + 3b). \tag{45}$$

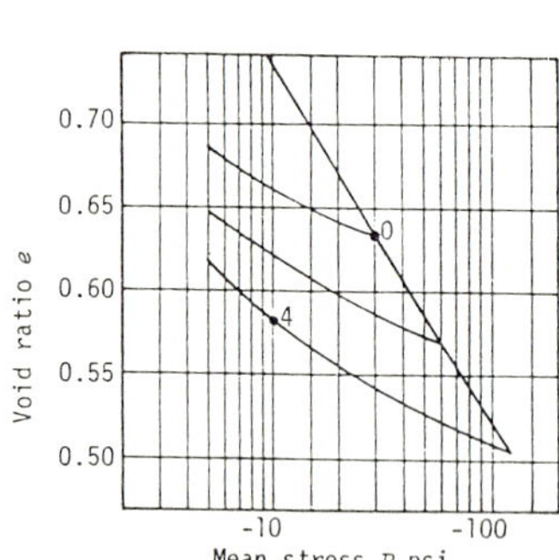

Fig.4 An experimental result on the isotropic loading of Weald clay (After Henkel (ref.6))

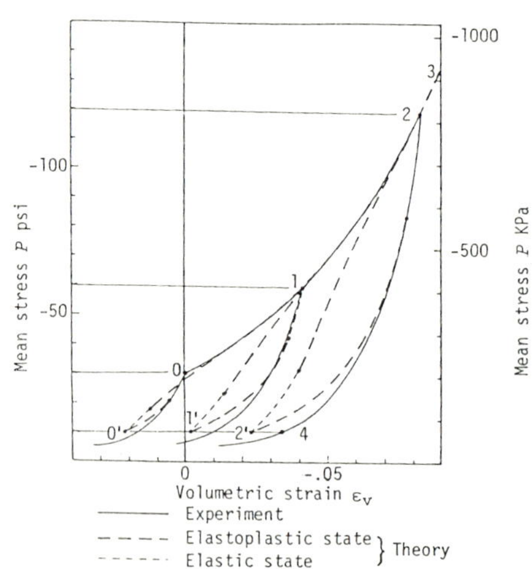

Fig.5 Comparison between theory and experiment on the isotropic loading of Weald clay

By loading from the initial state, i.e., the point 0, we obtain the curve 0123 coinciding with experimental curve. On the other hand, by reloadings after unloadings from -206.82 (-30), -413.64 (-60) and -827.28 (-120) to -689.48 KPa (-10 psi), we obtain the curves 00'3, 11'3 and 22'3 with histeresis loops, coinciding asymptotically with the first loading curve 0123.

Now, we compare the present theory with experimental data on the axisymmetric compression with a constant lateral stress of Weald clay measured by Parry (ref. 8). The experimental curves for the relations of axial stress σ_a and the volumetric strain ε_v to the axial strain ε_a for the loadings from the initial states 0 and 4 in Fig.4 or 5 are depicted by the solid lines in Fig.6 and 7, respectively.

On the other hand, theoretical curves calculated by (38)-(43) with material constants and functions prescribed in (45) are depicted by the broken lines in Fig.6 and 7. They are supplemented the elastic strain $\boldsymbol{\varepsilon}^e$ given by

$$\boldsymbol{\varepsilon}^e = \frac{1}{3}\gamma \ln\left(\frac{P}{P_o}\right)\mathbf{1} + \frac{1}{G}\boldsymbol{\sigma}^*, \tag{46}$$

where G is an elastic shear modulus, selecting as G = 2757.9 KPa (400 psi). Initial states for these calculations are the point 0 (P_o = -206.82 KPa (-30 psi), F_o = 103.42 KPa (15 psi)) in Fig.5 for Fig.6 and the point 2' (P_o = -68.95 KPa (-10 psi), F_o = 247.80 KPa (35.95 psi)) in Fig.5 for Fig.7.

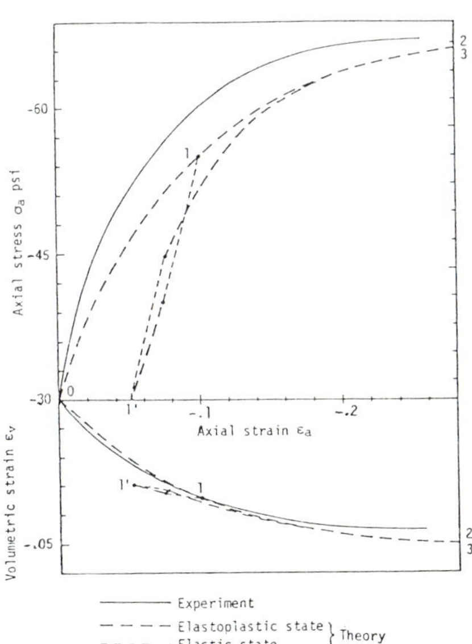

Fig.6 Comparison between theory and experiment on the axisymmetric loading of Weald clay in the normal-consolidation

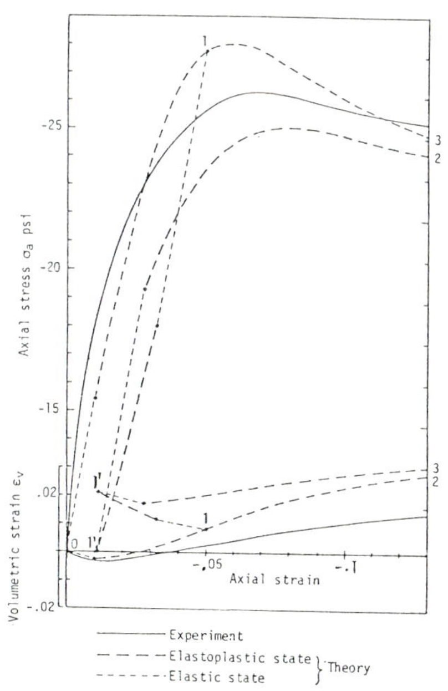

Fig.7 Comparison between theory and experiment on the axisymmetric loading of Weald clay in the over-consolidation

Good approximations of the experimental curves by the theory are observed in Fig.6 and 7. Besides, if we increase the axial strain from the point 1(ε_a = -0.1) to the point 1' (ε_a = -0.051, σ_a = 0) in Fig.6 and decrease it again, the theoretical curve 11'3 is depicted. Also, if we increase the axial strain from the point 1 (ε_a = -0.05) to the point 1' (ε_a = -0.011, σ_a = 0) in Fig.7 and decrease it again, the theoretical curve 11'3 is depicted.

CONCLUDING REMARKS

The constitutive equations of elastoplastic materials based on the two-surface theory are presented in this paper, which describe the anisotropic hardening and softening behaviors. Besides, by virture of the similarity of the distinct-yield and the subyield surfaces, they might be simple enough to be adopted in numerical analyses of practical problems in engineering.

REFERENCES

1 Mroz, Z., On the description of Anisotropic Workhardening, Journal of the Mechanics and Physics of Solids, Vol.15, 1967, pp.163-175.
2 Mroz, Z., Norris, V. A., and Zienkiewicz, O. C., An Anisotropic Hardening Model for Soils and its Application to Cyclic Loading, International Journal for Numerical Methods in Geomechanics, Vol.2, 1978, pp.203-211.
3 Krieg, R. D., A Practical Two Surface Plasticity Theory, Journal of Applied Mechanics, Transactions of the American Society of Mechanical Engineers, Vol. 42, 1975, pp.641-646.
4 Dafalias, Y. F., and Popov, E. P., A Model of Nonlinearly Hardening Materials Complex Loading, Acta Mechanica, Vol.21, 1975, pp.173-192.
5 Mroz, Z., Norris, V. A., and Zienkiewicz, O. C., Application of an Anisotropic Hardening Model in the Analysis of Elastoplastic Deformation of Solis, Geotechnique, Vol.29, No.1, pp.1-34
6 Henkel, D. J., The Effect of Overconsolidation on the Behavior of Clays during Shear, Goetechnique, Vol.6, 1956, pp139-150.
7 Hashiguchi, K., and Ueno, M., From Hvorslev's Failure Condition to Yield Condition, Proceedings of 9th International Conference on Soil Mechanics and Foundation Engineering, Vol.1, 1977, pp123-126.
8 Parry, R. H. G., Triaxial Compression and Extension Tests on Remoulded Saturated Clay, Geotechnique, Vol.10, No.4, 1959, pp166-180.

Mechanics of Granular Materials: New Models and Constitutive Relations, edited by
J.T. Jenkins and M. Satake, 1983
Elsevier Science Publishers B.V., Amsterdam — Printed in The Netherlands

MODELING OF MICROSCOPIC MECHANISMS IN GRANULAR MATERIAL

P.A. CUNDALL and O.D.L. STRACK
Department of Civil and Mineral Engineering, University of Minnesota, Minneapolis, Minnesota 55455, U.S.A.

ABSRACT
The computer program BALL has been used for the past several years to model and study the behavior of two-dimensional assemblies of discs. The main result of the study thus far, a qualitative description of the material behavior, is discussed. An attempt has been made to introduce measures that may be used in a future constitutive model. The main measures, a partitioned stress tensor, and a constraint ratio, which is related to the stability of the assembly, are introduced and illustrated for some numerical experiments with BALL.

INTRODUCTION

The computer program BALL models the movement and interaction of many discs, during simulated experiments on samples comprised of randomly-generated particles conforming to given size distributions. During such numerical experiments, a variety of boundary conditions and loading sequences may be specified, and measurements made of stresses, strains and internal parameters over circular contours within the samples. Descriptions of the method of calculation, the measurement and loading techniques, and the numerical experiments are contained in various papers and reports by Cundall and Strack (ref.1-4). Typical plots made by BALL are reproduced in Figures 1 and 4. In these plots, the lines drawn across contact points represent the contact forces in magnitude and direction, where the line thickness corresponds to magnitude.

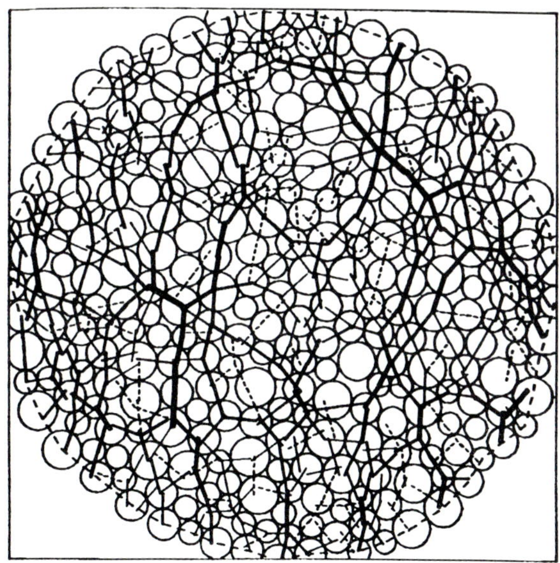

Fig. 1. 284-disc assembly, used to study effects of stress partitions.

REVIEW OF RESULTS FROM PROGRAM BALL

A catalog of mechanisms and effects observed from experiments with BALL is given by Cundall et al. (ref.5). The reader is urged to read that paper in conjunction with the present paper, but for convenience a summary of the observations is given below.

Microscopic Observations

1) Contact forces are concentrated in stiff chains of particles.

2) Slip almost never occurs in the stiff chains, but in the relatively unloaded regions between chains.

3) Observed macroscopic velocity discontinuities do not consist of contiguous lines of slipping contacts, but often correspond to 'hinge' regions, involving coherent particle rotations (spins).

4) Particle spins contribute largely to deformation in an assembly; the stiff columns collapse in a 'buckling' mode that involves spins of alternating sign.

5) During deviatoric loading, contacts with normals near the minor principal direction are broken preferentially, and are not fully recovered on unloading.

6) After a sample undergoes some sequence of loading and unloading, 'locked-in' shear forces persist at contacts even though the boundary deviatoric stress is zero.

Macroscopic Effects

1) The incremental dilatation rate is insensitive to changes in contact friction.

2) Sample strength is insensitive to contact stiffness.

3) The strength and deformability of a sample is dependent on boundary conditions (i.e. whether the sample is stress- or strain- controlled); Small samples are most sensitive to boundary conditions.

4) Large, apparently random, fluctuations in principal stress directions are observed in small samples at high strains.

QUALITATIVE DESCRIPTION OF SOIL BEHAVIOR

The picture that emerges from observations on many numerical experiments is as follows. Large forces are carried by chains of particles that are more or less aligned, by chance, in the direction of maximum compressive strainrate. These chains, or columns, become shorter and more kinked as increasing strain is applied: the distortion is similar to buckling or folding, and involves neighboring particles rotating in opposite directions, without sliding. Although the stiff columns dissipate no energy in 'buckling', their distortion is resisted by the surrounding groups of particles that are lightly loaded, and have many sliding contacts.

Simply stated, a granular assembly appears to consist of two phases. Phase A, made up of stiff columns, behaves like a pin-jointed structure that dissipates no energy and imposes local deformation patterns on Phase B particles. Phase B acts like a truly plastic material that dissipates energy at many contacts and serves as a restraint on Phase A.

As the applied strain increases, certain changes take place in the material structure. Columns become progressively more kinked, and therefore less able to carry load. Load is transferred to new columns that develop nearby, as increasing deformation in the major principal direction brings new particles into contact. When steady-state conditions are reached during continuous deformation, new columns are created at the same rate that old columns collapse and vanish: the total number of contacts remains more or less constant. However, during initial loading, contacts are lost in the direction of minor principal strainrate. This change in fabric causes anisotropy of moduli, and dilatation during shear.

The anisotropic fabric partly determines the history-dependence of granular materials, since fabric evolves gradually as contacts are added or deleted, wheras stresses may change instantaneously. A phenomenon connected with history dependence is the existence of locked-in shear forces after a change in direction of applied loading. The progressive deformation observed by Arthur (ref.6) during periodic variation in principal axes may also depend on fabric. An anisotropic fabric is established corresponding to the direction of applied loading. If this direction is changed, a new fabric is developed. Specifically, contacts are created in the new direction of maximum compressive stress to replace contacts that were lost when the maximum stress was in another direction. In order to develop new contacts, some movement must occur in order to bring particles together: this movement corresponds to a compressive strain in the mean direction of cyclic loading.

SOME SIMPLE DERIVATIONS

A constitutive model is currently being developed that attempts to reproduce the effects noted above. This model is incomplete at the time of writing but a number of analyses have been made of limited aspects of behavior. Two of these are presented below. The first explores a way in which history-dependence, or memory, may be represented by means of internal partitions of stress that cannot be measured at the surface of an assembly; these partitions can only be fully determined by taking the internal geometry into account. The second analysis explores the consequences of treating a granular material as a structural assembly.

Stress Partitions

When a granular material is loaded and then unloaded, certain changes take place internally that will cause a marked difference in material response to stress paths that are identical to those applied previously. An adequate constitutive model should keep track of internal changes in the material, and use them when predicting future material behavior. It is suggested here that the observed stress tensor can be decomposed into several 'partitions', each of which is associated with a different aspect of behavior. Stress partitions constitute at least some of the internal variables that are responsible for memory and path-dependence.

The first separation in the stress tensor is between stresses that derive from normal forces and stresses that derive from shear forces at contacts. It is shown that shear forces contribute only to the deviatoric components of the stress tensor; the isotropic part derives solely from normal forces at contacts. The isotropic stress is related to the average normal contact force for each particle, and is independent of the distribution of forces around the particle. The second split consists of separating, for each particle, the average normal force from the remaining normal forces. The latter forces constitute a partition of the stress tensor that is related to the angular variation of normal forces. The average normal force gives rise to yet two other partitions, one being associated with the isotropic stress and the other with the fabric.

These ideas are developed below in mathematical form for arbitrary assemblies of discs or spheres. Rather than interrupting the analysis with the frequent introduction of symbols, a list of these is provided below. Indices refer to vector or tensor components in a Cartesian coordinate system.

superscripts

(s) shear
(f) fabric
(nv) normal variation
(n) normal

Scalars

m number of contacts on particle p
N number of particles in V
$\overset{p}{R}$ radius of particle p
S surface area of assembly
σ_0 isotropic stress
V volume of assembly
$\overset{p}{V}$ volume of particle p

Vectors

$\overset{c}{F}_i$ force at contact c

$\overset{c}{n}_i$ unit normal vector at contact c

p_i stress vector

$\overset{c}{x}_i$ position vector of contact c

$\overset{p}{X}_i$ position vector of centroid, particle p

Tensors

σ_{ij} stress tensor

$\bar{\sigma}_{ij}$ average stress tensor

$\overset{p}{\sigma}_{ij}$ average stress tensor in particle p

The average stress tensor in a volume V of material is defined as

$$\bar{\sigma}_{ij} = \frac{1}{V} \int_V \sigma_{ij} dV \tag{1}$$

The integral becomes a summation for a particulate material because stresses exist only within particles:

$$\bar{\sigma}_{ij} = \frac{1}{V} \sum_{p=1}^{N} \overset{p}{\sigma}_{ij} \overset{p}{V} \tag{2}$$

In the same way the average stress tensor within particle p is defined as

$$\overset{p}{\sigma}_{ij} = \frac{1}{\overset{p}{V}} \int_{\overset{p}{V}} \sigma_{ij} d\overset{p}{V} \tag{3}$$

Applying Gauss' divergence theorem to (3) the following is obtained

$$\overset{p}{\sigma}_{ij} = \frac{1}{\overset{p}{V}} \int_S x_i p_j ds \tag{4}$$

which becomes a sum, noting that the surface forces are discrete

$$\overset{p}{\sigma}_{ij} = \frac{1}{\overset{p}{V}} \sum_{c=1}^{m} \overset{c}{x}_i \overset{c}{F}_j \tag{5}$$

The combination of (2) and (5) yields

$$\bar{\sigma}_{ij} = \frac{1}{V} \sum_{p=1}^{N} \sum_{c=1}^{m} \overset{c}{x}_i \overset{c}{F}_j \tag{6}$$

For discs and spheres, $\overset{c}{x}_i = \overset{p}{X}_i + \overset{p}{R} \overset{c}{n}_i$, and (6) becomes

$$\bar{\sigma}_{ij} = \frac{1}{V} \sum_{p=1}^{N} \overset{p}{R} \sum_{c=1}^{m} \overset{c}{n}_i \overset{c}{F}_j \tag{7}$$

where it is noted that for equilibrium

$$\sum_{c=1}^{m} \overset{p}{X}_i \overset{c}{F}_j = \overset{p}{X}_i \sum_{c=1}^{m} \overset{c}{F}_j = 0$$

The stress tensor may be split into two parts, that due to normal forces $(\bar{\sigma}_{ij}{}^{(n)})$ and that due to shear forces $(\bar{\sigma}_{ij}{}^{(s)})$. With

$$F_i = F_i{}^{(n)} + F_i{}^{(s)}$$

and

$$F_i{}^{(n)} = F_k n_k n_i$$

these two parts of the stress tensor become

$$\bar{\sigma}_{ij}{}^{(n)} = \frac{1}{V} \sum_{p=1}^{N} \overset{p}{R} \sum_{c=1}^{m} \overset{c}{F}_k \overset{c}{n}_k \overset{c}{n}_i \overset{c}{n}_j \tag{8}$$

$$\bar{\sigma}_{ij}{}^{(s)} = \frac{1}{V} \sum_{p=1}^{N} \overset{p}{R} \sum_{c=1}^{m} \left[\overset{c}{n}_i \overset{c}{F}_j - \overset{c}{F}_k \overset{c}{n}_k \overset{c}{n}_i \overset{c}{n}_j \right] \tag{9}$$

It may be verified that the expression for $\bar{\sigma}_{ij}{}^{(s)}$ vanishes if i = j in (9): hence the shear partition has no isotropic component.

The stress tensor due to normal forces may be decomposed further by separating from the normal force at each contact a constant part:

$$\overset{c}{F}_i{}^{(n)} = \overset{p}{F}{}^{(av)} \overset{c}{n}_i + \overset{c}{F}_i{}^{(nv)} \tag{10}$$

where $\overset{c}{F}_i{}^{(nv)}$ is the remaining part; (nv) stands for 'normal variation'. $\overset{p}{F}{}^{(av)}$ is the average normal force for a particle, given by

$$\overset{p}{F}{}^{(av)} = \frac{1}{m} \sum_{c=1}^{m} \overset{c}{F}{}^{(n)} = \frac{1}{m} \sum_{c=1}^{m} \overset{c}{F}_k \overset{c}{n}_k \tag{11}$$

The (nv) partition is then

$$\bar{\sigma}_{ij}{}^{(nv)} = \frac{1}{V} \sum_{p=1}^{N} \overset{p}{R} \sum_{c=1}^{m} \overset{c}{n}_i \overset{c}{F}_j{}^{(nv)}$$

$$= \frac{1}{V} \sum_{p=1}^{N} \overset{p}{R} \sum_{c=1}^{m} \left[\overset{c}{n}_i \overset{c}{F}_j{}^{(n)} - \overset{c}{n}_i \overset{p}{F}{}^{(av)} \overset{c}{n}_j \right]$$

$$= \frac{1}{V} \sum_{p=1}^{N} \overset{p}{R} \sum_{c=1}^{m} \left[\overset{c}{F}_k \overset{c}{n}_k \overset{c}{n}_i \overset{c}{n}_j - \overset{p}{F}{}^{(av)} \overset{c}{n}_i \overset{c}{n}_j \right] \tag{12}$$

Finally the remaining part of the stress tensor, equal to minus the second term in (12), may be split into the isotropic component and a 'fabric' partition, the latter depending only on geometry and the isotropic average particle stress. The isotropic average stress for the domain is

$$\bar{\sigma}_0 = \frac{\bar{\sigma}_{ii}}{\delta_{kk}} = \frac{1}{V\delta_{kk}} \sum_{p=1}^{N} \overset{p}{R} \sum_{c=1}^{m} \overset{c}{n}_k \overset{c}{F}_k$$

or, with (11),

$$\bar{\sigma}_0 = \frac{1}{V\delta_{kk}} \sum_{p=1}^{N} \overset{p}{R} m \overset{p}{F}(av) \tag{13}$$

Introduction of the average isotropic stress of particle p, $\overset{p}{\sigma}_0$, with

$$\overset{p}{\sigma}_0 = \frac{m\overset{p}{R}\overset{p}{F}(av)}{\overset{p}{V}\delta_{kk}} \tag{14}$$

gives

$$\bar{\sigma}_0 = \frac{1}{V} \sum_{p=1}^{N} \overset{p}{V}\overset{p}{\sigma}_0 \tag{15}$$

The fabric partition is found by subtracting the isotropic stress from the stress corresponding to the first term of (10),

$$\begin{aligned} \bar{\sigma}_{ij}(f) &= \frac{1}{V} \sum_{p=1}^{N} \overset{p}{R} \sum_{c=1}^{m} \overset{p}{F}(av)\, \overset{c}{n}_i \overset{c}{n}_j - \bar{\sigma}_0 \delta_{ij} \\ &= \frac{1}{V} \sum_{p=1}^{N} \overset{p}{V}\overset{p}{\sigma}_0 \left[\frac{\delta_{kk}}{m} \sum_{c=1}^{m} \overset{c}{n}_i \overset{c}{n}_j - \delta_{ij} \right] \end{aligned} \tag{16}$$

In summary, the average stress tensor can be decomposed into the following partitions:

$$\bar{\sigma}_{ij} = \bar{\sigma}_{ij}(s) + \bar{\sigma}_{ij}(f) + \bar{\sigma}_{ij}(nv) + \bar{\sigma}_0 \delta_{ij} \tag{17}$$

where the partitions are defined as follows in terms of the contact forces and contact normals

shear:

$$\bar{\sigma}_{ij}(s) = \frac{1}{V} \sum_{p=1}^{N} \overset{p}{R} \sum_{c=1}^{m} \left[\overset{c}{n}_i \overset{c}{F}_j - \overset{c}{F}_k \overset{c}{n}_k \overset{c}{n}_i \overset{c}{n}_j \right] \tag{18}$$

fabric:

$$\bar{\sigma}_{ij}(f) = \frac{1}{V} \sum_{p=1}^{N} \overset{p}{V}\overset{p}{\sigma}_0 \left[\frac{\delta_{kk}}{m} \sum_{c=1}^{m} \overset{c}{n}_i \overset{c}{n}_j - \delta_{ij} \right] \tag{19}$$

normal variation:

$$\bar{\sigma}_{ij}(nv) = \frac{1}{V} \sum_{p=1}^{N} \left[\overset{p}{R} \sum_{c=1}^{m} \overset{c}{F}_k \overset{c}{n}_k \overset{c}{n}_i \overset{c}{n}_j - \frac{\overset{p}{V}\overset{p}{\sigma}_0 \delta_{kk}}{m} \sum_{c=1}^{m} \overset{c}{n}_i \overset{c}{n}_j \right] \tag{20}$$

isotropic:

$$\bar{\sigma}_0 = \frac{1}{V} \sum_{p=1}^{N} \overset{p}{V}\overset{p}{\sigma}_0 \tag{21}$$

where

$$\overset{p}{\sigma}_o = \frac{m\overset{p}{R}\overset{p}{F}(av)}{\overset{p}{V}\delta_{kk}} = \frac{\overset{p}{R}}{\overset{p}{V}\delta_{kk}} \sum_{c=1}^{m} \overset{c}{F}_k \overset{c}{n}_k \tag{22}$$

The partitions defined by (18), (19) and (20) have only deviatoric components, and seem to correspond to three distinct characteristics of granular material:

$\bar{\sigma}_{ij}{}^{(s)}$ corresponds to mobilized shear forces; at any stage in a test this partition is related to the tendency of contacts to slide, and is thereby related to the dissipation of energy.

$\bar{\sigma}_{ij}{}^{(f)}$ corresponds to the angular distribution of contacts; this partition is directly related to the current state of anisotropy. Since it is proportional to the isotropic particle stress multiplied by a geometrical term, it is recovered even if an assembly is completely unloaded and then loaded again. The geometrical part of the fabric partition is similar in form to that proposed by Oda et al. (ref.7).

$\bar{\sigma}_{ij}{}^{(nv)}$ corresponds to variation in the magnitude of normal forces with angle; this partition reflects the eccentricity of forces acting on the particles, and may be related to the tendency of columns to buckle.

The property of memory observed in granular materials is implied by the partitioned nature of the stress tensor because the partitions may take any set of values for the same overall stress tensor provided that their sum is equal to the overall tensor. The internal state of the granular assembly is represented by the relative magnitudes and directions of the three deviatoric partitions, independently of the applied stress.

The three deviatoric partitions were measured during a loading and unloading test performed by A. Drescher, and described by Cundall et al. (ref.5). Results are recorded in Fig. 2, from which the following observations may be made. The (nv) partition follows the applied stress, but at a lower level; it returns to zero when the boundary stress returns to zero. The fabric and shear partitions also follow the applied stress on loading but the fabric does not return to zero on unloading. To compensate, the shear partition becomes negative, which corresponds to locked-in shear forces.

These data illustrate that the stress partitions record the effects of a changing internal geometry. Stress partitions may therefore form a useful basis for a continuum model.

STRUCTURAL INSTABILITY

Consider the assembly of discs shown in Fig. 3a: the assembly is infinite but consists of periodically repeating groups. The curious thing about this structure of discs is that it offers no resistance to deformation. In comparison, the assembly shown in Fig. 3b is less dense than that of Fig. 3a, but presents a greater resistance to deformation because sliding must occur. These are special cases, but nonetheless illustrate that the structure may have a profound influence on strength. Having observed that a dense assembly may be weaker than a loose assembly, it is tempting to postulate that strength depends not so much on density (or porosity) but on the particular arrangement of particles. The following analysis is an attempt to quantify the effect of structural arrangement.

At any stage in a test on granular material the assembly of particles may be regarded as a structural network consisting of a set of movable points connected by stiffnesses (or springs). In common with any structural network it is possible to make general statements concerning possible modes of behavior, by comparing the number of degrees of freedom with the number of constraints. A 'constraint', in this context, is a physical entity, such as a spring, that contributes an equation relating an increment in force to the incremental displacements of one or more degrees of freedom. The term 'degree of freedom' applies to any independent motion in the system, such as the rotation or translation of one particle. In two dimensions, each particle has three degrees of freedom, while each contact supplies two constraints (one normal, one shear) if no slip occurs. However, if slip occurs the increment in shear force is zero for any set of incremental displacements (provided that the normal force is constant). Hence the equation for incremental shear force vanishes at slipping contacts, reducing the number of constraints to one.

For a complete assembly,

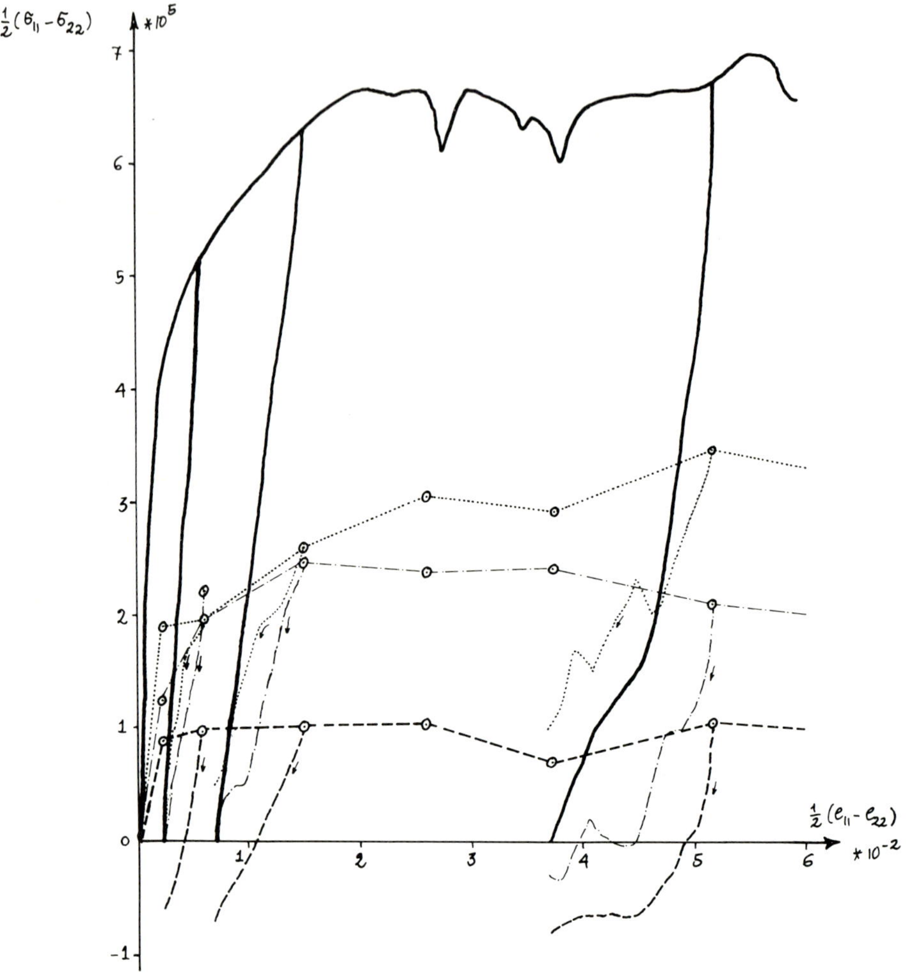

Fig. 2. Results from loading and unloading tests on 284-disc assembly, showing responses of individual stress partitions.

N_d = number of degrees of freedom = 3N

N_c = number of constraints = C(2 − S)

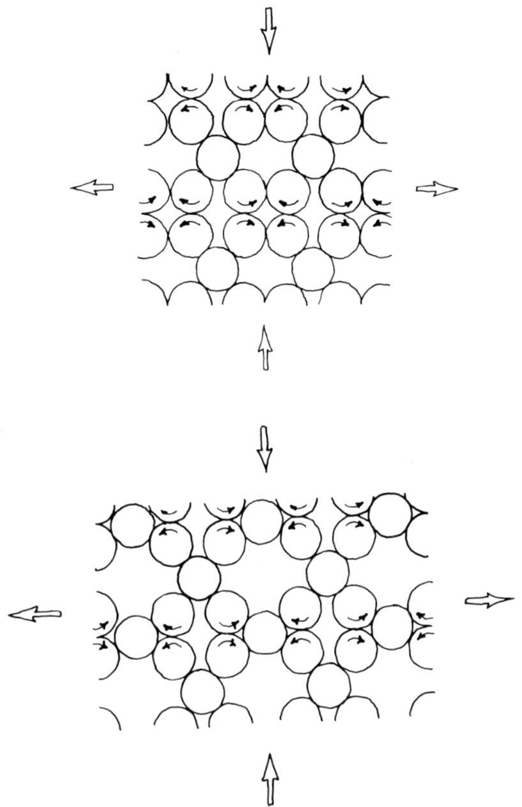

Fig. 3. Special arrangements of particles used to demonstrate the effect of structure on material strength: a) high density; low strength, b) lower density; higher strength.

where

N = number of particles
C = number of contacts
S = fraction of contacts that are slipping
$(0 \leq S \leq 1)$

The character of behavior changes according to the ratio of N_c to N_d (subject to certain conditions, given later). Defining R as the 'constraint ratio',

$$R = \frac{N_c}{N_d} = \frac{C(2-S)}{3N} = \frac{r}{3}(2-S) \tag{23}$$

$$\text{where} \quad r = \frac{\text{number of contacts}}{\text{number of particles}}$$

then the following observations may be made:

If $R > 1$, the system is overconstrained, or statically indeterminate; the assembly has a positive resistance to deformation which depends on the values of contact stiffnesses.

If $R < 1$, there are more degrees of freedom than constraints; therefore certain modes of deformation may take place for zero or negative increment in applied loading.

The state $R = 1$ is of fundamental importance for granular assemblies because it corresponds to the limiting stress that can be carried; at this point the assembly becomes a 'mechanism', in the sense that deformation proceeds in a particular mode with no increase in stress.

During a test under increasing deviator stress, the value of R decreases for two reasons: firstly, S increases as the proportion of sliding contacts increases with increasing obliquity of contact forces; secondly, r decreases as contacts are progressively broken in the direction of minor principal strainrate. It is postulated that peak strength is achieved when $R = 1$.

The above analysis rests on the following symplifying assumptions:

1) contact stiffnesses are incrementally linear;

2) the assembly of particles is irregular - i.e. all degrees of freedom are distinct and uncoupled. The analysis is not valid for assemblies with less than 3N degrees of freedom, such as regular assemblies and symmetrical assemblies.

3) the number of particles is large, so that the three rigid-body modes of the whole assembly are insignificant compared to the other modes of deformation;

4) the number of constraints or suppressed degrees of freedom caused by boundaries is small compared to the total number of constraints or degrees of freedom;

5) the assembly is two-dimensional.

Example 1

Although equation (23) is not strictly applicable to the assemblies of Fig. 3, since they are periodic, it is interesting to evaluate R for both cases.

For 3a, $R = 1.07$

For 3b, $R = 1.11$

Hence R seems to be more closely related to strength than density or porosity, since it indicates that 3b is stronger than 3a ,which is true, while the reverse is suggested by considering relative porosities.

Example 2

Cundall et al. (ref.5) report an experiment in which an 89-disc assembly was tested under three conditions:

1) strain-controlled boundary
2) stress-controlled boundary
3) sample embedded in larger assembly (1000 discs)

It was found that conditions 1 and 3 gave similar results but under condition 2 the sample behaved in a completely different way, producing strains of about ten times those of the other two tests. The constraint ratio, R, may be evaluated directly for all three samples, since the numbers of contacts and particles are known in all cases. These data are as follows, for the sample shown in Fig. 4.

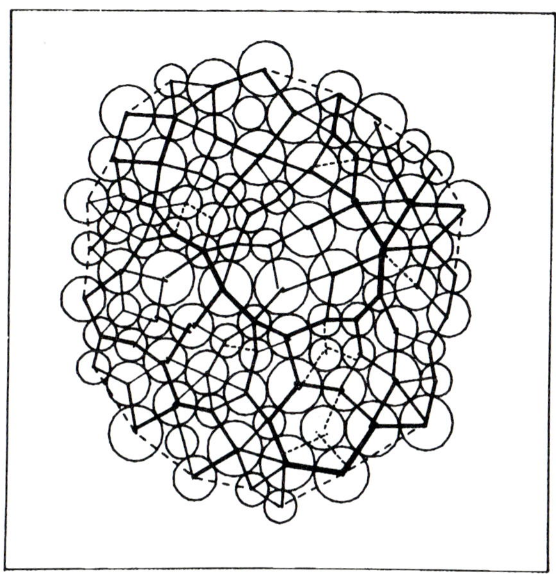

Fig. 4. 89-disc assembly used to investigate the effects of different boundary conditions.

89 total number of particles
25 boundary particles
42 contacts between boundary and larger assembly
12 contacts between boundary particles
39 contacts between interior and boundary particles
109 interior contacts

The three tests are analysed as follows, for the case of no sliding ($S = 0$) :

1) Strain-controlled boundary

The boundary particles contribute no degrees of freedom, and the contacts between boundary particles are inactive.

$$N = 89 - 25 = 64$$

$$C = 109 + 39 = 148$$

$$R = \frac{2 * 148}{3 * 64} = \underline{1.54}$$

2) Stress-controlled boundary

The boundary particles are all active, but there are no contacts outside the boundary.

$$N = 89$$

$$C = 109 + 39 + 12 = 160$$

$$R = \frac{2 * 160}{3 * 89} = \underline{1.20}$$

3) Sample embedded in larger assembly

In this case, more contacts are added between the sample boundary and the larger assembly.

$$N = 89$$

$$C = 109 + 39 + 12 + 42 = 202$$

$$R = \frac{2 * 202}{3 * 89} = \underline{1.51}$$

These analyses confirm the experimental observations: that a strain-controlled boundary produces conditions that are similar to those found when the sample is embedded in a larger assembly ($R = 1.54$ compared to $R = 1.51$). However the constraint ratio is much smaller ($R = 1.2$) for the sample with the stress-controlled boundary, indicating that it is less stable.

Another way to present the same results is to calculate the fraction of contacts that must slip in order to bring the assembly to failure ($R = 1$):

$$S = 2 - 3\frac{N}{C}$$

Case 1: $S = 0.70$

Case 2: $S = 0.33$

Case 3: $S = 0.68$

The same conclusion is evident: case 2 (stress- controlled boundary) will fail with far fewer sliding contacts, and is therefore more unstable. The strain-controlled sample is very similar to the sample embedded in a larger assembly.

Example 3 - general case

Since the stress-controlled boundary has considerable influence on the behavior of a sample it is interesting to derive a general formula that estimates the constraint ratio for a sample containing an arbitrary number of particles. Firstly it is necessary to estimate how many particles form the boundary of a circular sample. Assume that the number of particles (N) in a sample is proportional to sample area, where R is the radius.

$$N \propto \pi R^2$$

By proportion, one particle (and its associated void space) will, on average, occupy an area A, where

$$A = \frac{\pi R^2}{N}$$

Assume that the boundary layer consists of square cells of dimension d, where

$$d^2 = A = \frac{\pi R^2}{N} \tag{24}$$

The mid-point of each cell lies at radius $R - \frac{1}{2}d$. Hence the number of cells (and therefore particles) in the boundary layer is

$$n_b = \frac{2\pi (R - \frac{d}{2})}{d}$$

$$n_b = 2\sqrt{\pi N} - \pi \tag{25}$$

Consider a stress-controlled sample of N particles, with a contact to particle ratio of r. Each particle will, on average, be surrounded by 2r contacts, since each contact joins two particles. Each boundary particle under stress-control has three degrees of freedom, but lacks contacts outside the boundary. It is difficult to determine the proportion of contacts that are missing, but a conservative estimate may be made if we continue to regard the boundary layer as composed of square cells. One edge of each cell (i.e. one quarter of the cell boundary) faces the outer region; hence the number of missing contacts is

$$C_m = \left(\frac{1}{4}\right) 2r * n_b \tag{26}$$

Writing (23), with the missing contacts subtracted from the total number of contacts, the constraint ratio for the stress-controlled sample becomes

$$R_\sigma = \frac{(rN - C_m)(2 - S)}{3N}$$

$$= \frac{r(N - \frac{1}{2}n_b)(2 - S)}{3N} \tag{27}$$

In order to compare the sample response with the response of an infinite assembly, consider the state of incipient failure, $R = 1$, and assume the proportion of slipping contacts, S, to be 0.2, which is a value commonly observed in tests with BALL. Then

$$r = \frac{5}{3}$$

Using this value in (27), with $R_\sigma = 1$, Table 1 is obtained for the proportion of contacts that must slip in order to bring the stress-controlled sample to failure.

Even with a million particles, the effect of stress-control is noticeable, and for one thousand particles only half the number of sliding contacts is necessary for failure. It is concluded that the stress-controlled boundary makes a sample much more unstable than the same assembly of particles embedded in an infinite medium. Boundary conditions are seen to have a considerable influence on the behavior of granular assemblies, even for assemblies of many particles. It may be noted that the above boundary effect is underestimated because of the conservative value used for the number of missing boundary contacts.

TABLE 1. Proportion of slipping contacts as a function of the sample size.

N	S
∞	0.200
1,000,000	0.197
100,000	0.190
10,000	0.168
1,000	0.096
282	0.000

REFERENCES.

1 O.D.L. Strack and P.A. Cundall, The distinct element method as a tool for research in granular media, Report to NSF concerning grant ENG 76-20711, Part I, Dept. Civ. Min. Engng. Univ. Minnesota, 1978.

2 P.A. Cundall and O.D.L. Strack, The distinct element method as a tool for research in granular media, Report to NSF concerning grant ENG 76-20711, Part II, Dept. Civ. Min. Engng. Univ. Minnesota, 1979.

3 P.A. Cundall and O.D.L. Strack, A discrete numerical model for granular assemblies, Geotechnique, 29 (1979) 47-65.

4 P.A. Cundall and O.D.L. Strack. The development of constitutive laws for soil using the distinct element method, in: Numerical Methods in Geomechanics, Aachen, ed. W.Wittke, 1979, 289-298.

5 P.A. Cundall, A. Drescher and O.D.L. Strack, Numerical experiments on granular assemblies; measurements and observations. IUTAM Symposium on Deformation and Failure of Granular Materials, Delft, 1982, 355-370.

6 J.R.F. Arthur, K.S. Chua, T. Dunstan and J.I. Rodriguez del C , Principal stress rotation: a missing parameter, Proc. A.S.C.E, V106, No.GT4 (1980) 419-433.

7 M. Oda, S. Nemat-Nasser and M.M. Mehrabadi, A statistical study of fabric in a random assembly of spherical granules, Int. J. Num. Anal. Meth. Geomechanics 6, (1982) 77-94.

Mechanics of Granular Materials: New Models and Constitutive Relations, edited by
J.T. Jenkins and M. Satake, 1983
Elsevier Science Publishers B.V., Amsterdam — Printed in The Netherlands

NEW DISCRETE MODELS AND THEIR APPLICATION TO MECHANICS OF GRANULAR MATERIALS

T. KAWAI[1] and N. TAKEUCHI[2]

[1]Institute of Industrial Science, University of Tokyo

[2]Kyowa Consultants Company

ABSTRACT

Based on the experimental evidence of solids under the ultimate state of loading, the first author proposed a family of new discrete models. These models consist of rigid bodies and two types of connection springs, one of which resists the dilatational deformation, while the other the shear deformation. Theoretical basis of the models are described first and the application of these models to mechanics of granular materials is discussed with some simple examples.

INTRODUCTION

At the present moment the finite element method has become one indispensable tool for analysis of soil and rock mechanics problems. It is, however, common recognition that the soil and rock foundations are essentially discontinuous structures and the hysteresis is another important characteristic of these materials.

Therefore it is definitely needed to develop a new method of analysis in which effects of such discontinuous characteristics are fully taken into account. So far a limited number of works have been carried in the field of numerical analysis on the mechanics of granular materials. Considering this is the state, the present authors proposed a new discrete method of limit analysis recently. In this paper outline of the proposed method will be introduced briefly with some examples.

THEORETICAL BASIS OF NEW DISCRETE MODELS (ref. 1, 2)

Generally speaking all materials are composed of very large number of particles and therefore it is obvious that their deformation may be controled not only by the strength of the particles themselves but also by the intergranular strength. From the stand point of the solid mechanics, however, solids are generally considered as continua and their intergranular strength is usually neglected. But it is well known from ample experimental evidence that essential feature of plastic deformation is the slip movement and at the ultimate state of loading solids or structures may turn into link mechanisms composed of rigid blocks. The first author believed that at this point there must be a limitation in application of the existing theories to deformation of the structures where relative slip movement among particles play an important role.

Basing on such consideration he proposed a new discrete model entitled the "Rigid Bodies-Spring Model" as follows. Consider a set of three dimensional rigid bodies of arbitrary shape as shown in Fig. 1.

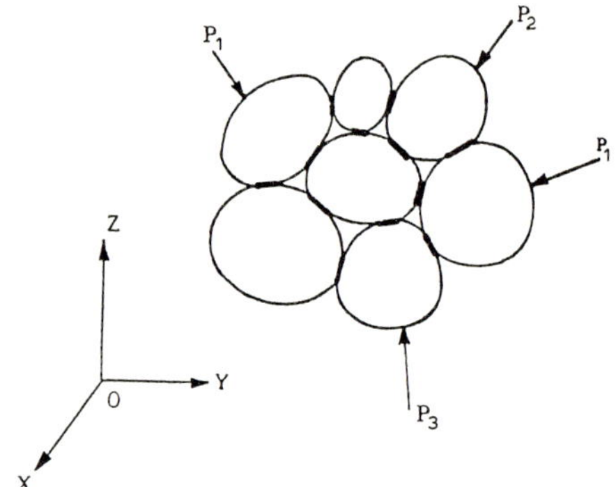

Fig. 1 Many discrete rigid bodies under contact

They are assumed to be in equilibrium with external loads, and reaction forces are produced by the spring system which is distributed over the contact surface of two adjacent bodies. For further development of new element models, it will be assumed that the contact area is known and fixed. It should be mentioned here that in actual contact problem, the contact surface is not known a priori and therefore it can be determined only in iterative way. Taking two such rigid bodies under contact,

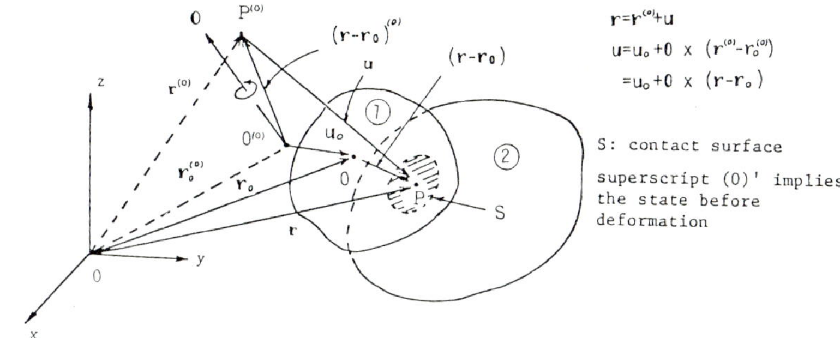

Fig. 2 Definition of rigid bodies-spring system.

an infinitesimal deformation of the spring system is considered (Fig. 2). The displacement $\mathbf{u}$ of an arbitrary point in a rigid body can be given by the following vector equation:

$$\mathbf{u} = \mathbf{u}_0 + \mathbf{0} \times (\mathbf{r}-\mathbf{r}_0) \tag{1}$$

Where $\mathbf{u}_0$ is the displacement vector of the centroid, $\mathbf{0}$ is the infinitesimal rotation vector and $(\mathbf{r}-\mathbf{r}_0)$ is the position vector of an arbitrary point with respect to the centroid before deformation,

$$\mathbf{u}_0 = (u_0, v_0, w_0) \qquad \mathbf{0} = (\theta, \phi, \chi) \tag{2}$$

Denoting the displacement vectors of an arbitrary point $P(x,y,z)$ in body ① or ② by $\mathbf{u}'$, $\mathbf{u}''$, respectively, the relative displacement vector of the point P can be defined as follows:

$$\overrightarrow{P'P''} = \mathbf{u}''-\mathbf{u}' = \boldsymbol{\delta} = (\delta_x, \delta_y, \delta_z)$$

Denoting the unit normal drawn outward to the contract surface at the point P by $\mathbf{n}$ (See Fig. 3).

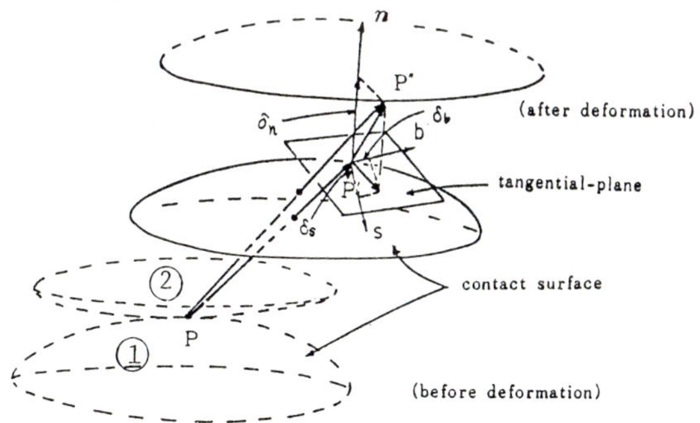

Fig. 3 Relative displacements of P on the contact surfaces in bodies.

$\boldsymbol{\delta}$ is resolved into δ_n and δ_s as follows:

$$\left.\begin{aligned} &\boldsymbol{\delta} = (\delta_n, \delta_s) \\ &\text{where} \\ &\delta_n = (\boldsymbol{\delta} \circ \mathbf{n}), \qquad \delta_s^2 = \| \mathbf{n} \times \boldsymbol{\delta} \|^2 \end{aligned}\right\} \tag{4}$$

Based on the above preliminaries, the strain energy due to the relative displacements (δ_n, δ_s) of the spring system distributed over the contact surface S can be given by the following equation:

$$V = \frac{1}{2} \iint_S (k_n \delta_n{}^2 + k_s \delta_s{}^2) dS \tag{5}$$

In view of equations (4) the strain energy V is a quadratic function of the displacement vector $\mathbf{u}$ of the centroids in bodies ① and ② as follows:

$$V(u) = \frac{1}{2} \iint_S \boldsymbol{\delta}^T \mathbf{D} \, \boldsymbol{\delta} \, dS = \frac{1}{2} \mathbf{u}^T \mathbf{k} \mathbf{u} \tag{6a}$$

where

$$\mathbf{k} = \iint_S \mathbf{B}^T \mathbf{D} \, \mathbf{B} \, dS, \qquad \boldsymbol{\delta} = \mathbf{B}\mathbf{u}$$

$$\mathbf{u}^T = \lfloor u_1, v_1, w_1, \theta_1, \phi_1, \chi_1; u_2, v_2, w_2, \theta_2, \phi_2, \chi_2 \rfloor \tag{6b}$$

It may not be necessary to mention that $\mathbf{k}$ is the (12 x 12) stiffness matrix of the RBSM 3D elements. From it the beam, plate and shell elements can be derived without any difficulty. The details of this can be seen in the literature listed at the end of this paper (ref. 4-9).

In case of a large displacement, eq. (1) has to be replaced by the following equation:

$$\mathbf{u} = \mathbf{u}_0 + (\mathbf{T} - \mathbf{I})(\mathbf{r} - \mathbf{r}_0) \tag{7}$$

where $\mathbf{T}$ is a transformation matrix of local coordinates attached to the centroid between before and after deformation, and $\mathbf{I}$ is an unit matrix. Here $\mathbf{u}_0$ is displacement vector of the centroid given by

$$\mathbf{u}_0 = (u_0, v_0, w_0) \tag{8}$$

and $(\mathbf{r}-\mathbf{r}_0)$ is a position vector of an arbitrary point with respect to the centroid before deformation.

Recently Toi has successfully derived an exact formula for components of **T** matrix in the series form of angular displacement (θ, ϕ, χ), a sketch of which is given briefly in Appendix.

Here only a formula of the second approximation is given:

$$[T] = \begin{bmatrix} 1-\frac{1}{2}(\phi^2+\chi^2) & -\chi+\frac{1}{2}\phi\theta & \phi+\frac{1}{2}\chi\theta \\ \chi+\frac{1}{2}\theta\phi & 1-\frac{1}{2}(\chi^2+\theta^2) & -\theta+\frac{1}{2}\chi\phi \\ -\phi+\frac{1}{2}\theta\chi & \theta+\frac{1}{2}\phi\chi & 1-\frac{1}{2}(\theta^2+\phi^2) \end{bmatrix} \tag{9}$$

Using eq.(9), x component of the rigid body displacement field given by eq.(7) is now obtained as follows:

$$U(x,y,z) = u_0 - \frac{1}{2}(x-x_0)(\phi^2+\chi^2) + (y-y_0)(-\chi+\frac{1}{2}\phi\theta) + (z-z_0)(\phi+\frac{1}{2}\chi\theta) \tag{10}$$

Similar formulae for the other components $V(x,y,z)$, $W(x,y,z)$ can be easily derived.

Therefore increments of the displacement $\Delta\mathbf{u}$ is given by:

$$\Delta u = (\mathbf{u}^{(0)} + \Delta\mathbf{u}) - \mathbf{u}^{(0)} = \Delta\mathbf{u}^{(1)} + \Delta\mathbf{u}^{(2)} \tag{11}$$

where superscript (0) implies the initial state, (1) linear components, (2) nonlinear components. Again formulae are given here for x component only.

(i) $\Delta U^{(1)}$ (linear component)

$$\Delta U^{(1)} = u_0 + (z-z_0)\Delta\phi - (y-y_0)\Delta\chi - (x-x_0)(\phi^{(0)}\Delta\phi + \chi^{(0)}\Delta\chi) + \frac{1}{2}(y-y_0)(\phi^{(0)}\Delta\theta + \theta^{(0)}\Delta\phi) + \frac{1}{2}(z-z_0)(\chi^{(0)}\Delta\theta + \theta^{(0)}\Delta\chi) \tag{12}$$

(ii) $\Delta U^{(2)}$ (nonlinear component)

$$\Delta U^{(2)} = -\frac{1}{2}(x-x_0)\{(\Delta\phi)^2 + (\Delta\chi)^2\} + \frac{1}{2}(y-y_0)(\Delta\phi)(\Delta\theta) + \frac{1}{2}(z-z_0)(\Delta\chi)(\Delta\theta) \tag{13}$$

And therefore increments of the relative displacment vector $\boldsymbol{\delta} = \overrightarrow{P'P''}$ can be expressed as follows: (refer to Fig. 3)

$$\boldsymbol{\delta} = \boldsymbol{\delta}^{(1)} + \boldsymbol{\delta}^{(2)} \tag{14}$$

where $\boldsymbol{\delta}^{(i)} = (\delta_x^{(i)}, \delta_y^{(i)}, \delta_z^{(i)}) \quad (i=1,2)$

For x component $\quad \boldsymbol{\delta}_x = \boldsymbol{\delta}_x^{(1)} + \boldsymbol{\delta}_x^{(2)}$

(i) $\Delta\delta_x^{(1)} = \Delta U_1^{(1)} - \Delta U_2^{(1)}$

where the incremental displacement of two adjacent elements is denoted by $\Delta U_i^{(i)}$ (i=1,2)

$$\begin{aligned}\Delta\delta_x^{(1)} =& (\Delta u_2 - \Delta u_1) + \{(z-z_2)\Delta\phi_2 - (z-z_1)\Delta\phi_1\} - \{(y-y_2)\Delta\chi_2 \\ & -(y-y_1)\Delta\chi_1\} - \{(x-x_2)(\phi_2^{(0)}\Delta\phi_2 + \chi_2{}^{(0)}\Delta\chi_2) - (x-x_1)\times \\ & (\phi_1^{(0)}\Delta\phi_1 + \chi_1{}^{(0)}\Delta\chi_1)\} + \frac{1}{2}\{(y-y_2)(\phi_2^{(0)}\Delta\theta_2 + \theta_2{}^{(0)}\Delta\phi_2) \\ & -(y-y_1)(\phi_1^{(0)}\Delta\theta_1 + \theta_1^{(0)}\Delta\phi_1)\} + \frac{1}{2}\{(z-z_2)(\chi_2^{(0)}\Delta\theta_2 + \\ & \theta_2^{(0)}\Delta\chi_2) - (z-z_1)(\chi_1^{(0)}\Delta\theta_1 + \theta_1^{(0)}\Delta\chi_1)\}\end{aligned} \tag{15}$$

(ii) $\Delta\delta_x^{(2)} = \Delta U_2^{(2)} - \Delta U_1^{(2)}$

$$\begin{aligned}\Delta\delta_x^{(2)} =& -\frac{1}{2}[(x-x_2)\{(\Delta\phi_2)^2 + (\Delta\chi_2)^2\} - (x-x_1)\{(\Delta\phi_1)^2 + (\Delta\chi_1)^2\}] \\ & +\frac{1}{2}\{(y-y_2)(\Delta\phi_2)(\Delta\theta_2) - (y-y_1)(\Delta\phi_1)(\Delta\theta_1)\} \\ & +\frac{1}{2}\{(z-z_2)(\Delta\chi_2)(\Delta\theta_2) - (z-z_1)(\Delta\chi_1)(\Delta\theta_1)\}\end{aligned} \tag{16}$$

In the similar way $\Delta\delta_y^{(i)}$, $\Delta\delta_z^{(i)}$ (i=1,2) can be obtained.

Next, the tangential plane at P is considered and local Cartesian coordinates (s,b,n) are defined as shown in Fig. 3.

On the contact boundary surface of two elements ① and ②, a set of spring system is considered whose force-displacement relationship is assumed to be given by the following equation:

$$\begin{Bmatrix} \Delta f_s \\ \hline \Delta f_b \\ \hline \Delta f_n \end{Bmatrix} = \begin{bmatrix} k_{sn} & k_{ss} & k_{sb} \\ k_{bn} & k_{bs} & k_{bb} \\ k_{nn} & k_{ns} & k_{nb} \end{bmatrix} \begin{Bmatrix} \Delta\delta_s \\ \hline \Delta\delta_b \\ \hline \Delta\delta_n \end{Bmatrix} \tag{17}$$

$$\Vert \qquad\qquad \Vert \qquad\qquad \Vert$$

$$\Delta \mathbf{f}_0 \qquad\qquad \mathbf{D}_0 \qquad\qquad \Delta\boldsymbol{\delta}_0$$

or

$$\Delta \mathbf{f}_0 = \mathbf{D}_0 \Delta \boldsymbol{\delta}_0 \tag{18}$$

where suffix (0) implies the interelement boundary surface and each quantity is defined with respect to the local coordinate (s, b, n).

It is standard practice to obtain the force-relative displacements with respect to the global coordinates by the following equation:

$$\Delta \mathbf{f} = \mathbf{D} \Delta \boldsymbol{\delta}$$

where $\mathbf{f} = \lfloor f_x, f_y, f_z \rfloor$, $\mathbf{D} = \mathbf{A}^t \mathbf{D}_0 \mathbf{A}$ (19)

$\mathbf{A}$ is the coordinate transformation matrix defined by the following equation:

$$\begin{Bmatrix} \delta_s \\ \hline \delta_b \\ \hline \delta_n \end{Bmatrix} = \begin{bmatrix} \ell_1 & m_1 & n_1 \\ \ell_2 & m_2 & n_2 \\ \ell_3 & m_3 & n_3 \end{bmatrix} \begin{Bmatrix} \delta_x \\ \hline \delta_y \\ \hline \delta_z \end{Bmatrix} \tag{20}$$

or

$$\boldsymbol{\delta} = \mathbf{A}^t \boldsymbol{\delta}_0 \tag{21}$$

With such preliminaries, the following incremental form of the virtual work equation can be derived:

$$\left.\begin{aligned} &\Sigma\iint \delta(\Delta\boldsymbol{\delta}^{(1)})^t\ (\mathbf{f}^{(0)}+\Delta\mathbf{f})ds + \Sigma\iint \delta(\Delta\boldsymbol{\delta}^{(2)})^t\ \mathbf{f}^{(0)}ds \\ &- \Sigma\iiint \delta(\Delta\mathbf{u}^{(1)})^t\ (\bar{\mathbf{p}}^{(0)}+\Delta\bar{\mathbf{p}})dv - \Sigma\iiint \delta(\Delta\mathbf{u}^{(2)})^t\ (\bar{\mathbf{p}}^{(0)}+\Delta\bar{\mathbf{p}})dv \\ &- \Sigma\iint \delta(\Delta\mathbf{u}^{(1)})^t (\bar{\mathbf{f}}^{(0)}+\Delta\bar{\mathbf{f}})ds - \Sigma\iint \delta(\Delta\mathbf{u}^{(2)})^t(\bar{\mathbf{f}}^{(0)}+\Delta\bar{\mathbf{f}})ds = 0 \end{aligned}\right\} \tag{22}$$

In derivation of eq.(22), the terms of higher order products of the incremental displacements were neglected.

Eq.(22) is further transformed into the following equation:

$$\Sigma \iint \delta(\Delta\boldsymbol{\delta}^{(1)})^t \mathbf{D}\Delta\delta^{(1)} ds + \Sigma \iint \delta(\Delta\boldsymbol{\delta}^{(2)})^t \mathbf{f}^{(0)} ds$$

$$-\Sigma \iiint \delta(\Delta\mathbf{u}^{(1)})^t \Delta\bar{\mathbf{p}}\, dv - \Sigma \iiint \delta(\Delta\mathbf{u}^{(2)})^t (\bar{\mathbf{p}}^{(0)} + \Delta\bar{\mathbf{p}}) dv$$

$$-\Sigma \iint \delta(\Delta\mathbf{u}^{(1)})^t \Delta\bar{\mathbf{f}}\, ds - \Sigma \iint \delta(\Delta\mathbf{u}^{(2)})^t (\bar{\mathbf{f}}^{(0)} + \Delta\bar{\mathbf{f}})\, ds = \mathbf{R} \qquad (23)$$

where $\mathbf{R} = -\Sigma \iint \delta(\Delta\boldsymbol{\delta}^{(1)})^t \mathbf{f}^{(0)} ds - \Sigma \iiint \delta(\Delta\mathbf{u}^{(1)})^t \bar{\mathbf{p}}^{(0)} dv$

$$- \Sigma \iint \delta(\Delta\mathbf{u}^{(1)})^t \bar{\mathbf{f}}^{(0)} ds$$

It is not so difficult to derive the following matrix equation in incremental form after some calculations.

$$\mathbf{K}_T \Delta\mathbf{u} = \Delta\bar{\mathbf{F}} - \bar{\mathbf{F}}_r \qquad (24)$$

where $\mathbf{K}_T$ is the total tangential stiffness matrix given by

$$\mathbf{K}_T = \mathbf{K} + \mathbf{K}_0 + \mathbf{K}_G \qquad (25)$$

$\mathbf{K}$ is the original stiffness matrix, $\mathbf{K}_0$ the initial displacement matrix and $\mathbf{K}_G$ corresponds to the initial stress (geometric) matrix in the well known finite element methods.

It should be mentioned here that $\bar{\mathbf{F}}_r$ is an unbalance force due to manipulation error in previous stage of loading.

For determination of the spring constants in plastic range, Yamada's method[3] is adopted.

First it is assumed that plastic yielding will occur if stresses in these spring system satisfy the following condition:

$$f(\boldsymbol{\sigma}) = 0 \quad (\boldsymbol{\sigma}^t = \lfloor \sigma_n, \tau_{ns}, \tau_{nb} \rfloor) \qquad (26)$$

where $f(\boldsymbol{\sigma})$ is the plastic potential in the flow theory of plasticity. Based on the plastic flow rule, the relation between stress increments

$\Delta\sigma$ and strain increments $\Delta\varepsilon$ can be finally obtained in the following form:

$$\Delta\sigma = \left[K^e - \frac{(\frac{\partial f}{\partial \sigma})^t K^{(e)} \cdot K^{(e)} \frac{\partial f}{\partial \sigma}}{(\frac{\partial f}{\partial \sigma})^t K^{(e)} \frac{\partial f}{\partial \sigma}} \right] \Delta\varepsilon \qquad (27)$$

It should be mentioned here that the method described was proposed for the rigid-plastic materials and therefore the effect of strain hardening is not taken into account.

It is needless to say that a pertinent yield condition should be selected depending upon the material to be used.

Crack initiation or contact problems can be also handled easily by using the spring characteristics as shown in Figs. 4a and 4b respectively.

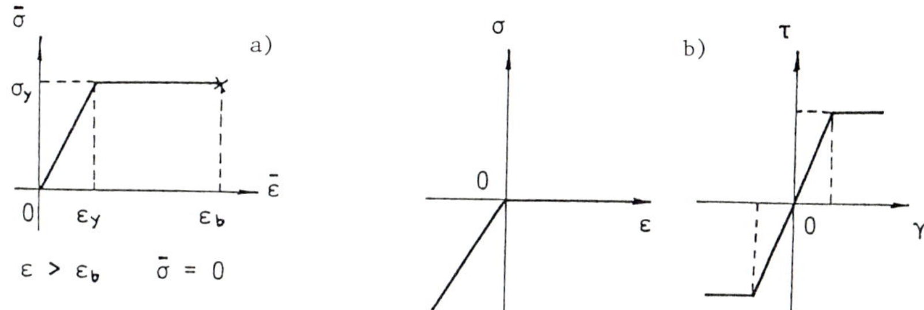

Fig. 4 a) A condition of the crack initiation on the interelement boundary; b) Characteristics of the interelement springs in contact problems.

CONVERGENCE OF THE RBSM SOLUTIONS AND DETERMINATION OF THE SPRING CONSTANTS

Generally speaking the elastic deformation of solids is characterized by the deformation of the finite elements without slip on the interelement boundaries.

The finite element displacement models can satisfy this displacement continuity on the interelement boundaries so that convergency of the FEM elastic solutions can be usually guaranteed.

However, convergence of the FEM inelastic solutions up to the limit

loading stage can hardly be guaranteed because the FEM analysis may always require such displacement continuity along the interelement boundaries even beyond the elastic limit. This might cause substantial error of the calculated results from the true solutions.
On the other hand, in the RBSM models the elements are assumed to be rigid and therefore the element deformations are not considered, but the interelement displacement can be taken into account by introducing two types of spring connecting two adjacent elements.
Consequently it is extremely difficult to represent the elastic deformation by using the RBSM models, and then convergence of the RBSM elastic solutions can not be guaranteed in general except in the cases of some elements such as beam element, ring element for the axisymmetric deformation of shells of revolution.
However, once the structural deformation reaches the stage of the large scale yielding, deformation due to slip on the intergranular boundaries may become predominant and consequently the elastic deformation of the particles may become negligibly small. Therefore, if the real collapse mechanism is included in the assumed mesh pattern, the incremental RBSM solutions always yield the true collapse loads. Or even if this is not the case, it can be proved that such incremental method can always give the best upper bound solutions.
From the practical point of view, however, it is still not sufficient to guarantee accuracy of the RBSM solutions unless the corresponding lower bound solutions can be derived.
Kawai has considered that a lower bound solution may be obtained from RBSM solutions by following the diagram as shown in Fig. 5
The practical program to obtain such lower bound solutions has been developed recently by Takeuchi and he verified the effectiveness of his method by solving several simple examples, one of which is shown in Fig.6.
In this way a reliable method for determination of collapse loads of solids has been developed by using the RBSM models. The rigorous

mathematical proof, however, has not been completed, because exact determination of the stress distribution in individual elements is extremely difficult so that approximate methods of solution such as the least square method are obliged to use.

Nevertheless this method may be acceptable from the limit analysis point of view. In near future attempt will be made to extend application of this method to the dynamic collapse analysis of solids.

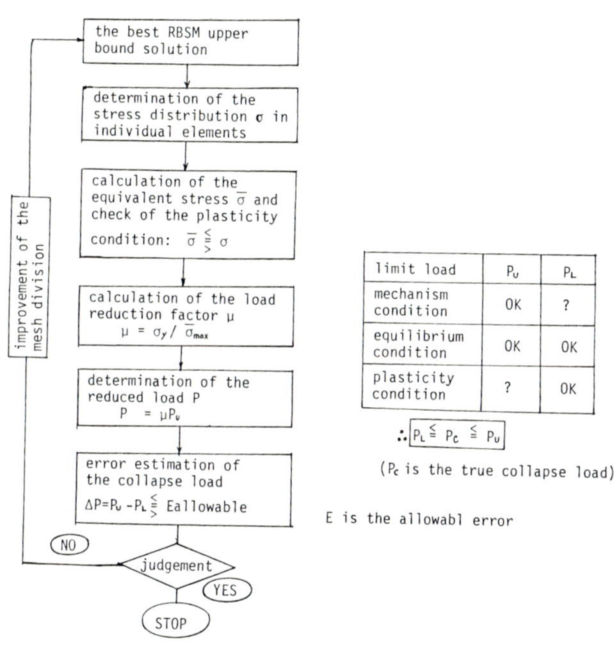

limit load	P_U	P_L
mechanism condition	OK	?
equilibrium condition	OK	OK
plasticity condition	?	OK

Fig. 5. Flow diagram for evaluation of the reliable limit loads.

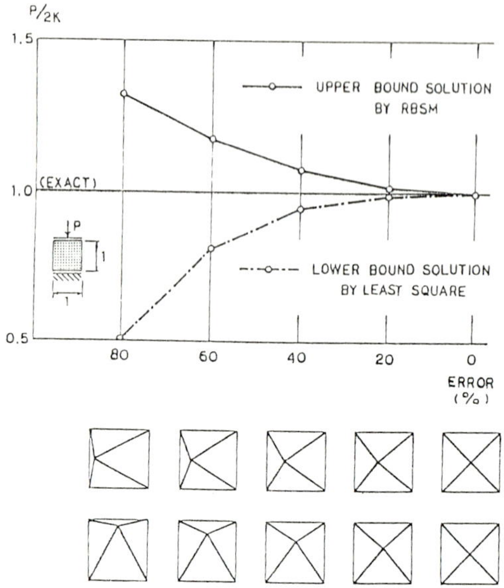

Fig. 6. Determination of the upper and lower bound solutions in the present limit analysis.

NO-TENSION AND TENSION CRACK ANALYSIS OF SOIL AND ROCK FOUNDATIONS BY MEANS OF NEWLY PROPOSED DISCRETE ELEMENTS

Zienkiewicz and other proposed a method of "No-Tension Analysis" to study behavior of the granular materials, Fig. 7 shows the flow diagram of their method (ref. 10).

In the above-mentioned method of analysis the initial stiffness matrix can be used for the entire process of calculation, and therefore reduction of computing time may be possible, but the accuracy of displacement obtained by this method may not be guaranteed if the crack initiates somewhere. Especially in the case of analysis of granular materials, actual behavior may be completely different from those predicted by this no-tension analysis and to make it worse, the contact problem can not be handled easily by this method. In the above mentioned method tensile residual forces are released at every stage. During the stress-releasing process possibility of the re-contact of adjacent elements is considered so that the algorithm becomes somewhat complicated, and computing time may be increased due to various checking schemes. In this computational scheme introduction of the crack initiation, consideration of initial

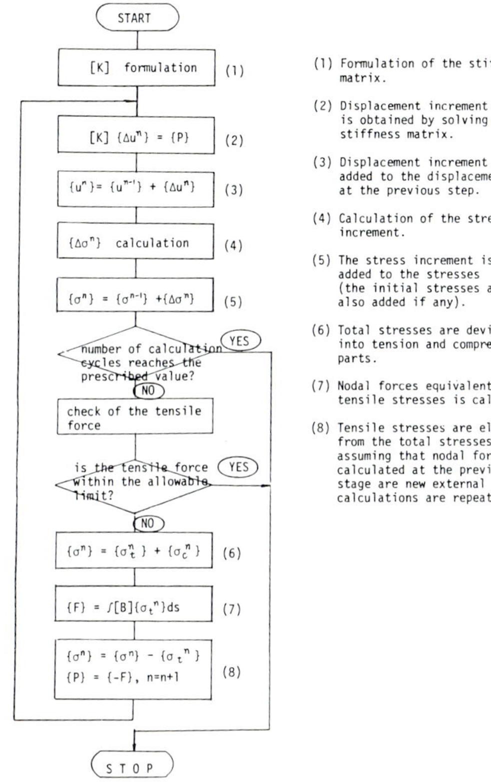

Fig. 7 Flow chart for no-tension analysis.

stresses and application to the contact as well as elasto-plastic analyses may be possible and these advantages may fully compensate the above-mentioned drawback. From such consideration, Takeuchi attempted to modify the scheme of no-tension crack analysis by using RBSM models. He developed a scheme for tension-crack analysis as shown in Fig. 8.

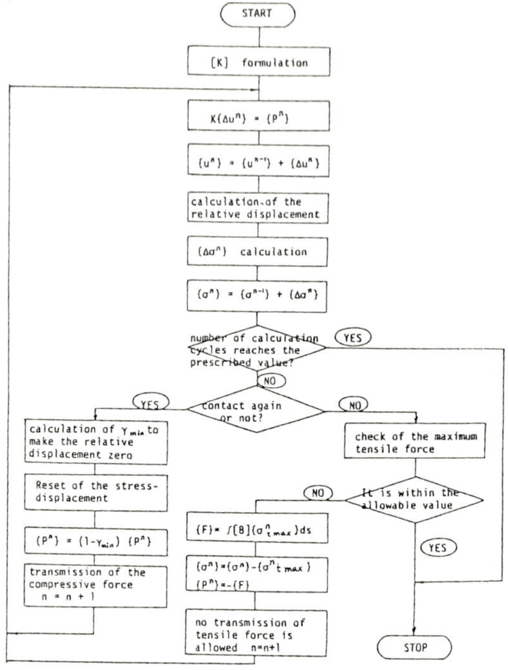

Fig. 8 Flow chart for tension-crack analysis.

SOME NUMERICAL EXAMPLES

Computer Simulation on the Load Transfer Mechanism of the Fuel Block System in the High Temperature Gas Reactor (HTGR) (ref. 11)

It has been revealed from the two dimensional photoelastic study that when the reactor core structure of the HTGR as shown in Fig. 9(a) is subjected to external loads, they are transmitted to the fuel blocks lying along certain lines because each block is independent and therefore internal forces are not distributed like in the continuum.
Since the individual block is considered to slide freely each other though they are in contact, the reactor core structure is idealized as a set of the RBSM elements without the shear springs.

Fig. 9 (a) shows the result of elastic analysis in which the arch action is not built up and tensile forces are almost diffused into the whole structure.

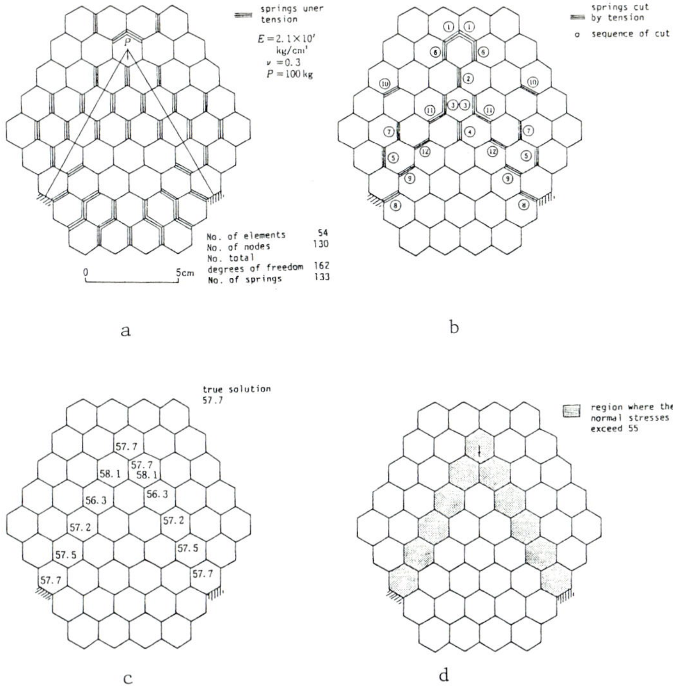

Fig. 9 Stress analysis of the HTGR fuel core structures: a) Elastic analysis; b) Collapse load; c) Stress distribution; d) Stress transfer mechanisms.

In the real structure, however, such tensile forces may not be observed because they are at most under contact condition. Therefore calculation was carried out by releasing such fictitious tensile forces in the step by step manner.

In Fig. 9 (b) the sequence number of spring cut and locations are shown. The obtained stress distribution is also shown in Fig. 9(c).

The result of the tension-crack analysis revealed that the external load is transmitted in some blocks which are lined up in certain directions and practically no stress is distributed in the rest of the structure as shown in Fig. 9 (d).

Computer Simulation of the Microscopic Behavior of the Polycrystalline Structures (ref. 9)

It can be concluded from this study that stress analysis of a set of discrete bodies may be possible by using the present algorithm of the tension crack analysis.

Assuming individual crystals to be rigid and representing their bonding forces on the grain boundary by the normal and shear spring systems, computer simulation of the microscopic behavior of the metal structures can be made without any trouble.

A preliminary study has been made on the computer simulation of the "superplasticity problem" to which a keen attention has been focused recently among engineers in metallurgy and material sciences.

Fig. 10 shows the discrete model of the microscopis structure of the Magnesium (Mg-6%, Zr-0.5%, Zn alloy) alloy specimen which is produced from the picture of the electromicroscope.

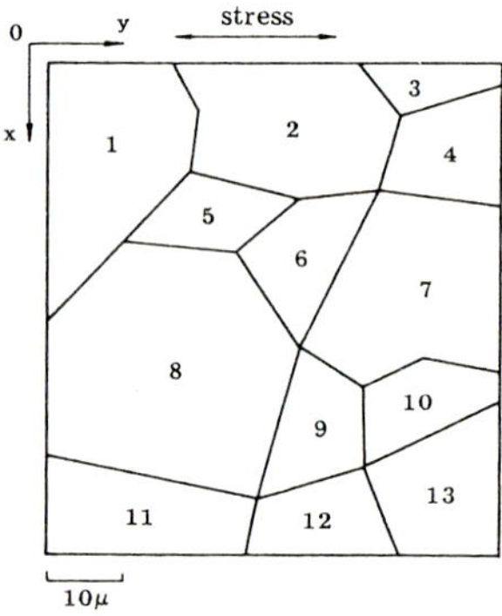

Fig. 10 Idealization of the microscopic structure of a Magnesium alloy specimen. The element 1 is clamped and the displacement in y direction and angular displacement of the element 11 are restrained.

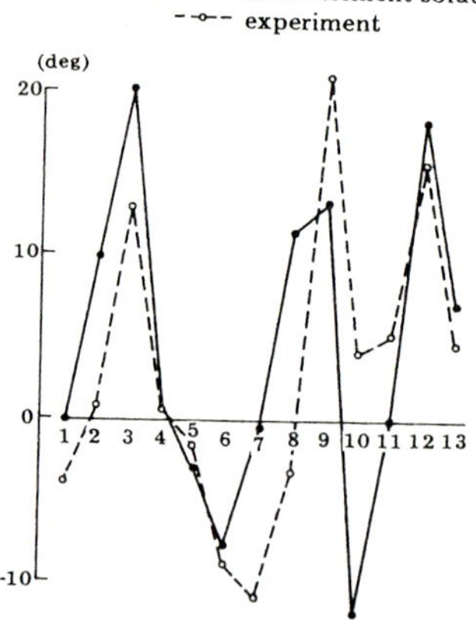

Fig. 11 Distribution of the angular displacement of the coarse crystals.

Figure 11 shows the calculated distribution of the angular displacement of individual coarse grains to be compared with the experimental data directly measured from the picture of the electron microscope.

In this analysis each grain is idealized by the four element rheological model.

It can be concluded that there exists a significant correlationship between the calculation and experiment and successful application of such computer simulation to microscopic behavior of polycrystalline structures will be expected in near future.

6. CONCLUSION

Outline of a new discrete method of limit analysis proposed by Kawai in 1976 was described briefly and then methods of no-tension as well as tension-crack analyses proposed by Takeuchi were introduced. Usefulness of the tension-crack analysis on the mechanics problems of granular materials was verified by showing two numerical examples. In near future extensive application of this method to analysis of static as well as dynamic behavior of the granular materials will be anticipated.

REFERENCES

1. T. KAWAI, "New Discrete Structural Models and Generalization of the Method of Limit Analysis", International Conference on Finite Elements in Nonlinear Solid and Structural Mechanics, Vol. 2, Geilo, Norway, pp. G04.1-G04.20, 1977.

2. T. KAWAI, "Some Considerations on the Finite Element Method", International Journal of Numerical Methods in Engineering, Vol. 16, PP. 81-120, 1980.

3. Y. YAMADA, N. YOSHIMURA and T. SAKURAI, "Plastic Stress-Strain Matrix and its Application for the Solution of Elastic-Plastic Problems by the Finite Element Method", Int. J. Mech. Sci., Vol. 10, pp. 343-354, 1968.

4. T. KAWAI, "New Discrete Models and Their Application to Seismic Response Analysis of Structures", Nuclear Engineering and Design 48, North-Holland Publishing Company, pp. 207-229, 1978.

5. T. KAWAI, "New Element Models in Discrete Structural Analysis", Journal of the Society of Naval Architects of Japan, Vol. 141,

6. Y. TOI and T. KAWAI, "A New Discrete Analysis on Dynamic Collapse of Structures", Journal of the Society of Naval Architects of Japan, Vol. 143, pp. 275-281, May, 1978.

7. T. KAWAI, U. KAWABATA and K. KUMAGAI, "A New Discrete Model for Analysis of Solid Mechanics Problems", Proc. of the First Int. Conf. on Numerical Methods in Fracture Mechanics (editors A.R. Luxmoore, and D.R.J. Owen, Swansea), 1978.

8. K. KONDOU, "Basic Studies on the Discrete Analysis of Beam and Plate Elements by Using the Lower Order Shape Functions", Doctorate Thesis submitted to the University of Tokyo, (in Japanese) Dec. 1977.

9. Y. TOI, "Studies on the Discrete Analysis of Plate and Shell Structures by using the Simplified Element Models", Doctorate Thesis submitted to the University of Tokyo, (in Japanese), Dec. 1978.

10. O.C. ZIENKIEWICZ, The Finite Element Method, McGraw-Hill Book Co. (U.K.), 3rd ed. 1977.

11. N. TAKEUCHI, "Fundamental Investigation for Limit Analysis of Soil and Rock Foundations by Means of New Discrete Models", Doctorate Thesis submitted to the University of Tokyo, (in Japanese), July 1981.

12. T.KAWAI, "A New Approach to Soil Mechanics and Geotechnical Engineering", The Keynote Address presented the Third International Conference in Australia on Finite Element Methods, University of New South Wales, Sydney, July 1979.

13. T. KAWAI, N. TAKEUCHI and Y. KAMIO, "A New Discrete Limit Analysis of Underground Structures", ASME, A Century 2 Publication PVP-43, August 12-15, 1980, San Francisco.

14. T. KAWAI and N. TAKEUCHI, "A Discrete Method of Limit Analysis with Simplified Elements", ASCE, International Conference on Computing in Civil Engineering, New York, May 12-14, 1981.

15. T. KAWAI, N. TAKEUCHI and T. KUMETA, "New Discrete Models and their Application to Rock Mechanics", Proc. of the International Symposium Weak Rock, Tokyo, Sept. 21-24, 1981.

Appendix

Derivation of the Exact Rigid Body Displacement Field in Three Dimensional Space

The exact rigid body displacement function in three dimensional space can be expressed in the following form by using Eulerian angles θ, ϕ, ψ as defined in Fig. A-1.

$$\left.\begin{aligned} u(x,y,z) &= u_0 + \ell_\xi x + \ell_\eta y + \ell_\zeta z - x \\ v(x,y,z) &= v_0 + m_\xi x + m_\eta y + m_\zeta z - y \\ w(x,y,z) &= w_0 + n_\xi x + n_\eta y + n_\zeta z - z \end{aligned}\right\} \quad \text{(A-1)}$$

where

$$\left.\begin{aligned} \ell_\xi &= \cos\theta \cos\phi \cos\psi - \sin\phi \sin\psi \\ \ell_\eta &= -\cos\theta \cos\phi \sin\psi + \sin\phi \cos\psi \\ \ell_\zeta &= \sin\theta \cos\phi \\ m_\xi &= \cos\theta \sin\phi \cos\psi + \cos\phi \sin\psi \\ m_\eta &= -\cos\theta \sin\phi \sin\psi + \cos\phi \cos\psi \\ m_\zeta &= \sin\theta \sin\phi \\ n_\xi &= -\sin\theta \cos\psi \\ n_\eta &= \sin\theta \sin\psi \\ n_\zeta &= \cos\theta \end{aligned}\right\} \quad \text{(A-2)}$$

ℓ_ξ, ℓ_η ..... n_ζ are direction cosines between the coordinate axes (x,y,z) fixed to the body before displacement and the ones after displacement (ξ,η,ζ). u_0, v_0, w_0 are components of the translational displacement of the origin.

When Eulerian angles θ, ϕ, ψ as shown in Fig. A-1 are used, the condition of no rotation before and after displacement can be expressed as follows:

$$\theta = 0, \qquad \phi + \psi = 0 \quad \text{(A-3)}$$

In view of this, the following new rotational parameters are introduced:

$$\begin{aligned} \phi_1 &= \phi + \psi \\ \phi_2 &= \phi - \psi \end{aligned} \qquad \text{(A-4)}$$

Using new variables θ, ϕ_1 and ϕ_2, the direction cosines given by eq. (A-2) can be expressed by the following equation:

$$\left.\begin{aligned}
\ell_\xi &= \frac{1}{2}\cos\theta(\cos\phi_1 + \cos\phi_2) - \frac{1}{2}(\cos\phi_2 - \cos\phi_1) \\
\ell_\eta &= -\frac{1}{2}\cos\theta(\sin\phi_1 - \sin\phi_2) + \frac{1}{2}(\sin\phi_1 + \sin\phi_2) \\
\ell_\zeta &= \frac{1}{2}\sin\theta\,\cos\{(\phi_1 + \phi_2)\} \\
m_\xi &= \frac{1}{2}\cos\theta(\sin\phi_1 + \sin\phi_2) + \frac{1}{2}(\sin\phi_1 - \sin\phi_2) \\
m_\eta &= -\frac{1}{2}\cos\theta(\cos\phi_1 - \cos\phi_2) + \frac{1}{2}(\cos\phi_1 + \cos\phi_2) \\
m_\zeta &= \frac{1}{2}\sin\theta\,\sin\{(\phi_1 + \phi_2)\} \\
n_\xi &= -\frac{1}{2}\sin\theta\,\cos\{(\phi_1 + \phi_2)\} \\
n_\eta &= \frac{1}{2}\sin\theta\,\cos\{(\phi_1 + \phi_2)\} \\
n_\zeta &= \cos\theta
\end{aligned}\right\} \qquad \text{(A-5)}$$

Now expanding eq.(A-5) into Taylor series of θ, ϕ_1, ϕ_2, around θ=0 and ϕ_1=0 and taking up to the second order terms the following approximate expression of the direction cosines can be obtained.

$$\begin{aligned}
\ell_\xi &= 1-\frac{\phi_1{}^2}{2} - \frac{\theta^2}{4}(1 + \cos\phi_2) \\
\ell_\eta &= -\phi_1 - \frac{1}{4}(\theta^2\sin\phi_2) \\
\ell_\zeta &= \theta\cos(\frac{\phi_2}{2}) - \frac{1}{2}\{\theta\phi_1\sin(\frac{\phi_2}{2})\}
\end{aligned}$$

$$
\left.
\begin{aligned}
m_\xi &= \phi_1 - \frac{1}{4}(\theta^2 \sin\phi_2) \\
m_\eta &= 1 - \frac{\phi_1{}^2}{2} - \frac{\theta^2}{4}(1-\cos\phi_2) \\
m_\zeta &= \theta\sin(\frac{\phi_2}{2}) + \frac{1}{2}\{\theta\phi_1\cos(\frac{\phi_2}{2})\} \\
n_\xi &= -\theta\cos(\frac{\phi_2}{2}) - \frac{1}{2}\{\theta\phi_1\sin(\frac{\phi_2}{2})\} \\
n_\eta &= -\theta\sin(\frac{\phi_2}{2}) + \frac{1}{2}\{\theta\phi_1\cos(\frac{\phi_2}{2})\}
\end{aligned}
\right\} \qquad \text{(A-6)}
$$

$$
n_\zeta = 1 - \frac{\theta^2}{2}
$$

Now let's introduce the following new notations:

$$
\left.
\begin{aligned}
a_1 &= -\theta\sin(\frac{\phi_2}{2}) \\
a_2 &= \theta\cos(\frac{\phi_2}{2}) \\
a_3 &= \phi_1
\end{aligned}
\right\} \qquad \text{(A-7)}
$$

Then eq.(A-6) can be written by the following equations:

$$
\left.
\begin{aligned}
\ell_\xi &= 1 - \frac{1}{2}\{(a_2)^2 + (a_3)^2\} \\
\ell_\eta &= -a_3 + \frac{1}{2}(a_1 a_2) \\
\ell_\zeta &= a_2 + \frac{1}{2}(a_1 a_3) \\
m_\xi &= a_3 + \frac{1}{2}(a_2 a_1) \\
m_\eta &= 1 - \frac{1}{2}\{(a_3)^2 + (a_1)^2\} \\
m_\zeta &= -a_1 + \frac{1}{2}(a_2 a_3) \\
n_\xi &= -a_2 + \frac{1}{2}(a_3 a_1) \\
n_\eta &= a_1 + \frac{1}{2}(a_3 a_2) \\
n_\zeta &= 1 - \frac{1}{2}\{(a_1)^2 + (a_2)^2\}
\end{aligned}
\right\} \qquad \text{(A-8)}
$$

Substituting eq.(A-8) into eq.(A-1).

the followng expressions can be finally obtained:

$$\left.\begin{aligned} u(x,y,z) &= a_4-a_3y+a_2z-\frac{1}{2}x\{(a_2)^2+(a_3)^2\}+\frac{1}{2}a_1a_2y+\frac{1}{2}a_1a_3z \\ v(x,y,z) &= a_5-a_1z+a_3x-\frac{1}{2}y\{(a_3)^2+(a_1)^2\}+\frac{1}{2}a_2a_3z+\frac{1}{2}a_2a_1x \\ w(x,y,z) &= a_6-a_2x+a_1y-\frac{1}{2}z\{(a_1)^2+(a_2)^2\}+\frac{1}{2}a_3a_1x+\frac{1}{2}a_3a_2y \end{aligned}\right\} \quad (A-9)$$

Eq.(A- 9) is the approximate rigid body displacement functions in which terms up to the second order are retained, and it can be clearly seen that a_1, a_2 and a_3 are rotational displacements around x, y and z axes respectively, while a_4, a_5, a_6 are componts of translational displacement in x, y and z directions respectively. It should be mentioned here that angular displacements a_1, a_2 and a_3 do not depend on the order of rotation and their second order terms are symmetric as can be seen from eq.(A-8), while Eulerian angles θ, ϕ, ψ depend on the sequence of rotation.

And it is not too difficult to show that all the finite strain components vanish if eq.(A- 9) is substituted into the three dimensional strain-displacement relations where displacement terms up to the second order are retained.

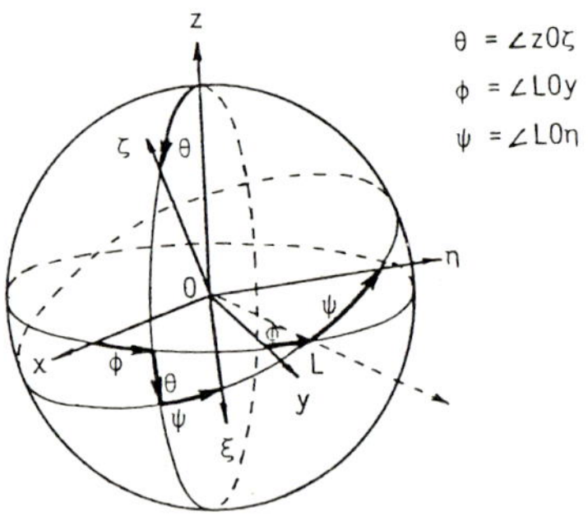

Fig. A-1 Definition of Eulerian angles.

Mechanics of Granular Materials: New Models and Constitutive Relations, edited by
J.T. Jenkins and M. Satake, 1983
Elsevier Science Publishers B.V., Amsterdam — Printed in The Netherlands

FREQUENCY DISTRIBUTION OF VOIDS IN RANDOMLY PACKED MONOGRANULAR LAYERS

M. SHAHINPOOR
Granular Materials Research Laboratory and
Department of Mechanical and Industrial Engineering
Clarkson College of Technology, Potsdam, NY 13676 (USA)

ABSTRACT

An optical scanning and recording technique is applied to randomly packed monolayers of equal spheres to obtain the frequency distributions of voids in their fabrics. The monogranular layers are randomly packed by rotation and finite shear motion of grains in encapsulated circular cylindrical cells. The resulting frequency distributions exhibit a biased Maxwellian trend favored towards the population of denser "Voronoi Cells."

INTRODUCTION

This paper briefly reports some recent experimental results on the frequency distribution of voids, characterized as "Voronoi Cells" (ref. 1-6) (see Fig. 1 and Table 1), in monolayers of randomly packed equal spheres. The spheres are made up of 1/8" to 3/16" in diameter steel, glass, nylon, and teflon balls. The frequency distribution density is a normalized pdf defined as a ratio. This is the ratio of the number of "Voronoi Cells" whose area void ratio is within

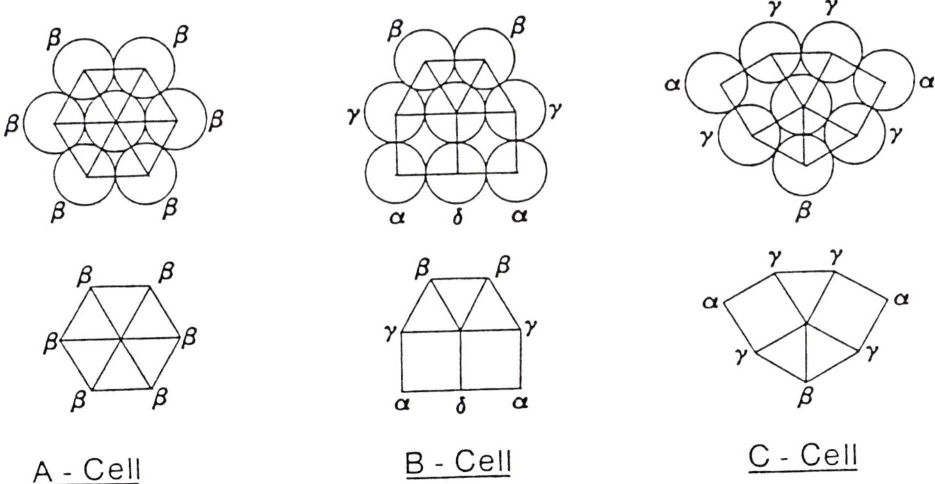

Fig. 1. Some typical "Voronoi Cells" in a random packing of monogranular layers

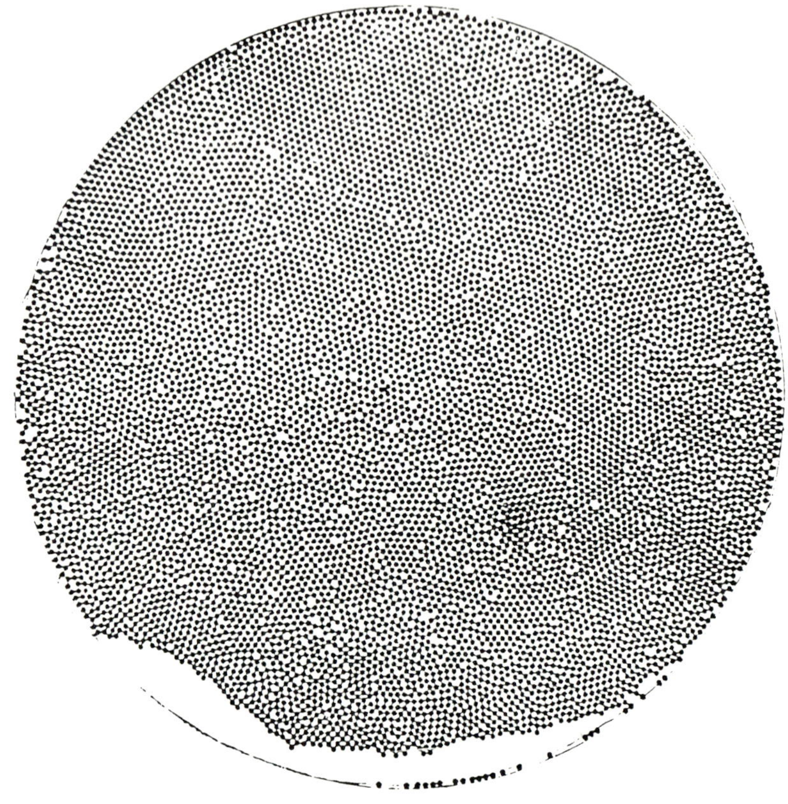

Fig. 2. Random monogranular layer of nylon balls of 1/8" diameter.

a narrow band (e, e+Δe) of void ratio space $\{e_m, e_M\}$ to the total number of cells. Here e_m and e_M are, respectively, the minimum and the maximum void ratios of all "Voronoi Cells" within the random monogranular assembly. The void ratio space is spanned by the void ratios of a group of "Voronoi Cells" (Fig. 1). For every random monogranular packing the values of e_m and e_M are first measured by direct visual inspection of their clear photographs (Figs. 2, 3, 4, 5).

The actual random two-dimensional packing of equal spheres (monogranular layers) are produced by sandwiching them between two transparent glass plates in a space with circular geometry (Figs. 2, 3, 4, 5). The setup is such that it can be rotated about the perpendicular central axis while slightly oriented (α~5-10°) with respect to the gravitational axis. While rotating the monogranular layer, random shear deformation and flow gives rise to many regions of critical density (Fig. 2). Critical density is a term referred to the

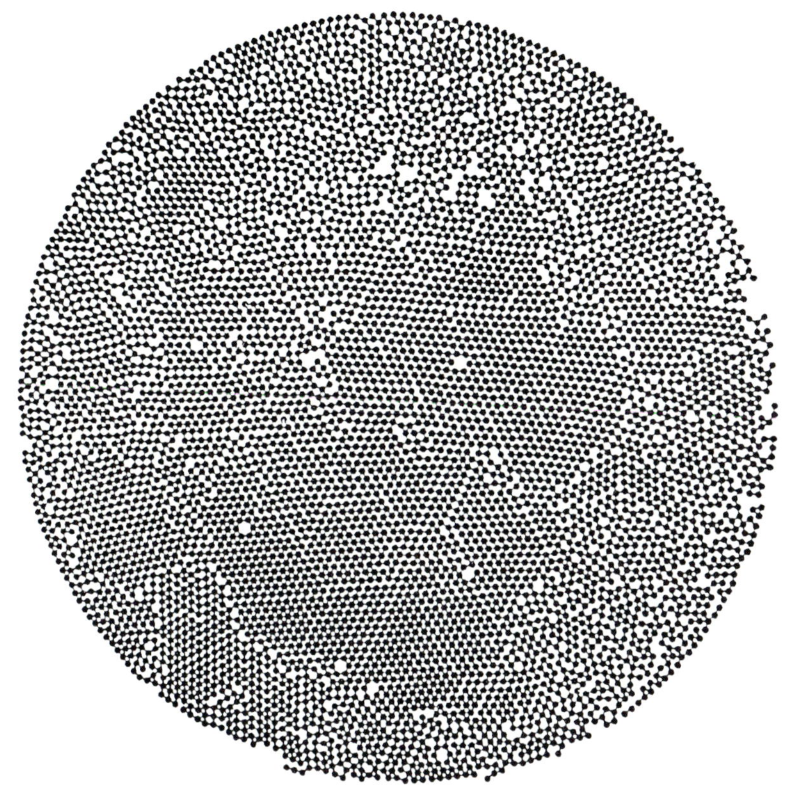

Fig. 3. Random monogranular layer of steel balls of 1/8" in diameter.

density of granular assemblies which are sufficiently sheared under a given confining pressure (ref. 1-6). The critical state in this case possesses a uniform frequency distribution of voids. This is due to the fact that sufficient shear motion statistically creates equal probabilities of formation of all possible (stable) "Voronoi Cells," Table 1. In our experiments, however, we have not observed such uniform distributions. Most of the frequency distributions obtained are biased positively towards the population of denser "Voronoi Cells," (Figs. 6, 7, 8, 9, 10, 11, 12, 13).

The experimental set-up consists of a wand type optical scanner with LED which scans the "Voronoi Cells" at a uniform speed in a random manner and sends out voltage signals (voltage of 250 mv) which are then fed to a strip-chart recorder. The signals from the strip chart recorder are then accordingly calibrated (Figs. 14, 15, 16, 17). The signals are then further amplified through operational amplifiers to 3-5 volts and are then fed to an AIM 65 Microcomputer capable of sorting out various signals and plotting the relevant

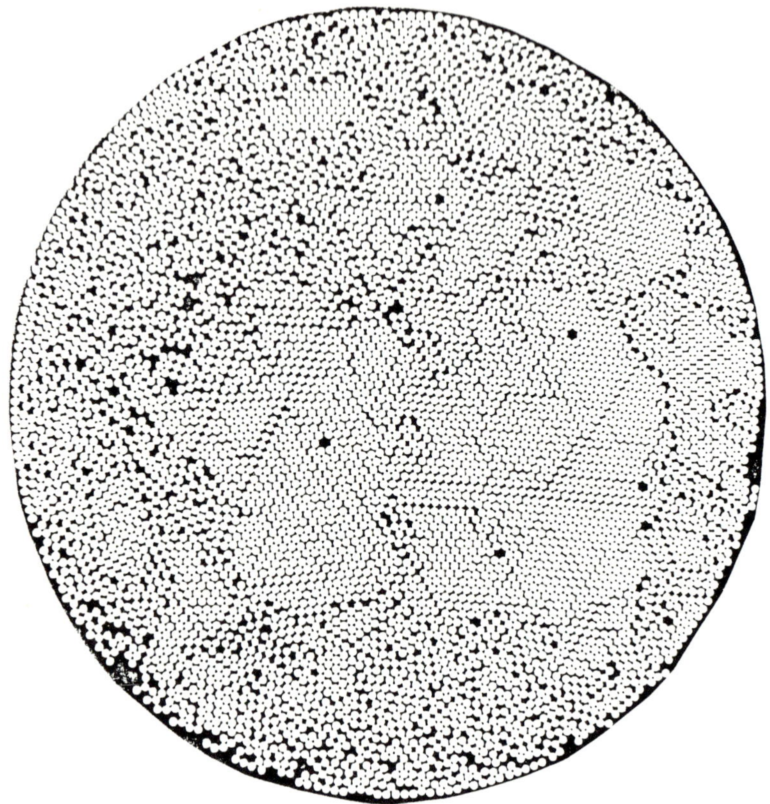

Fig. 4. Random monogranular layer of glass marbles of 3/16" diameter.

histograms. The RAM microprocessor is instructed to pick out only signals that are identified and calibrated corresponding to known "Voronoi Cells."

This procedure has been carried out for random monogranular assemblies of steel, glass, nylon and teflon spherical balls. However, it appears fully applicable to other monogranular fabric arrangement of real materials. Such statistical characterizations of two dimensional fabrics may be generalized to three dimensional fabrics through standard stereological theorems (see Santalo (ref. 7).

The general trend of all frequency distributions obtained exhibits a biased Maxwellian trend with the Maxwellian tail favored towards the population of denser "Voronoi Cells." This appears to be in complete agreement with the statistical mechanical results obtained by Shahinpoor (ref. 1,2) which suggest that

$$p(e) = \left[\frac{\lambda \exp.(-\lambda e)}{\exp.(-\lambda e_m) - \exp.(-\lambda e_M)}\right] , \qquad (1)$$

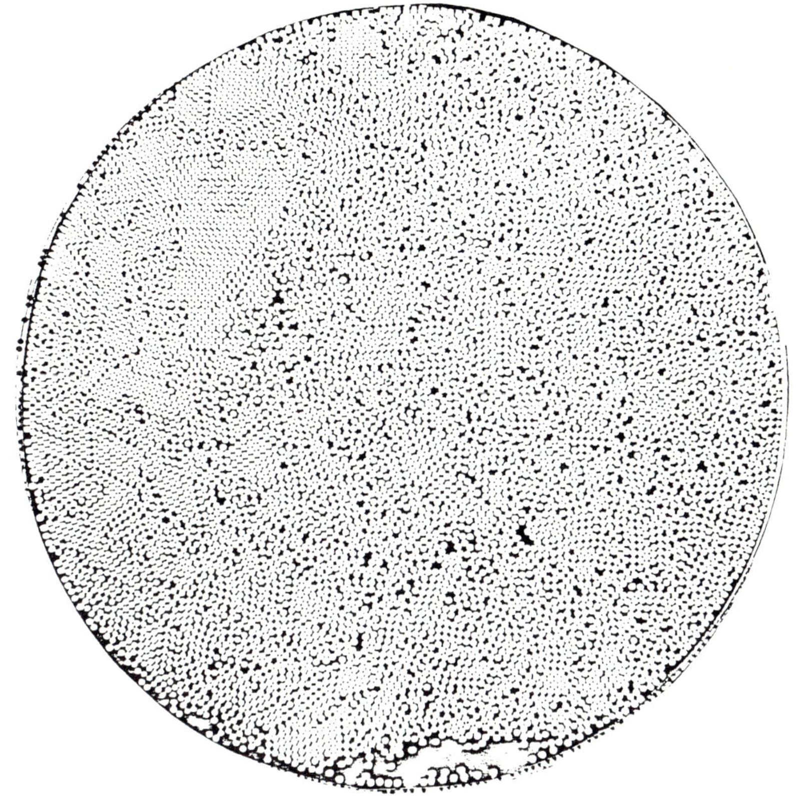

Fig. 5. Random monogranular layer of teflon balls of 1/8" diameter.

where λ is obtained from the following transcendental equation:

$$\bar{e} = \lambda^{-1} + \left[\frac{e_m \exp.(-\lambda e_m) - e_M \exp.(-\lambda e_M)}{\exp.(-\lambda e_m) - \exp.(-\lambda e_M)}\right] , \qquad (2)$$

where e_m, e_M, and e are, respectively, the minimum, the maximum and the mean void ratio. Figure (18) shows the general trend of equation (1) and Figure (19) displays the general solutions of equation (2) for λ for various values of $\bar{e}$.

The continuation of the present research is underway to improve the data collection and sampling, the creation of more realistic random aggregates of monogranular layers, as well as studying their fabrics interference patterns (Fig. 20).

ACKNOWLEDGMENTS

This material is based upon work supported by the National Science Foundation Geotechnical Engineering Program Grant No. CME-8021032.

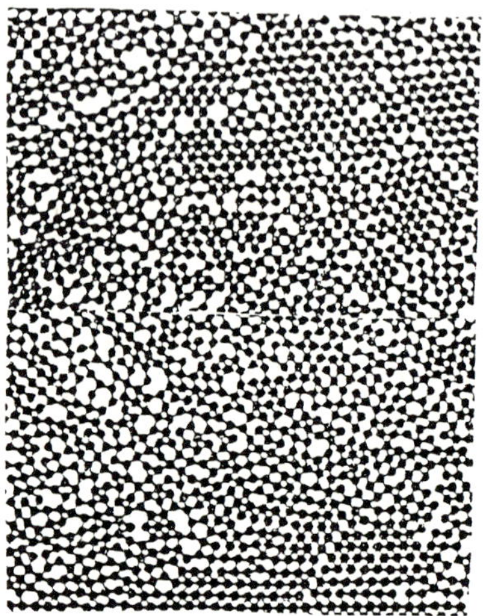

Fig. 6. Random packing of monogranular layers of equal spherical steel balls of 1/8" diameter.

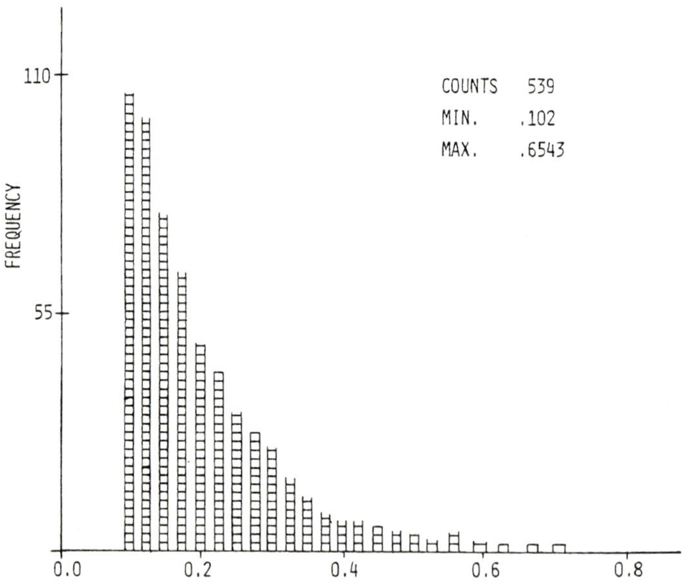

Fig. 7. Histogram generated by optical scanning of Figure 6.

Fig. 8. Random packing of monogranular layers of equal spherical nylon balls of 1/8" diameter.

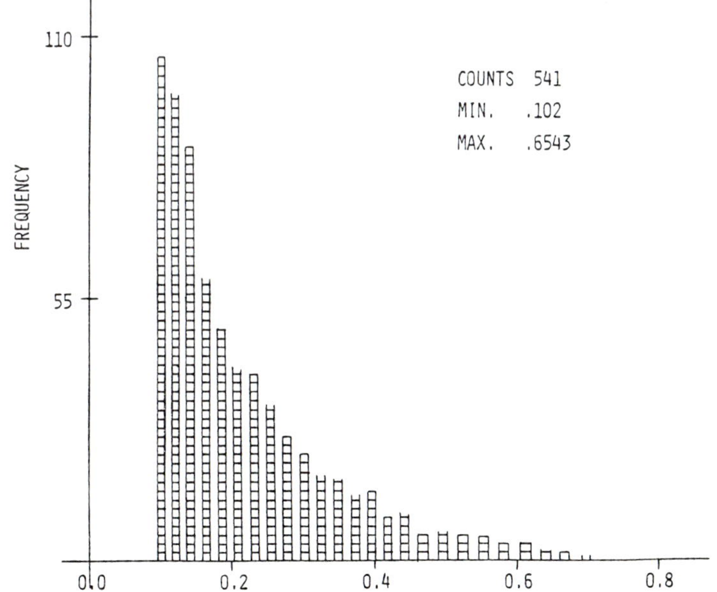

Fig. 9. Histogram generated by optical scanning of Figure 8.

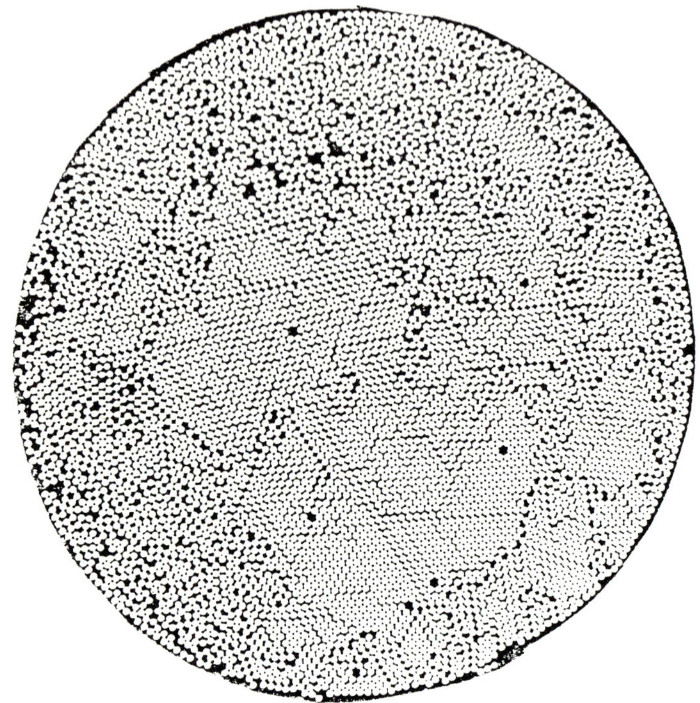

Fig. 10. Random packing of monogranular layers of equal glass marbles of 3/16" diameter.

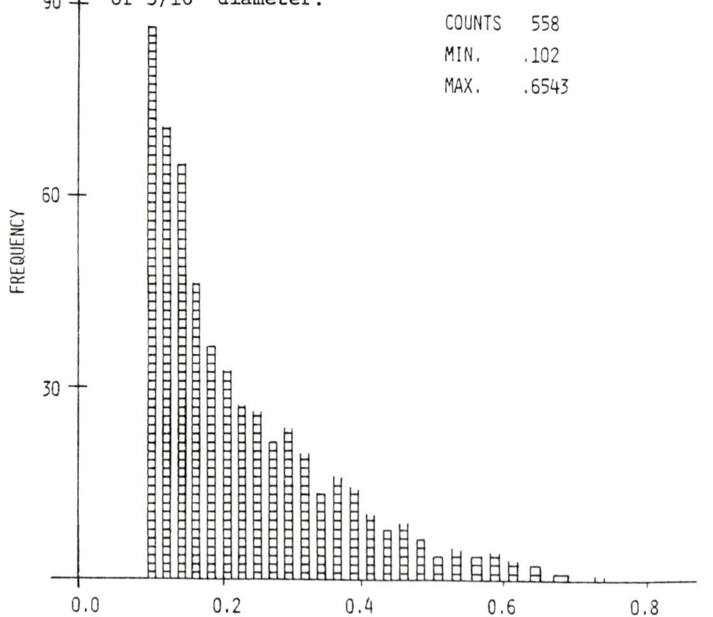

Fig. 11. Histogram generated by optical scanning of Fig. 10.

Fig. 12. Random packing of monogranular layers of equal teflon balls of 1/8" diameter.

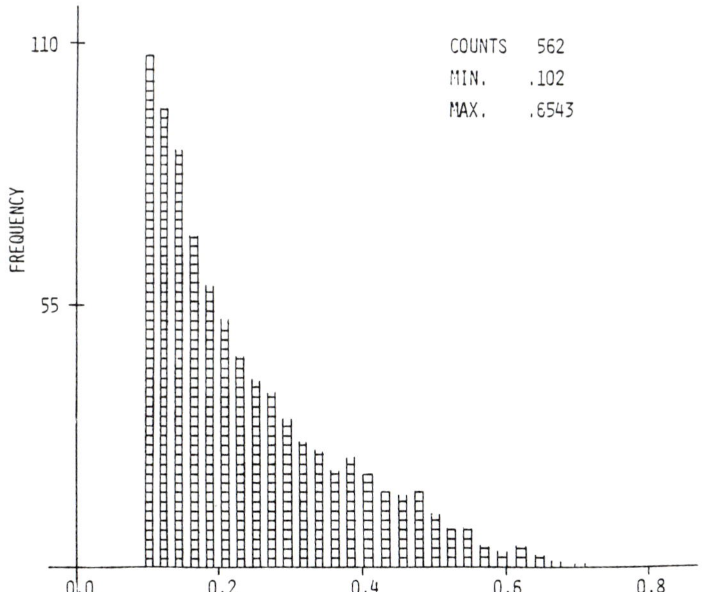

Fig. 13. Histogram generated by optical scanning of Fig. 12.

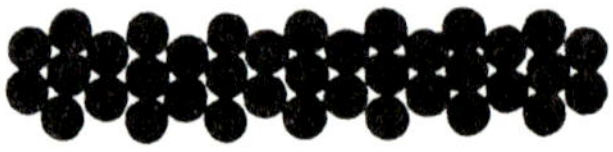

Figure 14. The densest possible "Voronoi Cells" possessing a void ratio of 0.1027.

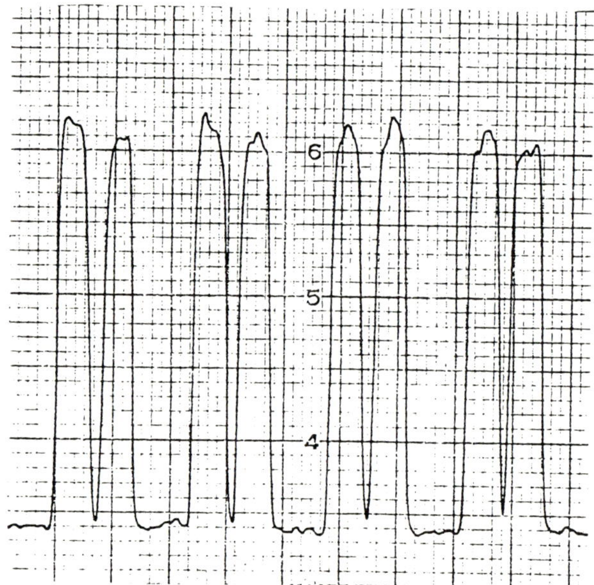

Figure 15. Output of the optical scanner corresponding to Figure 14.

Figure 16. The second densest possible "Voronoi Cells" possessing a void ratio of 0.2732.

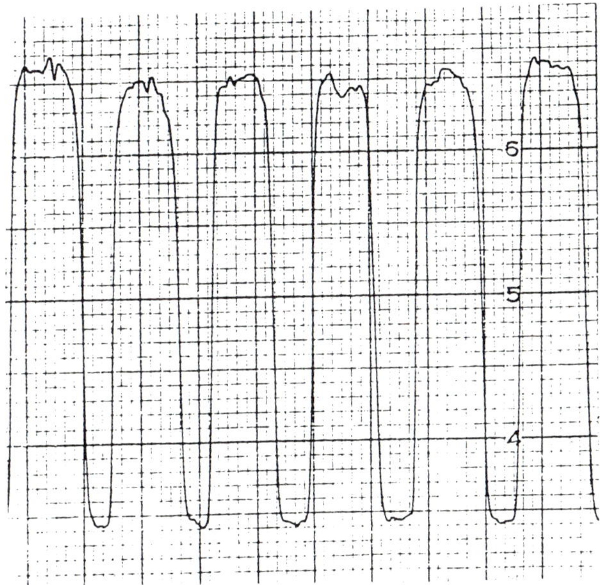

Figure 17. Output of the optical scanner corresponding to Figure 16.

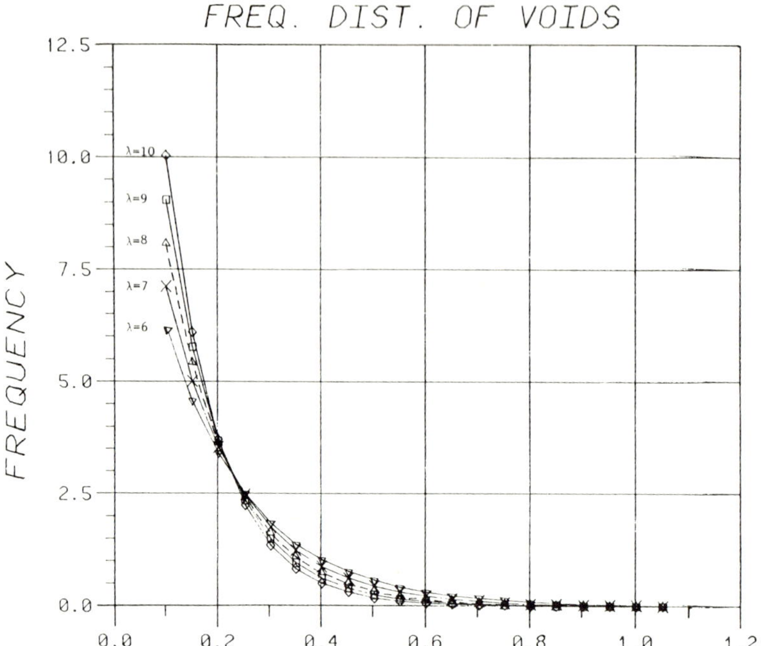

Fig. 18. Variation of Frequency Distribution of Voids for Different Values of λ.

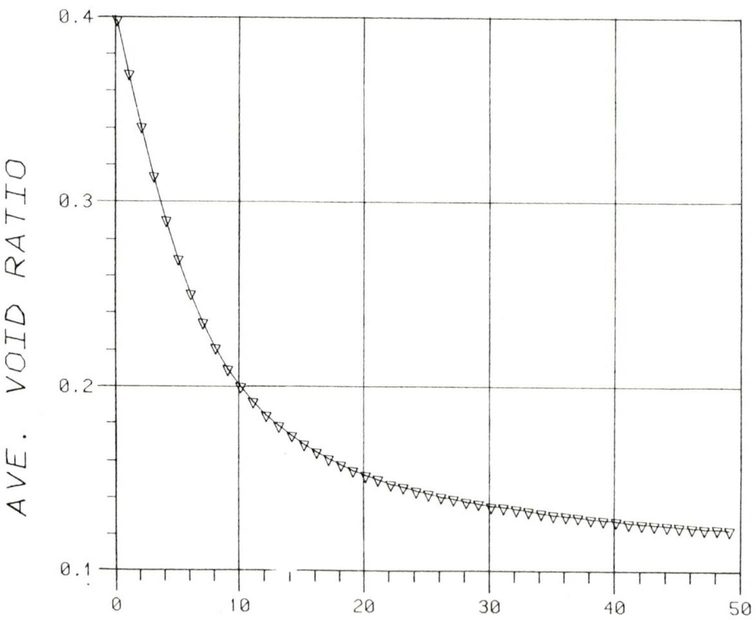

Fig. 19. Variation of Average Void Ratio $\bar{e}$ with λ .

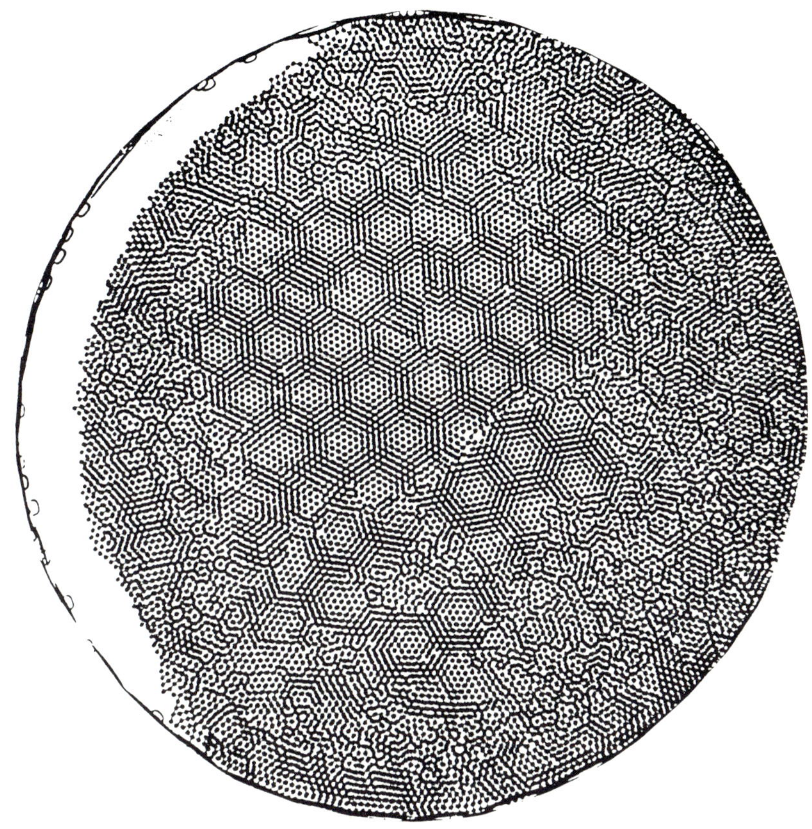

Fig. 20. An Interference Pattern for Monogranular Layer.

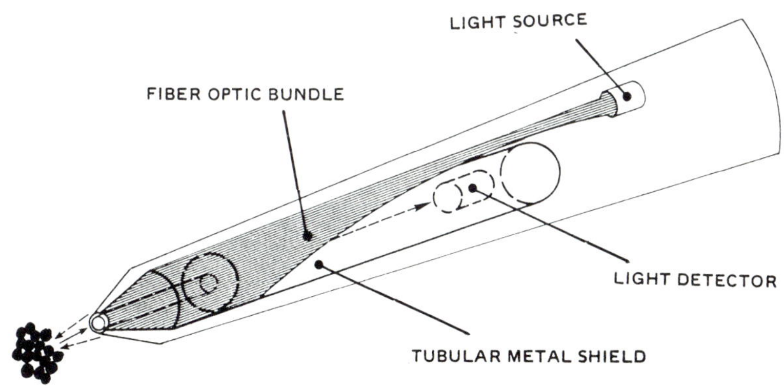

Fig. 21. The Optical Scanner.

Table 1

Some Typical Two-Dimensional Bulk "Voronoi Cells"

Cellular Structure	Coordination Number	Porosity	Void Ratio	Cell Number
	6	0.0931	0.1027	1
	5	0.1582	0.1879	2
	5	0.1582	0.1879	3
	4	0.2146	0.2732	4
	3	0.3954	0.6540	5

REFERENCES

1. M. Shahinpoor, "Statistical Mechanical Consideration on Random Packing of Granular Materials," J. Powder Tech., 25 (1979), pp. 163-176
2. M. Shahinpoor, "Statistical Mechanical Consideration on Storing Bulk Solids," Bulk Solids. Handl. Int. J., 1, 1 (1981), pp. 31-36.
3. A. Shahrpass, "Theoretical and Experimental Analysis on the Frequency Distribution of Voids in Random Two-Dimensional Packing of Granular Materials," M.Sc. Thesis, Clarkson College of Technology, August (1981), Also Res. Rep. MIE-071, August (1981).
4. M. Shahinpoor and A. Shahrpass, "Determining the Frequency Distribution of Voids in Randomly Packed Two-Dimensional Assemblies of Granular Materials," MIE-CCT Res. Rep. No. MIE-072, September (1981).
5. M. Shahinpoor and A. Shahrpass, "Frequency Distribution of Voids in Mono-layers of Randomly Packed Equal Spheres," Bulk Solids Handl. Int. J., 3, 3 (1982), pp.
6. M. Shahinpoor, "Crystallization of monomolecular Layers on Contracting Surfaces," J. Colloid and Interface Sci., 85, 1 (1982), pp. 227-234.
7. L.A. Santalo, "Integral Geometry and Geometric Probability," Addison-Wesley Pub. Comp., London, Amsterdam, Tokyo (1976).

Mechanics of Granular Materials: New Models and Constitutive Relations, edited by
J.T. Jenkins and M. Satake, 1983
Elsevier Science Publishers B.V., Amsterdam — Printed in The Netherlands

THE ENTROPY OF GRANULAR MATERIALS IN SHEARING DEFORMATIONS

N.MOROTO
Civil Engineering Dept.,Hachinohe Institute of Technology
Hachinohe,Aomori,031,(Japan)

ABSTRACT

The plastic work done during the shearing deformation of granular materials such as sand depends largely on stress paths. A new parameter defines as

$$\int (\text{increment of plastic work done/mean pressure})$$

is independent of stress paths. This parameter is called 'entropy' and it plays an important role in the study of plastic deformation of granular materials

INTRODUCTION

The deformation of granular materials is composed of two deformation mechanism:shear and consolidation. In this paper, the author directs his attention to shearing deformation of granular materials. In this case,(1)he reviews the new state function S_s called 'entropy' of granular materials from the phenomenological point of view and (2)considers the shearing deformation rule by means of entropy of the information theory. Symbols used here are as follows:

σ'_1: major principal stress σ'_3: minor principal stress
σ'_a: axial stress σ'_r: radial stress
$\sigma'_1=\sigma'_a, \sigma'_3=\sigma'_r$ for triaxial compression test
$\sigma'_1=\sigma'_r, \sigma'_3=\sigma'_a$ for triaxial extension test

$p=(\sigma'_a + 2\sigma'_r)/3$: mean principal stress(mean pressure)

$q=\sigma'_1-\sigma'_3$: deviator stress

$\eta=q/p$: stress ratio

ε_1: major principal strain ε_a: axial strain
ε_3: minor pricipal strain ε_r: radial strain

$\varepsilon_1=\varepsilon_a, \varepsilon_3=\varepsilon_r$ for triaxial compression test

$\varepsilon_1=\varepsilon_r, \varepsilon_3=\varepsilon_a$ for triaxial extension test

$v=\varepsilon_a + 2\varepsilon_r$: volumetric strain

$\gamma = \varepsilon_1 - \varepsilon_3$: deviator strain $\quad \epsilon = 2\gamma/3$

dv: increment of v $\quad$ dγ: increment of γ

$d\epsilon = 2d\gamma/3$

Compression is taken to be positive for both stress and strain. The increment of plastic work done due to shearing force can be written as

$$dw_s^p = pdv_d^p + \frac{2}{3} qd\gamma^p \quad (1)$$

where dv_d^p: increment of plastic volumetric strain due to dilatancy, $d\gamma^p$:increment of plastic deviator strain.

The discussions are limited to the following conditions:

1) isotropically consolidated samples,
2) monotonously and smoothly increasing stress paths,and
3) the virginal shear deformation

REVIEW OF NEW STATE FUNCTION

By increasing the shearing stress under different stress paths, the stress state of the specimens of granular materials changed from the same initial 'isotropic consolidation state' to the final condition of 'failure'. During the deformation from the initial state to the terminal,the plastic work done W_s^p carried out by shearing clearly depends on stress path. For instance, the amount of plastic work done along the stress path of p=3.0 kgf/cm^2 is almost three times as great as that along the stress path of p=1.0 kgf/cm^2. Therefore, the author introduced the new parameter defined as

$$S_s = \int dW_s^p/p \quad (2)$$

(refs.1,2). Performing the triaxial tests on sand, Tatsuoka(ref.3) showed that the deviator strain γ was determined almost exclusively by the stress condition of (p,q) irrespective of the stress paths of p-constant,σ_r-constant and σ_a-constant.

In Figs.1,2,3 and 4, the value of S_s are plotted against mean principal stress p for the given values of $\gamma(\doteqdot\gamma^p)$. The set of points in S_s -γ graphs shows a unique relationship which is almost parallel to the p-axis for both loosely and densely prepared samples. This is true for both cases of triaxial compression and triaxial extension(ref.1). These facts confirm the postulate that parameter S_s is an index for representing the state variable for shearing deformation. We may call this state function S_s as

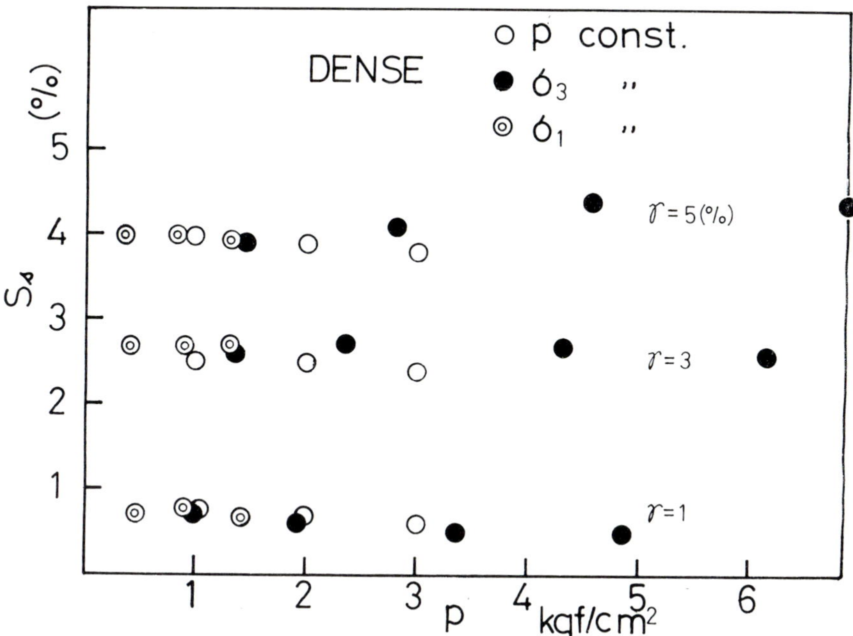

Fig.1. Value of S_s for dense sand
(Sagami River sand, compression)

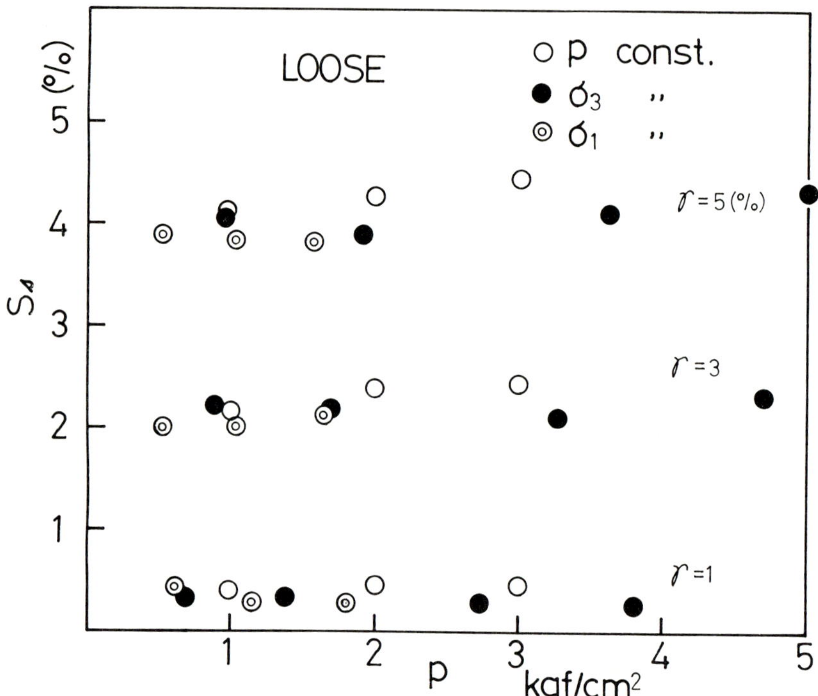

Fig.2. Value of S_s for loose sand
(Sagami River sand, compression)

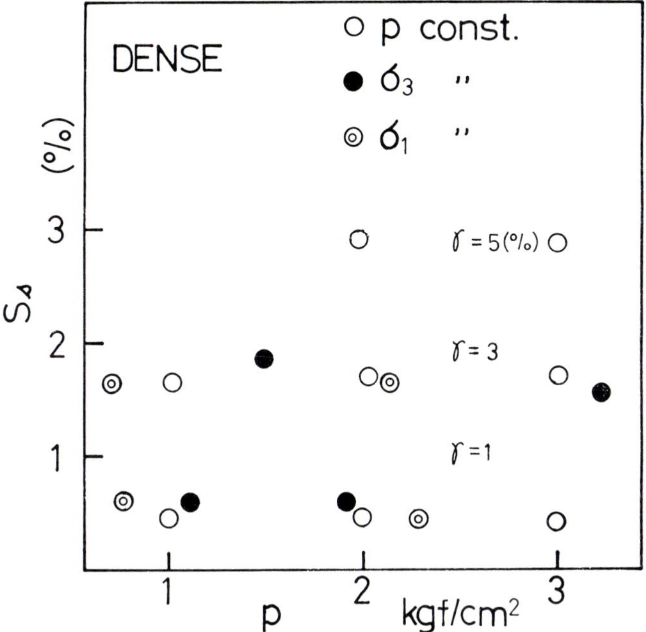

Fig.3. Value of S_s for dense sand(Sagami River sand, extension)

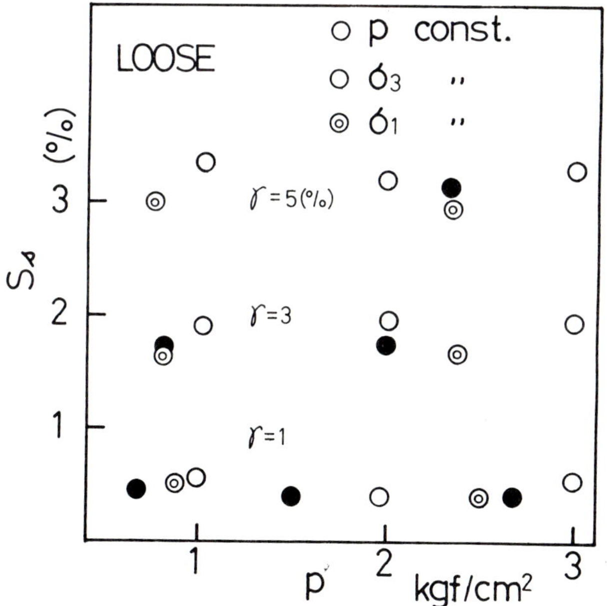

Fig.4. Value of S_s for loose sand(Sagami River sand,extension)

'entropy' of granular materials from a phenomenological point of view. The graphs of S_s and γ has an almost linear relationship (ref.1). Then, we write

$$dS_s = M\, d\epsilon^p \qquad M : \text{constant} \tag{3}$$

From Eq.(1),the increment of S_s can be written as

$$dS_s = dv_d^p + \eta d\epsilon^p \tag{4}$$

$$\text{and} \quad M = dv_d^p/d\epsilon^p + \eta \tag{5}$$

ENTROPY OF INFORMATION THEORY

If the granular material composed of the cells of each micro-state is randomly arranged, the microstate can be stated by the following expression(ref.4)

$$\begin{pmatrix} e_1 & e_2 & & e_i & & e_n \\ V_1 & V_2 & & V_i & & V_n \\ N_1 & N_2 & & N_i & & N_n \end{pmatrix} \tag{6}$$

where e_i : the void ratio in the i-th cell
V_i : the volume of the i-th cell
N_i : the number of spheres in the i-th cell

and $N = N_1 + N_2 + \quad + N_i + \quad + N_n$

$V = V_1 + V_2 + \quad + V_i + \quad + V_n$

$V_i = (1 + e_i)\, v_0\, N_i$

N : the total number of spheres
V : the total volume
v_0: the volume of each sphere

The expression given by Eq.(6) can be transformed to

$$\begin{pmatrix} f_1 & f_2 & & f_i & & f_n \\ p_1 & p_2 & & p_i & & p_n \end{pmatrix} \tag{7}$$

where $f_i = 1 + e_i$

$p_i = N_i/N$

Volume parameter $f = 1 + e$ is called the specific volume and p_i is the probability.

The Shannon's measurement of uncertainty(entropy) associated with the number of grains is given by

$$H = -\Sigma p_i \ln p_i \quad (8)$$

Maximizing Eq.(8) are subject to

$$\Sigma p_i = 1$$

$$\Sigma f_i p_i = f = 1 + e$$

Solving this problem, we can obtain

$$dH = -\lambda_2 df \quad (9)$$

where λ_2 is the Lagrangian multiplier. Eq.(9) can be written

$$dH = \lambda_2 f\, dv_d^p = \theta\, dv_d^p \quad (10)$$

where $\theta = \lambda_2 f$

Substituting Eq.(10) into (4), we obtain

$$dS_s = dH/\theta + \eta d\epsilon^p \quad (11)$$

$$\text{Then} \quad dH/d\epsilon^p = (dS_s/d\epsilon^p - \eta) = (M - \eta) \quad (12)$$

According to the dilatancy, the state of void ratio can be classified into

(1) denser than critical(dry state) $e < e_{cr}$

(2) at the critical state $e = e_{cr}$ (a)

(3) looser than critical(wet state) $e > e_{cr}$

where e_{cr} is the critical void ratio.

Now, we have an reasonable assumption that

dry state $dv_d^p/d\epsilon^p < 0$ (dilatancy)

critical state $dv_d^p/d\epsilon^p = 0$ (b)

wet state $dv_d^p/d\epsilon^p > 0$ (contractancy)

From Eq.(5) and the above relationships, we obtain

dry state $M - \eta < 0$

critical state $M - \eta = 0$ (c)

wet state $M - \eta > 0$

Kanatani's theory(ref.5) tells about that

dry state $\theta < 0$

critical state $\theta = 0$ (d)

wet state $\theta > 0$

Combining these relationships given by (b),(c) and (d), we can finally obtain the following entropy production rule

$$dH/d\epsilon^{p} \geq 0 \quad (13)$$

REFERENCES

1 N.Moroto, State Function of Sand Deformation,Proceedings JSCE, 229 (1974),77-86(in Japanese)
2 N.Moroto, A New Parameter to Measure Degree of Shear Deformation of Granular Material in Triaxial Compression Tests, Soils and Foundations, Vol.20, No.4,1976, pp.1-10
3 F.Tatsuoka, Fundamental Study on Deformation of Sand in Triaxial Apparatus, Doctor Thesis of Tokyo Unversity,1973
4 T.Mogami, A Statistical Approach to the Mechanics of Granular Materials, Soils and Foundations, Vol.V, No.2.1965, pp.26-36
5 K.Kanatani, The Use of Entropy in the Description of Granular Materials, Powder Technology, 30(1981), pp.217-223

Mechanics of Granular Materials: New Models and Constitutive Relations, edited by
J.T. Jenkins and M. Satake, 1983
Elsevier Science Publishers B.V., Amsterdam — Printed in The Netherlands

AN EFFECTIVE STRESS PRINCIPLE FOR PARTIALLY SATURATED GRANULAR MEDIA

D. F. MCTIGUE, R. K. WILSON, J. W. NUNZIATO, Fluid Mechanics and Heat Transfer Division I, Sandia National Laboratories Albuquerque, New Mexico

ABSTRACT

In this paper, we use the continuum theory of mixtures to extend the classical notion of effective stress to partially saturated granular media. Our analysis recovers previously proposed phenomenological representations for the effective stress in terms of the capillary pressure. Using this result, we consider the one-dimensional consolidation of a half-space with the granular medium described by linear poroelasticity. The solution is contrasted with the fully-saturated case. The influence of grain fabric is also briefly discussed.

INTRODUCTION

The presence of one or more fluid phases in the pore space of a granular or porous material can have a strong effect on the mechanical behavior of the medium. In classical soil mechanics, this observation has led to the concept of "effective stress," in which compressive normal stresses are diminished in magnitude by the interstitial fluid pressure. It has been shown recently that this concept emerges quite naturally from a continuum theory for a mixture of an ideal fluid and a matrix of arbitrary constitutive behavior (McTigue and Passman [1]). Thus, a well established phenomenological law appears to be embodied in mixture theory and, in turn, mixture theory offers a framework within which to generalize and extend the effective stress concept to multiphase systems.

In this paper we consider a granular or porous medium with two fluid phases in the pores. Motivation stems primarily from concern for fluid flow and matrix deformation in partially saturated media, a typical example being a soil or rock containing interstitial water and air. The heart of the following development lies in the choice of constitutive equations for the forces of interaction between the various components of the mixture. In particular we draw upon thermodynamic arguments to motivate the form of the interaction terms. As a result, a concept of effective stress emerges which agrees with previously proposed models (Bishop and Blight [2], and Raats [3]), and capillary effects are identified clearly in a thermodynamic context.

We illustrate the resulting theory by examining several simple problems in partially saturated media. First, we specialize to the case of a linear elastic matrix and show how capillary suction reduces the porosity. Second, we consider the problem of one-dimensional consolidation of a partially saturated half-space in which the air is free to escape. The problem for the fluid pressure reduces to a nonlinear diffusion problem, similar to that for horizontal infiltration [4], with a saturation front propagating downward into the medium. No water drains through the surface. The results for the fluid pressure indicate that the granular medium undergoes larger elastic deformation than in the saturated case. We also comment briefly on the effects of porosity gradients on consolidation. This provides some insight into the importance of the fabric of the granular medium.

BASIC CONCEPTS FOR MULTIPHASE MIXTURES

In this section we outline the kinematics for a mixture of three constituents, liquid, gas and granular solid, and give the set of mechanical balance laws for the individual constituents along with the summation rules necessary to achieve the correct balance laws for the mixture. In a companion article [5], we show how these equations can be generalized for N constituents.

Kinematics

Consider a mixture of three constituents, liquid, gas and granular solid denoted by the subscripts ℓ, g, and s, respectively. Let ϕ_a represent the volume fraction of each constituent and let $n = 1 - \phi_s$ denote the connected porosity. Dimensionally, ϕ_a is the ratio of the volume of constituent a to the total volume. The saturation s_a of constituent a is then defined by

$$s_a = \phi_a/n \quad , \qquad a = \ell, g \tag{1}$$

and, since all of the pore volume is occupied,

$$s_\ell + s_g = 1 \quad . \tag{2}$$

Associated with each phase are two densities, a partial density ρ_a which is the mass of phase a per unit volume of the bulk material and a material density γ_a which is the mass of phase a per unit volume of phase a. They are related by

$$\rho_a = \phi_a \gamma_a \quad . \tag{3}$$

The motion of each phase is described in terms of a deformation gradient $\underset{\sim}{F}_a$, which is related to the velocity field $\underset{\sim}{v}_a$ by

$$\grave{\underset{\sim}{F}}_a = (\nabla \underset{\sim}{v}_a)\, \underset{\sim}{F}_a . \tag{4}$$

Here ($\grave{}$) denotes the material time derivative of $\underset{\sim}{F}_a$ following the motion of a,

$$\grave{\underset{\sim}{F}}_a = \partial \underset{\sim}{F}_a/\partial t + (\nabla \underset{\sim}{F}_a)v_a \tag{5}$$

and "∇" denotes the gradient with respect to the spatial coordinates $\underset{\sim}{x}$.

The barycentric motion of the mixture is described by the velocity

$$\underset{\sim}{v} = (\rho_\ell \underset{\sim}{v}_\ell + \rho_g \underset{\sim}{v}_g + \rho_s \underset{\sim}{v}_s)/\rho \tag{6}$$

where

$$\rho = \rho_\ell + \rho_g + \rho_s \ . \tag{7}$$

A diffusion velocity is also defined for the motion of any phase relative to the motion of the mixture:

$$\underset{\sim}{u}_a = \underset{\sim}{v}_a - \underset{\sim}{v} \ . \tag{8}$$

Balance Laws

Our primary interest here is in the response of a partially saturated granular medium in the absence of thermal effects. Thus, we need only require that the porous medium and the mobile constituents satisfy equations for balance of mass and momentum. For each phase, these balance laws are expressed in the form

$$\partial \rho_a/\partial t + \nabla \cdot (\rho_a \underset{\sim}{v}_a) = 0 \ , \tag{9}$$

$$\nabla \cdot \underset{\sim}{\sigma}_a + \rho_a \underset{\sim}{b}_a + \overset{+}{\underset{\sim}{m}}_a = \underset{\sim}{0} \ , \tag{10}$$

$$\underset{\sim}{\sigma}_a = \underset{\sim}{\sigma}_a^T \ , \tag{11}$$

where $\underset{\sim}{\sigma}_a$ is the partial stress of phase a, $\underset{\sim}{b}_a$ is the external body force due to gravity, and $\overset{+}{\underset{\sim}{m}}_a$ is the momentum exchange vector between the mobile constituents and between the constituents and the granular medium. In writing (9) and (10), we have assumed that there is no exchange of mass between constituents and that the inertial forces are negligible. Of course, whatever momentum is lost by one constituent must be gained by another and thus $\underset{\sim}{m}_a$ must satisfy the summation rule

$$\overset{+}{\underset{\sim}{m}}_s = -(\overset{+}{\underset{\sim}{m}}_\ell + \overset{+}{\underset{\sim}{m}}_g). \tag{12}$$

CONSTITUTIVE ASSUMPTIONS - THE MOMENTUM EXCHANGE VECTOR

An expression for the effective stress of a partially saturated granular medium follows from a consideration of the momentum balance equations for the constituents along with constitutive equations for the momentum exchange terms $\underset{\sim}{m}_a^+$. In general, the constitutive equations for $\underset{\sim}{m}_a^+$ are motivated by thermodynamic results, the details of which are given elsewhere [5]. Here we state the result for the isothermal case:

$$\underset{\sim}{m}_a^+ = (p_\ell - \beta_\ell)\nabla\phi_a - \Sigma\, \underset{\sim}{R}_{ab}(\underset{\sim}{u}_a - \underset{\sim}{u}_b) \;, \quad a = \ell, g \tag{13}$$

where $\underset{\sim}{R}_{ab}$ is a tensor whose components are the resistivities associated with the drag between the constituents; for example, $\underset{\sim}{R}_{as}$ is a tensor whose components are the resistivities associated with the drag between mobile constituent a and the granular solid, and the summation is over the constituents b = ℓ, g, s.

In order to give the first term of (13) a physical interpretation, we need the following thermodynamic relationship [5]:

$$p_a - p_\ell = \beta_a - \beta_\ell \tag{14}$$

where

$$p_a = -(\mathrm{tr}\, \underset{\sim}{\sigma}_a)/3\phi_a \tag{15}$$

is the hydrostatic pressure of each constituent and

$$\beta_a = p_a + \rho_a(\partial\psi_a/\partial\phi_a) \tag{16}$$

has the dimensions of pressure and represents the contribution to momentum exchange that arises from changes in free energy with respect to the volume fraction. Physically, the difference between β_g for the gas phase and β_ℓ for the liquid phase is related to the effects of surface tension. This identification leads one to a definition for capillary pressure P_c in the form

$$P_c = \beta_g - \beta_\ell \tag{17}$$

so that (14) becomes

$$P_c = p_g - p_\ell \;; \tag{18}$$

thus we see that the usual definition of capillary pressure arises naturally in the context of thermodynamic arguments. It is also physically reasonable to define the difference between the pressure in the solid p_s and the total fluid pressure p_f as the the compaction pressure S_c:

$$S_c = p_s - p_f \tag{19}$$

where

$$p_f = s_g p_g + s_\ell p_\ell \tag{20}$$

$$= p_g - s_\ell p_c \quad .$$

Clearly, the compaction pressure S_c can be related to the changes in the free energy ψ_a with respect to the ϕ_a; that is,

$$S_c = \beta_s - \beta_g + s_\ell(\beta_g - \beta_\ell) \ . \tag{21}$$

Hence it represents the static compaction behavior of the granular matrix as a result of intergranular contact forces.

With these results, the momentum exchange vector $\underset{\sim}{m}_a^+$ can be written as

$$\underset{\sim}{m}_a^+ = (p_a - \beta_a)\nabla\phi_a - \Sigma\ \dot{\underset{\sim}{R}}_{ab}(\underset{\sim}{u}_a - \underset{\sim}{u}_b) \ , \ a = \ell, \ g. \tag{22}$$

It is now evident that the first term of (22) represents the buoyancy forces acting on phase a as a result of the other phases present. It is important to note that on purely mechanical grounds Raats [3] arrived at a slightly different expression for $\underset{\sim}{m}_a^+$ of the form

$$\underset{\sim}{m}_a^+ = p_a\nabla\phi_a - \Sigma\ \underset{\sim}{R}_{ab}(\underset{\sim}{u}_a - \underset{\sim}{u}_b) \quad .$$

This result is based on a generalization of single phase flow and does not address capillary effects, which enter through the terms involving β_a in (22).

A DEFINITION OF EFFECTIVE STRESS

In this section, we explore some of the effects of the momentum exchange (22) on the behavior of the individual phases and the mixture as a whole. We start with a consideration of the familiar case of a granular medium with two phases, liquid and gas, in the pores. Classical results for the effective stress in a partially saturated medium are recovered, and additional body force terms that do not arise in classical theory emerge as well. Extension of this development to an arbitrary number of fluid phases follows in a straightforward fashion [5].

Consider the momentum balance for the matrix (10); using (12) and (22), we obtain

$$\nabla\cdot\underset{\sim}{\sigma}_s - (p_\ell - \beta_\ell)\nabla\phi_\ell - (p_g - \beta_g)\nabla\phi_g - \underset{\sim}{D}_\ell - \underset{\sim}{D}_g + \rho_s\underset{\sim}{b}_s = \underset{\sim}{0} \tag{23}$$

where, for convenience, we have let

$$\underset{\sim}{D}_a = -\Sigma\, \underset{\sim}{R}_{ab}(\underset{\sim}{u}_a - \underset{\sim}{u}_b) \tag{24}$$

represent the drag force on each phase. Using (1) to eliminate ϕ_ℓ and ϕ_g, equation (23) can be rearranged to read

$$\begin{aligned}\nabla\cdot[\underset{\sim}{\sigma}_s &+ (1-n)(s_\ell p_\ell + s_g p_g)\underset{\sim}{1}] - (1-n)\nabla(s_\ell p_\ell + s_g p_g)\\ &+ n(p_g - p_\ell)\nabla s_\ell + \beta_\ell\nabla(ns_\ell) + \beta_g\nabla(ns_g) - \underset{\sim}{D}_\ell - \underset{\sim}{D}_g + \rho_s\underset{\sim}{b}_s = \underset{\sim}{0}\end{aligned} \tag{25}$$

where we have used (2) to eliminate s_g in the third term. Defining the effective stress $\underset{\sim}{\sigma}_{eff}$

$$\underset{\sim}{\sigma}_{eff} = \underset{\sim}{\sigma}_s + (1-n)p_f\,\underset{\sim}{1} \quad , \tag{26}$$

and using (20), (14), and (18), equation (25) can be written as

$$\nabla\cdot\underset{\sim}{\sigma}_{eff} - (1-n)\nabla p_f - (s_\ell P_c - \beta g)\nabla n - \underset{\sim}{D}_\ell - \underset{\sim}{D}_g + \rho_s\underset{\sim}{b}_s = \underset{\sim}{0} \tag{27}$$

The first three terms in (27) represent forces on the matrix associated with tractions due to the effective stress, a body force due to the total fluid pressure gradient, and a body force due to capillary pressure. The expression (27) agrees with the result obtained by Raats [3] in the context of the purely mechanical theory except for the form of the body force due to the capillary pressure. This difference is a consequence of including the terms β_ℓ and β_g that emerge from thermodynamic arguments. Note that this extra force vanishes in a homogeneous matrix, for which $\nabla n \equiv 0$.

Turning now to the momentum balance for the liquid and gas, (10); using (1), (14), and (18), we have

$$- ns_\ell\nabla p_\ell - (\beta_g - P_c)\nabla(ns_\ell) + \rho_\ell\underset{\sim}{b}_\ell + \underset{\sim}{D}_\ell = \underset{\sim}{0} \quad , \tag{28}$$

$$- ns_g\nabla p_g - \beta_g\nabla(ns_g) + \rho_g\underset{\sim}{b}_g + \underset{\sim}{D}_g = \underset{\sim}{0} \quad . \tag{29}$$

Adding (28) and (29) results in a momentum balance for the total fluid:

$$-n\nabla p_f + (s_\ell p_c - \beta_g)\nabla n + \rho_\ell \underset{\sim}{b}_\ell + \rho_g \underset{\sim}{b}_g + \underset{\sim}{D}_\ell + \underset{\sim}{D}_g = \underset{\sim}{0} \ . \tag{30}$$

From (28) and (29) it seems that a body force associated with capillary pressure acts on the liquid phase, while forces associated with β_g act on both the liquid and gas. For the total fluid (30), we again obtain a term due to inhomogeneity of the matrix.

Finally, adding the matrix momentum balance (27) and the fluid balance (30), we obtain a balance for the mixture as a whole,

$$\nabla \cdot \underset{\sim}{\sigma}_{eff} - \nabla p_f + \rho \underset{\sim}{b} = \underset{\sim}{0} \tag{31}$$

in which

$$\underset{\sim}{b} = (1/\rho)(\rho_\ell \underset{\sim}{b}_\ell + \rho_g \underset{\sim}{b}_g + \rho_s \underset{\sim}{b}_s).$$

At this point, our results correspond exactly to the classical concept of the effective stress. Equation (31) states that a load applied to a porous granular medium is partitioned between the pressurized pore fluid and an effective stress on the matrix. The result for a partially saturated medium is identical to that for a saturated medium when a total fluid pressure, equal to the sum of the saturation-weighted phase pressures, is introduced. For each constituent, however, the partially saturated case introduces extra body forces related to capillarity.

The definition of the effective stress (26) is more easily interpreted when written in several alternative forms. First, we note that, neglecting inertia, the total stress $\underset{\sim}{\sigma}$ is defined by

$$\underset{\sim}{\sigma} = \underset{\sim}{\sigma}_s + \underset{\sim}{\sigma}_\ell + \underset{\sim}{\sigma}_g = \underset{\sim}{\sigma}_s - p_f \underset{\sim}{1} \ . \tag{32}$$

Then equation (26) can be written as

$$\underset{\sim}{\sigma}_{eff} = \underset{\sim}{\sigma} + p_f \underset{\sim}{1} \tag{33}$$

which is exactly the classical definition, and is implicit in (31). For partially saturated media, (33) is usually written, using (20), in the form

$$\underset{\sim}{\sigma}_{eff} = \underset{\sim}{\sigma} + p_g \underset{\sim}{1} - s_\ell p_c \underset{\sim}{1} \tag{34}$$

which emphasizes the role of the capillary pressure (cf. Bishop and Blight [2]). An alternative that emphasizes the partitioning of loads between the matrix and fluid (McTigue and Passman [1]) is

$$\sigma_{eff} = (1-n)(\tau_s + p_f \mathbf{1}) \tag{35}$$

where τ_s is interpreted as a force per unit area of the solid, $\sigma_s = (1-n)\tau_s$. In this view, we see the effective stress clearly related to the load carried by the matrix.

LINEAR POROELASTICITY AND CONSOLIDATION OF A HALFSPACE

Equation (31) shows that the classical definition of the effective stress emerges quite naturally from a continuum theory for mixtures, and it is easily extended to accommodate an arbitrary number of fluid components. We emphasize that the result is obtained without having specified anything about the constitutive behavior of the matrix.

The term $\nabla \cdot \sigma_{eff}$ in (31) suggests that the deformation of the matrix should be related to the effective stress rather than to the actual solid stress σ_s, and this procedure is in fact consistent with classical interpretations. The simplest case is to assume that σ_{eff} is related linearly to the Cauchy strain, E:

$$\sigma_{eff} = 2G\,[E + (\nu/1-2\nu)\,(\mathrm{tr}\,E)\,\mathbf{1}] \quad , \tag{36}$$

where G is the shear modulus, and ν is the drained Poisson's ratio. Equations (2), (9), (30), (31), and (36), along with an assumption about β_g, represent an extension of the theory of consolidation (e.g., Biot, [6]), to partially saturated media.

Notice that as a result of (15), (19), and (26), we can compute the form of the compaction pressure S_c corresponding to the constitutive assumption (36). That is, it follows that

$$S_c = -\,K\,(\mathrm{tr}\,E)/(1-n) \tag{37}$$

where K is the drained bulk modulus:

$$K = 2G(1+\nu)/3(1-2\nu) \quad . \tag{38}$$

If, in addition, we assume that each of the constituents is incompressible, then any volumetric strain of the medium is due entirely to changes in the pore volume:

$$n - n_o = \mathrm{tr}\,E \tag{39}$$

where n_o is the reference porosity. Thus, the compaction pressure S_c becomes a function only of the porosity,

$$S_c = - K\ (n-n_o)/(1-n)\ . \tag{40}$$

The effect of capillarity on changing the porosity is seen if we imagine a partially saturated medium under zero total stress ($\underset{\sim}{\sigma} = \underset{\sim}{0}$) and $|p_g| << |s_\ell P_c|$. Then (20), (26), (36), and (39) give

$$n-n_o = - s_\ell P_c/K\ . \tag{41}$$

Since the capillary pressure P_c is positive, the effect of partial saturation is to draw the grains closer together and reduce the porosity.

Another interesting problem to consider in the context of linear poro-elasticity for partially saturated media is the problem of one-dimensional consolidation. To begin with, we make the following simplifying assumptions:

a) $\beta_g = 0$,
b) $\underset{\sim}{R}_{\ell g} = \underset{\sim}{0}$,
c) $\underset{\sim}{R}_{gs} = [n(1-s_\ell)]^2 \mu_g \underset{\sim}{1}/k_{gs}$,
   $\underset{\sim}{R}_{\ell s} = (ns_\ell)^2 \mu_\ell \underset{\sim}{1}/k_{\ell s}$,
d) $|p_g| << |s_\ell P_c|$,

where μ_g and μ_ℓ are the viscosities of the gas and liquid, respectively. The first of these, (a), assumes that the free energy of the non-wetting gas (air) phase does not depend on the volume fraction of gas. The second, (b), implies that there is negligible resistance between the liquid and the gas. The third, (c), asserts that the porous medium is isotropic with the intrinsic permeabilities k_{gs}, $k_{\ell s}$ depending on the liquid saturation s_ℓ . Finally, with negligible resistance between the liquid and the gas, it is reasonable to assume the gas pressure is essentially atmospheric and (d) follows. In this case, (20) implies that the total fluid pressure p_f is

$$p_f = s_\ell p_\ell = - s_\ell P_c\ . \tag{42}$$

In considering consolidation of a granular medium, it is more convenient to employ mass balance for the mixture:

$$\nabla \cdot \underset{\sim}{v}_s = - \nabla \cdot [n(1-s_\ell)(\underset{\sim}{v}_g - \underset{\sim}{v}_s) + ns_\ell(\underset{\sim}{v}_\ell - \underset{\sim}{v}_s)] \tag{43}$$

which results from summing (9) and using (1), (2) and (3). In view of our assumptions (a)-(c), the gas and liquid fluxes, (28) and (29), reduce to

$$n(1-s_\ell)\,(\underset{\sim}{v}_g-\underset{\sim}{v}_s) = -\,(k_{gs}/\mu_g)\,\nabla p_g \quad , \tag{44}$$

$$ns_\ell(\underset{\sim}{v}_\ell-\underset{\sim}{v}_s) = -\,(k_{\ell s}/\mu_\ell s_\ell)\,[\nabla(s_\ell p_\ell) - p_g \nabla s_\ell] \quad ,$$

neglecting gravity and gradients of porosity. Combining (43), (44) and (45) yields

$$\nabla\cdot\underset{\sim}{v}_s = \nabla\cdot[(k_{gs}/\mu_g)\,\nabla p_g + (k_{\ell s}/\mu_\ell s_\ell)(\nabla(s_\ell p_\ell)-p_g\nabla s_\ell)] \;.$$

By further invoking assumption (d) and using (42), this becomes

$$\nabla\cdot\underset{\sim}{v}_s = \nabla\cdot[(k_{\ell s}/\mu_\ell s_\ell)\,\nabla p_f] \;. \tag{47}$$

Clearly, changes in the volume of the porous medium will be determined by the effective pressure gradients and the permeability $k_{\ell s}(s_\ell)$.

The problem of interest concerns the one-dimensional consolidation of a semi-infinite region (see Figure 1). Assume the surface is under a constant load, $\sigma_{xx} = -\,\sigma_0$ (constant), and that it is permeable for fluid drainage. By (36), the effective stress in one dimension is

$$-\,\sigma_0 + p_f = 2G[(1-\nu)/(1-2\nu)]\,E_{xx} \quad . \tag{48}$$

For small strains, time differentiation of (48) yields

$$\partial p_f/\partial t = 2G[(1-\nu)/(1-2\nu)]\,(\partial v_s/\partial x) \tag{49}$$

which combines with (47) to give

$$\partial p_f/\partial t = \partial[c(\partial p_f/\partial x)]/\partial x \tag{50}$$

where

$$c = 2G(1-\nu)k_{\ell s}/(1-2\nu)\,\mu_\ell s_\ell \tag{51}$$

is called the <u>coefficient</u> <u>of</u> <u>consolidation</u> [7]. Notice that since the capillary pressure P_c and the permeability $k_{\ell s}$ are functions of the liquid saturation s_ℓ , so also is the effective fluid pressure p_f and thus the coefficient of consolidation c can be viewed as a nonlinear function of the fluid pressure p_f. For example, in many applications, the permeability $k_{\ell s}$ can be represented by [8]

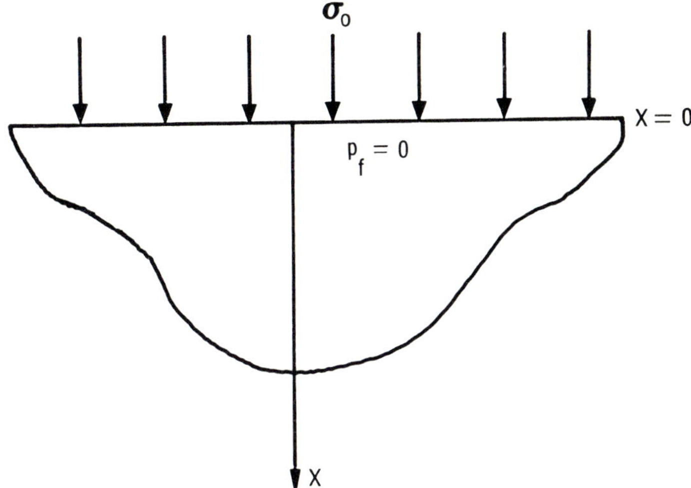

Figure 1. Geometry for one - dimensional consolidation of a half - space.

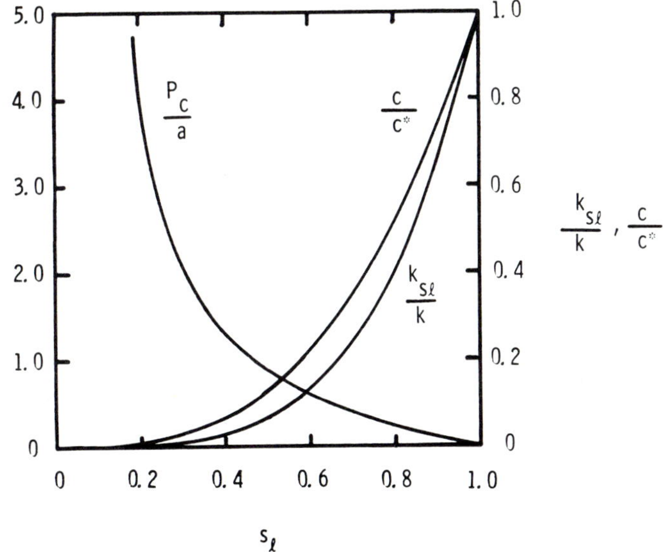

Figure 2. Dependence of capillary pressure P_c, permeability $k_{s\ell}$, and coefficient of consolidation c on liquid saturation s_ℓ.

$$k_{\ell s} = k s_\ell^{\,4} \ , \tag{52}$$

where k is the permeability at full saturation. Moreover, a reasonable representation for the capillary pressure P_c is

$$P_c(s_\ell) = a(1-s_\ell)/s_\ell \ , \tag{53}$$

(see Figure 2). In this case, with the aid of (42), the coefficient of consolidation c has the form

$$\begin{aligned} c &= 2G(1-\nu)k(1+p_f/a)^3/(1-2\nu)\mu_\ell \ , \\ &= c^* \ (1+p_f/a)^3 \ . \end{aligned} \tag{54}$$

This form should be appropriate for a wide range of saturation s_ℓ .

Equation (50) is a nonlinear diffusion equation which is assumed to satisfy the initial condition

$$p_f(x,0) = -\pi \tag{55}$$

and the boundary conditions

$$\begin{aligned} &p_f(0,t) = 0 \ , \\ &p_f(x,t) \rightarrow -\pi \ \text{ as } x \rightarrow \infty \ . \end{aligned} \tag{56}$$

Condition (55) asserts that initially the halfspace is at a uniform saturation corresponding to the constant fluid pressure $-\pi$; while (56) specifies that the upper surface is permeable and at atmospheric pressure. Boundary value problems of this type have been studied extensively in the literature [9]. Using the similarity variable

$$\alpha = x/2(c^*t)^{1/2} \tag{57}$$

and non-dimensionalizing the fluid pressure $p = 1+p_f/\pi$, the problem can be reformulated as a two-point boundary value problem with the general solution $\alpha(p)$ satisfying

$$c(p) = -\,2(d\alpha/dp)\int_0^p \alpha \; dp \tag{58}$$

where $\alpha(1) = 0$.

This solution is difficult to analyze in general. However, if we choose the initial fluid pressure such that π=a, the magnitude of the slope of the capillary pressure at $s_{\ell}=1$, then we can draw on the numerical results of Brutsaert and Weisman [10] to illustrate the main features of the solution. In particular, we graph the numerical results in Figure 3 as the curve labeled "partially saturated." For comparative purposes, we also show the solution for the fully saturated case [7], in which the normalizing pressure π is the boundary load σ . Notice that the partially saturated case is very much like an infiltration problem; that is, as the granular medium is consolidated, water is driven deeper into the half-space as a result of the pressure gradient, similar to the propagation of a saturation front. As a consequence of this, there is no flow out of the upper surface as one might expect physically. This is exactly opposite to the saturated case where the pressure gradient drives water to the surface.

From (39) and (48), it follows that the one-dimensional strain and the change in porosity are given by

$$E_{xx} = n-n_o = - (1-2\nu)/2G(1-\nu) \, [\sigma_o - p_f] \quad . \tag{59}$$

Using this result along with Figure 3, it is also possible to make some observations on the deformation of the granular medium during consolidation. Consider first the saturated case. As time increases and fluid is driven out of the granular medium, the fluid pressure decreases and thus the strain becomes more negative and the porosity decreases (i.e., the grains are compressed). On the other hand, in the unsaturated case, as time increases and fluid is driven into the granular medium, the tension pulling the grains together is relieved and thus the strain decreases and the porosity increases.

Effect of Porosity Gradients on Consolidation

In linear poroelasticity, the presence of the pore fluid is accounted for by relating the effective stress to the elastic deformation. However, the generality of the effective stress concept that emerges from the mixture theory suggests that we may consider any appropriate model for the drained deformation of the matrix. It is of considerable interest, then, to investigate consolidation problems for granular materials described by more elaborate constitutive equations.

Recent work on continuum theories for granular materials has sought to account for the effects of microstructure by including constitutive dependence on variables that represent some measure of local fabric. One such tensor

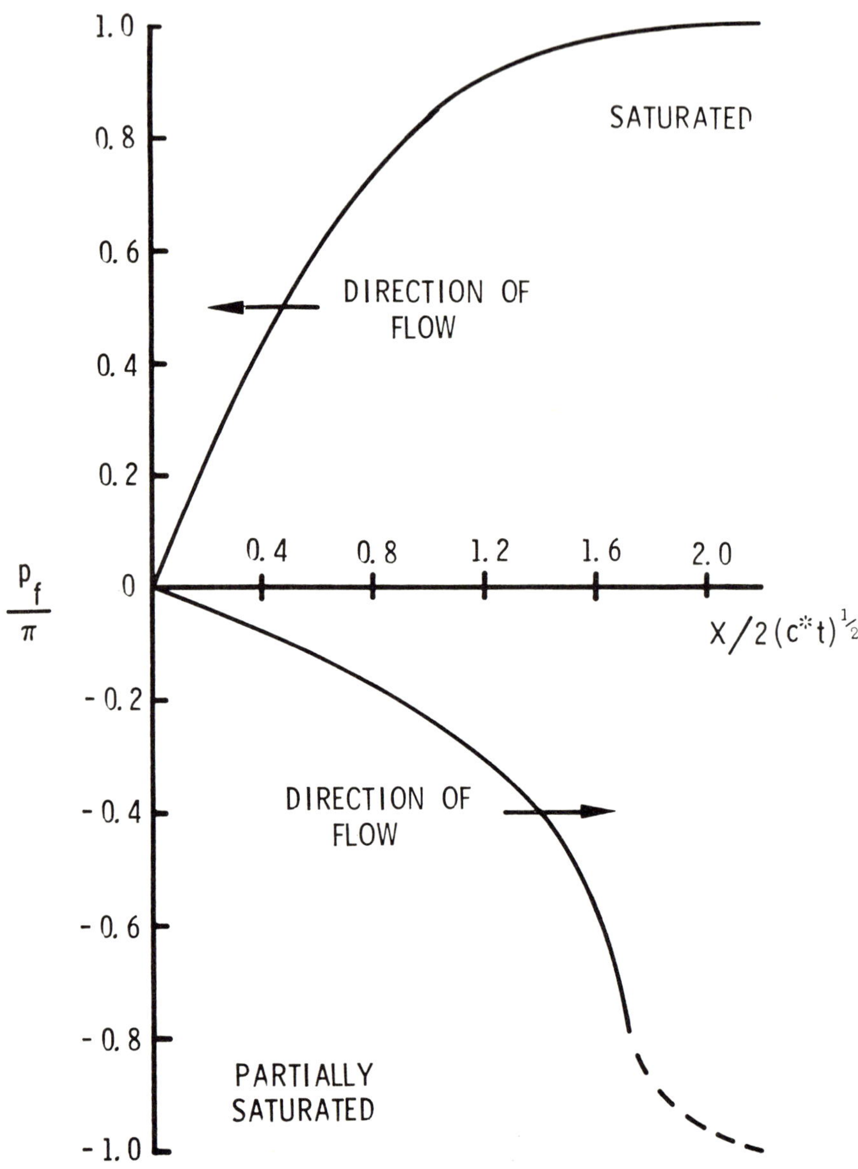

Figure 3. Fluid pressure profiles for consolidation of a saturated and a partially saturated half - space.

measure of microstructure, suggested by Cowin [11], is $\nabla\phi_a \otimes \nabla\phi_a$. A theory for elastic materials with voids that incorporates this dependence has been postulated by Nunziato and Cowin [12]. We consider here the effect of such microstructural stress terms on one-dimensional consolidation of saturated and unsaturated media. In particular, we assume that the effective stress is linear and isotropic in the infinitesimal strain, $\underset{\sim}{E}$, and the tensor $\underset{\sim}{M} \equiv \nabla\phi_s \otimes \nabla\phi_s = \nabla n \otimes \nabla n$:

$$\underset{\sim}{\sigma} + p_f \underset{\sim}{1} = 2G[\underset{\sim}{E} + (\nu/1-2\nu)(\mathrm{tr}\, \underset{\sim}{E})\underset{\sim}{1}] + \alpha_1 \underset{\sim}{M} + \alpha_2 (\mathrm{tr}\, \underset{\sim}{M})\, \underset{\sim}{1} \quad , \tag{60}$$

where α_1 and α_2 are material constants, the signs of which are not known *a priori*.

If the microstructural effects are relatively small, equation (5) indicates that the porosity gradient is, to leading order, proportional to the fluid pressure gradient. Thus, it is apparent by inspection of Figure 3 that the modification due to fabric is quite different in the saturated and unsaturated cases. Consolidation of a saturated halfspace results in a maximum porosity gradient at the drained surface ($x = 0$). This suggests that the fabric will only result in a "skin" effect near the drainage surface. In contrast, microstructural effects on the stress (60) in a partially saturated region are expected to be most important in the neighborhood of the fluid pressure front propagating into the medium. In this case, the fabric of the material affects the entire stress field. Analysis of these problems is the subject of ongoing work.

CONCLUSIONS

The continuum theory of mixtures serves as a basis for obtaining the governing equations for fluid flow and matrix deformation in partially saturated granular or porous media. Thermodynamic arguments lead to a constitutive equation for the forces of interaction between constituents, including the effects of capillarity. A definition of the effective stress in the solid matrix corresponding to classical usage emerges directly. The theory is illustrated by considering consolidation of a partially saturated poroelastic halfspace. Liquid flows away from the drained surface, increasing the saturation, decreasing the capillary pressure, and decreasing the effective compression of the region. The effect of fabric is expected to be important in the neighborhood of the steep fluid pressure front propagating into the medium.

ACKNOWLEDGEMENT

This work was performed at Sandia National Laboratories, supported by the U. S. Department of Energy under contract number DE-AC04-76-DP00789.

REFERENCES

1 D. F. McTigue and S. L. Passman, "The Effective Stress Principle and Mixture Theory," to appear.

2 A. W. Bishop and G. E. Blight, "Some Aspects of Effective Stress in Saturated and Partly Saturated Soils," Geotechnique, 13, 177-197 (1963).

3 P. A. C. Raats, "Forces Acting Upon the Solid Phase of a Porous Medium," ZAMP, 19, 606-613 (1968).

4 R. K. Wilson, D. F. McTigue and J. W. Nunziato, "Partially Saturated Flow Through Porous Media," Sandia National Laboratories, Report No. SAND82-2349 (1982).

5 D. F. McTigue, R. K. Wilson and J. W. Nunziato, "An Effective Stress Principle for Partially Saturated Media," Sandia National Laboratories Report No. SAND82-1977 (1982).

6 M. A. Biot, "General Theory of Three-Dimensional Consolidation," J. Appl. Phys., 12, 155-164 (1941).

7 T. W. Lambe and R. V. Whitman, Soil Mechanics, Wiley, New York (1979).

8 A. T. Corey, "Measurement of Water and Air Permeability in Unsaturated Soils," Proc. Soil Sci. Soc. Am., 21, 7-10 (1957).

9 J. Crank, The Mathematics of Diffusion, Clarendon, Oxford (1979).

10 W. Brutsaert and R. N. Weisman, "Comparison of Solutions of a Nonlinear Diffusion Equation," Water Resources Res., 6, 642-644 (1970).

11 S. C. Cowin, "Microstructural Continuum Models for Granular Materials," Proceedings of the U. S. - Japan Seminar on Continuum-Mechanical and Statistical Aproaches in the Mechanics of Granular Materials, Gakujutsu Bunken Fukyukai, Tokyo, Japan, 162-170 (1978).

12 J. W. Nunziato and S. C. Cowin, "A Nonlinear theory of Elastic Materials with Voids," Arch. Rat. Mech. Anal., 72, 175-201 (1979).

Mechanics of Granular Materials: New Models and Constitutive Relations, edited by
J.T. Jenkins and M. Satake, 1983
Elsevier Science Publishers B.V., Amsterdam — Printed in The Netherlands

DISCUSSIONS ON THE TRANSMISSION OF THE PARTICLE BULK DENSITY GENERATED IN A POWDER LAYER

K. MAKINO[1], S. ITOH[1], M. YAMADA[1], J. HIDAKA[2] and T.P. MELOY[3]
[1]Dept. of Chemical Engineering, Kyoto University, Kyoto 606 (Japan)
[2]Dept. of Chemical Engineering, Doshisha University, Kyoto 602 (Japan)
[3]Dept. of Minerals and Engineering, Univ.of West Virginia, West Virginia (U.S.A.)

ABSTRACT

The transmission characteristics of the particle bulk density relating to plastic deformation which is generated in a powder layer in a rupture state due to external forces are qualitatively discussed on the basis of motion.

The results show that it satisfies the wave equation and propagates as a kind of dynamic wave in the powder layer. It is pointed out that acoustic emission from a powder is closely related to the transmission of the particle bulk density.

INTRODUCTION

Basic and theoretical reports concerned with the particle bulk density built up in a powder layer have been presented by Kuno(1), Goodman and Cowin(2)(3), Kanatani(4) and the authors(5)(6)(7). On the other hand experimental reports including X-ray graphs of the distribution of the particle bulk density in a two-dimensional hopper were given by Bransby et al(8), Cowin et al(9); those around a plunger were given by the authors(10). The former paper pointed out various constitutive relations in a powder layer and the latter indicated the fundamental characteristics of the distribution of the particle bulk density built up in a powder layer such as the wave phenomena observed in the distribution of the particle bulk density which are generated around a plunger on its penetration into the powder layer(10). But the agreement between the theoretical studies and experimental ones is not yet good enough. By utilizing the results of the previous reports(7), in this report the equation of motion in a powder layer is described in a general form, and the transmission of the distribution of particle bulk density generated in it is investigated. The results make it clear that the transmission of the plastic deformation that accompanies compression satisfies the wave equation if the compressibility of a powder layer depends on the isotropic particle velocity vectors(rotation components). Furthermore transmission due to the turbulent components of particles is also shown to obey the wave equation.

TRANSMISSION OF THE PARTICLE BULK DENSITY GENERATED IN A POWDER LAYER

The stress tensor σ_{jk} is expressed by the following equations if the component equations obtained in the previous paper(7) are used:

$$\sigma_{jk} = -(p + \delta_d p)\delta_{jk} + \tilde{\sigma}_{jk} + \sigma_{[jk]}$$

$$= -(p + \delta_d p)\delta_{jk} + \frac{\sqrt{6}}{2}(\frac{\mu p}{\bar{w}} + \frac{aNc\rho}{3m\bar{w}})[\frac{3}{5}E_{jk} + R_{jk}]$$

$$= -(p + \delta_d p)\delta_{jk} + \frac{\sqrt{6}}{2}(\frac{\mu p}{\bar{w}} + \frac{aNc\rho}{3m\bar{w}})[\frac{3}{5}v_{j,k} - \frac{1}{5}v_{k,j} - \frac{1}{5}\delta_{jk}v_{1,1} - \omega_{jk}] \quad (1)$$

All symbols are identical with those in the previous paper(7). Here, the stress tensor σ_{jk} given by Eq.(1) is unsymmetrical. Then, speaking strictly, the balance of momentum has to be considered simultaneously with the equation of motion. However, in this paper, the behavior of dynamic particle density wave, which relates strongly to the equation of motion, is considered. From this standpoint, particle bulk density is qualitatively discussed on the basis of equation of motion only. Therefore, using Eq.(1), the equation of motion of a powder layer is given in a general form:

$$\rho[\frac{\partial(v_k + v'_k)}{\partial t} + (v_k + v'_k)_{,i}(v_i + v'_i)]$$

$$= -(p + \delta_d p)_{,k} + \frac{\sqrt{6}}{2}(\frac{\mu p}{\bar{w}} + \frac{aNc\rho}{3m\bar{w}})_{,j}[\frac{4}{5}v_{j,k} - \frac{1}{5}v_{k,j} - \frac{1}{5}\delta_{jk}v_{1,1} - \omega_{jk}]$$

$$+ \frac{\sqrt{6}}{2}(\frac{\mu p}{\bar{w}} + \frac{aNc\rho}{3m\bar{w}})[\frac{4}{5}v_{j,kj} - \frac{1}{5}v_{k,jj} - \frac{1}{5}\delta_{jk}v_{1,1j} - \omega_{jk,j}] + \rho b_k$$

$$= -(p + \delta_d p)_{,k} + \frac{\sqrt{6}}{2}(\frac{\mu p}{\bar{w}} + \frac{aNc\rho}{3m\bar{w}})_{,j}[\frac{4}{5}v_{j,k} - \frac{1}{5}v_{k,j} - \frac{1}{5}\delta_{jk}v_{1,1} - \omega_{jk}]$$

$$+ \frac{\sqrt{6}}{2}(\frac{\mu p}{\bar{w}} + \frac{aNc\rho}{3m\bar{w}})[\frac{3}{5}v_{j,kj} - \frac{1}{5}v_{k,jj} - \omega_{jk,j}] + \rho b_k \quad (2)$$

(Note 1)

In Eq.(2) the turbulent components(v_i', $\delta_d p$, etc.) of the particles are taken into account.

We consider a small displacement of a plunger in a powder layer as shown in Fig.1. More specifically the case in which changes in v_k and x_k in terms of time and position are very small is considered. In this case the second order terms(i.e. non-linear terms) are negligible compared with the first order ones. Therefore Eq.(2) reduces to

$$\rho\frac{\partial(v_k + v'_k)}{\partial t} = -(p + \delta_d p)_{,k} - \frac{\sqrt{6}}{2}(\frac{\mu p}{\bar{w}} + \frac{aNc\rho}{3m\bar{w}})_{,j}\omega_{jk}$$

$$+ \frac{\sqrt{6}}{2}\left(\frac{\mu p}{\bar{w}} + \frac{aNc\rho}{3m\bar{w}}\right)\left(\frac{3}{5}v_{j,kj} - \frac{1}{5}v_{k,jj} - \omega_{jk,j}\right) + \rho b_k \tag{3}$$

In Eq.(3) the turbulent components(v_k', $\delta_d p$, etc) and the nonturbulent components of particles are independent of each other. This means that Eq.(3) can be divided into the following two equations:

$$\rho \frac{\partial v_k}{\partial t} = -p_{,k} - \frac{\sqrt{6}}{2}\left(\frac{\mu p}{\bar{w}} + \frac{aNc\rho}{3m\bar{w}}\right)_{,j} \omega_{jk} + \frac{\sqrt{6}}{2}\left(\frac{\mu p}{\bar{w}} + \frac{aNc\rho}{3m\bar{w}}\right)\left(\frac{3}{5}v_{j,kj} - \frac{1}{5}v_{k,jj} - \omega_{jk,j}\right) + \rho b_k \tag{4}$$

$$\rho \frac{\partial v_k'}{\partial t} = - \delta_d p_{,k} \tag{5}$$

Eq.(4) is an equation of motion which takes both compressibility and plasticity of the powder into account. Therefore information regarding the transmission of the particle bulk density obtained from Eq.(4) can be regarded to be applicable to that of plastic deformation of particles accompanied with compression. On the other hand information from Eq.(5) should be applied to that of high frequency components of the particle bulk density.

<u>Transmission of plastic deformation accompanied by compression</u>

We assume that generally the rotation of particle ω_{jk} is, by definition, expressed in terms of the v_j by

$$2\omega_{jk} = v_{k,j} - v_{j,k} \tag{6}$$

Substitution of Eq.(6) into Eq.(4) leads to

$$\begin{aligned}\rho \frac{\partial v_k}{\partial t} &= -p_{,k} - \frac{\sqrt{6}}{4}\left(\frac{\mu p}{\bar{w}} + \frac{aNc\rho}{3m\bar{w}}\right)_{,j}(v_{k,j} - v_{j,k}) \\ &\quad + \frac{\sqrt{6}}{2}\left(\frac{\mu p}{\bar{w}} + \frac{aNc\rho}{3m\bar{w}}\right)\left(\frac{3}{5}v_{j,kj} - \frac{1}{5}v_{k,jj} - \frac{1}{2}v_{k,jj} + \frac{1}{2}v_{j,kj}\right) \\ &= -\frac{\sqrt{6}}{4}\left(\frac{\mu p}{\bar{w}} + \frac{aNc\rho}{3m\bar{w}}\right)_{,j}(v_{k,j} - v_{j,k}) - p_{,k} + \frac{\sqrt{6}}{2}\left(\frac{\mu p}{\bar{w}} + \frac{aNc\rho}{3m\bar{w}}\right) \times \left(\frac{11}{10}v_{j,kj} - \frac{7}{10}v_{k,jj}\right)\end{aligned} \tag{7}$$

Neglect of the second order terms in Eq.(7) gives

$$\rho \frac{\partial v_k}{\partial t} = \frac{\sqrt{6}}{2}\left(\frac{\mu p}{\bar{w}} + \frac{aNc\rho}{3m\bar{w}}\right)\left(\frac{11}{10}v_{j,kj} - \frac{7}{10}v_{k,jj}\right) - p_{,k} + \rho b_k \tag{8}$$

Eq.(8) can be differentiated with respect to x_k and treated in the same manner to give

$$\rho \frac{\partial}{\partial t}(v_{k,k}) = \frac{\sqrt{6}}{2}(\frac{\mu p}{\bar{w}} + \frac{aNc\rho}{3m\bar{w}})(\frac{11}{10}v_{j,kjk} - \frac{7}{10}v_{k,jjk}) - p_{,kk} + (\rho b_k)_{,k}$$
$$= \frac{\sqrt{6}}{2}(\frac{\mu p}{\bar{w}} + \frac{aNc\rho}{3m\bar{w}})(\frac{11}{10}v_{j,jkk} - \frac{7}{10}v_{j,jkk}) - p_{,kk} + (\rho b_k)_{,k}$$
$$= \frac{\sqrt{6}}{5}(\frac{\mu p}{\bar{w}} + \frac{aNc\rho}{3m\bar{w}})v_{j,jkk} - p_{,kk} + (\rho b_k)_{,k} \tag{9}$$

That is, Eq.(9) can be expressed as follows:

$$\rho\frac{\partial}{\partial t}(\mathrm{div}\,\mathbb{V}) = \frac{\sqrt{6}}{5}(\frac{\mu p}{\bar{w}} + \frac{aNc\rho}{3m\bar{w}})\Delta(\mathrm{div}\,\mathbb{V}) - \Delta p + (\rho b_k)_{,k} \tag{10}$$

As div $\mathbb{V}$ is equal to $-\dot{\rho}/\rho$ (where the symbol • stands for the time-derivative) because of the equation of continuity, and Eq.(10) takes account of plasticity, Eq.(10) can be regarded as the basic equation which describes the transmission of plastic deformation accompanied by compression generated in the powder layer. Simultaneously Eq.(10) means that the transmission of the particle bulk density depends on three factors, i.e. compressibility(div $\mathbb{V}$), powder pressure p and external forces(or the force of gravity) ρb_k. The intention of this report is to bring out the fundamental features of the transmission of the distribution of the particle bulk density, and only the compressibility of the powder enables the transmission of the particle bulk density. Therefore we will uncover the characteristics of the transmission mechanism of the particle bulk density given by the following equation which was derived from Eq.(10) by neglecting the second and third terms on the right-hand side:

$$\rho \frac{\partial}{\partial t}(\mathrm{div}\,\mathbb{V}) = \frac{\sqrt{6}}{5}(\frac{\mu p}{\bar{w}} + \frac{aNc\rho}{3m\bar{w}})\Delta(\mathrm{div}\,\mathbb{V}) \tag{11}$$

As a simplification, let Eq. (11) be written in the following form:

$$\frac{\partial}{\partial t}(\mathrm{div}\,\mathbb{V}) = K^2\Delta(\mathrm{div}\,\mathbb{V}) \tag{12}$$

where

$$K^2 = \frac{\sqrt{6}}{5}(\frac{\mu p}{\bar{w}} + \frac{aNc}{3m\bar{w}}) \tag{13}$$

It should be obvious from the above description that the motion of a powder layer in its continuous rupture process is to be focused on in this report. Therefore in the case where such a problem is discussed a rational study requires that the yield point, as a failure criterion, should be defined as the very point when a dynamic state changes into a static state. Particles in a powder layer have rotation ω and translation velocity v_j in its dynamic state. As stated above, the yield point of a powder layer can be regarded as the point of time when the rotation ω just reaches critical zero. This means that the component τ_i, which lies in the tangential plane, of the contact point force at an arbitrary contact point of the particle considered can act as a rotational moment having the same direction as that of the rotation ω.

Therefore the friction force and the cohesive force which act against the shear force τ_i have a negative rotation moment compared with the above one. Integration of the rotation moments around the whole surface of the considered particle gives the following equation:

$$\oint\{\tau_i a - (\mu'\nu + a^2c')a\}\mathbf{e}_w d\Omega = 0 \qquad (14)$$

Where, $\mathbf{e}_w$ is the unit vector having the direction of particle rotation. Eq.(14) can be expressed in a scalar form to give

$$\oint\tau_i\tau_i a^2\mathbf{e}_w\mathbf{e}_w d\Omega = \oint(\mu'\nu + a^2c')^2a^2\mathbf{e}_w\mathbf{e}_w d\Omega \qquad (15)$$

Considering that $\mathbf{e}_w\mathbf{e}_w = 1$, Eq.(14) reduces to

$$\oint[\tau_i\tau_i - (\mu'\nu + a^2c')^2]d\Omega = 0 \qquad (16)$$

Eq.(16) coincides with a special case of the basic postulate Kanatani used (4) when he derived the failure criterion of a powder layer. This means, therefore, the failure of a powder layer can be presumed to happen at the very time when elementary particles which constitute it begin to rotate. Therefore, from the viewpoint mentioned above, an actual powder layer is considered, in this report, to be composed of cohesive particles of various sizes. Let the particle velocity $\mathbb{V}$ in Eq.(12) be the sum of two components expressed as follows:

$$\mathbb{V} = \mathbb{V}_0(r,\theta,t) + \delta\mathbb{V}(r,\theta,t) \qquad (17)$$

Eq.(17) describes that the particle velocity $\mathbb{V}$ is the sum of the non-rotational component $\mathbb{V}_0(r,\theta,t)$ and the rotational one $\delta\mathbb{V}(r,\theta,t)$. We assume that the former has no relation with the above-mentioned cyclical rupture phenomenon,i.e. change in the particle bulk density, and only the latter is directly related to it.

Generally the latter should be described as the sum of various frequency components. But the studies that follow pick up only a typical rotation component expressed in the following equation and discuss

The reason why such an assumption does not make our discussion lose its generality is because of the linearity of Eq. (12). Write

$$\delta\mathbb{V} = \delta\mathbb{V}_0 e^{j\nu t} \qquad (18)$$

where $\delta\mathbb{V}_0$ is a function which is dependent only on position. An arbitrary velocity vector $\mathbb{V}$ can always be decomposed in such a way as Eq.(17). Substitutton of Eq.(17) into Eq.(12) gives

$$\frac{\partial}{\partial t}(\mathrm{div}\ \mathbb{V}_0 + \mathrm{div}\ \delta\mathbb{V}) = K^2\Delta(\mathrm{div}\ \mathbb{V}_0 + \mathrm{div}\ \delta V) \qquad (19)$$

As $\mathbb{V}_0$ has no relation with the particle bulk density, we obtain

$$\mathrm{div}\ \mathbb{V}_0 = 0 \qquad (20)$$

Substittution of Eq.(20) into Eq.(19) leads to

$$\frac{\partial}{\partial t}(\mathrm{div}\ \delta\mathbb{V}) = K^2\Delta(\mathrm{div}\ \delta\mathbb{V}) \tag{21}$$

On the other hand, the following equation can be obtained from Eq.(18) easily.

$$\delta\dot{\mathbb{V}} = j\nu\delta\mathbb{V} \tag{22}$$

Substitution of Eq.(22) into Eq.(21) followed by its modification gives

$$\frac{\partial}{\partial t}\left(\frac{1}{j\nu}\mathrm{div}\ \delta\dot{\mathbb{V}}\right) = \frac{1}{j\nu}\frac{\partial^2}{\partial t^2}(\mathrm{div}\ \delta\mathbb{V})$$

$$= K^2\Delta(\mathrm{div}\ \delta\mathbb{V}) \tag{23}$$

Further rearrangement gives the following equation:

$$\frac{\partial^2}{\partial t^2}(\mathrm{div}\ \delta\mathbb{V}) = j\nu K^2\Delta(\mathrm{div}\ \delta\mathbb{V}) \tag{24}$$

As mentioned below, div $\delta\mathbb{V}$ in Eq.(24) is directly related to the particle bulk density. Consequently Eq.(24) means that the particle bulk density propagates just like a wave having the velocity $K\sqrt{\nu}$ when the powder layer ruptures continuously under the influence of external forces.

The above discussions are concerned with the case in which the transmission of the particle bulk density in the powder layer is determined by only a unidirectional velocity component of the rotating particles. In actual cases, however, it may be reasonable to take some independent rotation components into account simultaneously. In such a case, Eq.(17) needs to be extended in a manner which follows:

$$\mathbb{V}(r,\theta,t) = \mathbb{V}_0(r,\theta,t) + \delta\mathbb{V}_1(r,\theta,t)$$

$$= \mathbb{V}_0(r,\theta,t) + \delta\mathbb{V}(r,\theta,t) + \delta\mathbb{V}^*(r,\theta,t) \tag{25}$$

where $\delta\mathbb{V}_1(r,\theta,t) = \delta\mathbb{V}(r,\theta,t) + \delta\mathbb{V}^*(r,\theta,t)$ and $\delta\mathbb{V}^*(r,\theta,t)$ is a conjugate vector of $\delta\mathbb{V}(r,\theta,t)$. Substituting Eq.(25) into Eq.(19), we obtain the following equation:

$$\frac{\partial}{\partial t}\{\mathrm{div}\ \mathbb{V}_0 + \mathrm{div}(\delta\mathbb{V}_1)\} = \frac{\partial}{\partial t}\{\mathrm{div}\ \mathbb{V}_0 + \mathrm{div}(\delta\mathbb{V}^*)$$

$$= K^2\delta\{\mathrm{div}\ \mathbb{V}_0 + \mathrm{div}(\delta\mathbb{V}) + \mathrm{div}(\delta\mathbb{V}^*) \tag{26}$$

Taking account of Eq.(20), Eq.(26) reduces to

$$\frac{\partial}{\partial t}\{\mathrm{div}(\delta\mathbb{V})\} = \frac{\partial}{\partial t}\{\mathrm{div}(\delta\mathbb{V}) + \mathrm{div}(\delta\mathbb{V}^*)\}$$

$$= K^2\{\mathrm{div}(\delta\mathbb{V}) + \mathrm{div}(\delta\mathbb{V}^*)\} \tag{27}$$

In the meantime let changes in the particle existing density corresponding to div $\delta\mathbb{V}_1$, div $\delta\mathbb{V}$ and div $\delta\mathbb{V}^*$ be $\delta\rho_1$, $\delta\rho$ and $\delta\rho^*$. Taking $\dot{\rho}_0 = 0$ into account, we obtain following equations:

$$\mathrm{div}\ \delta\mathbb{V}_1 \doteq -\overline{\dot{\ln\delta\rho_1}} \tag{28}$$

$$\mathrm{div}\ \delta\mathbb{V} \doteq -\overline{\dot{\ln\delta\rho}} \tag{29}$$

$$\mathrm{div}\ \delta\mathbb{V}^* \doteq -\overline{\dot{\ln\delta\rho^*}} \tag{30}$$

Therefore applying the relation $\delta\mathbb{V}_1 = \delta\mathbb{V} + \delta\mathbb{V}^*$ to Eqs.(28)-(30), we obtain

$$\overline{\dot{\ln\delta\rho_1}} = \overline{\dot{\ln\delta\rho}} + \overline{\dot{\ln\delta\rho^*}} = \overline{\dot{\ln\delta\rho \times \delta\rho^*}} \tag{31}$$

Eq.(31) gives

$$\delta\rho_1 = \delta\rho \times \delta\rho^* \tag{32}$$

Eq.(27) shows that div $\delta\mathbb{V}_1$ propagates in the powder layer as if it were a diffusion phenomenon with a diffusion coefficient K,i.e. div $\delta\mathbb{V}_1$ is a solution of the linear differential equation. This fact simultaneously means that div $\delta\mathbb{V}$ and div $\delta\mathbb{V}^*$ also satisfy Eq.(27). Eq.(18) represents that div $\delta\mathbb{V}$ and div $\delta\mathbb{V}^*$ are rotation components which are conjugate each other. Eventually it is recognized that each of them satisfies the following wave equation in the same way as mentioned above:

$$\frac{\partial^2}{\partial t^2}\{\mathrm{div}\ \delta\mathbb{V}\} = \pm j\nu K^2\Delta\{\mathrm{div}\ \delta\mathbb{V}\} \tag{33}$$

where the negative sign in the right-hand side means the case of div $\delta\mathbb{V}$. However, as stated before, it is evident that div $\delta\mathbb{V}$ and div $\delta\mathbb{V}^*$ are conjugate each other. Threfore if div $\delta\mathbb{V}$ satisfying the following equation is determined, div $\delta\mathbb{V}^*$ can be obtained immediately.

$$\frac{\partial^2}{\partial t^2}\{\mathrm{div}\ \delta\mathbb{V}\} = -j\nu K^2\Delta\{\mathrm{div}\ \delta\mathbb{V}\} \tag{34}$$

Introduction of the assumption $\rho_0 >> \delta\rho_1,\ \delta\rho,\ \delta\rho^*$ takes Eqs.(28)-(30) to the following expressions respectively:

$$\mathrm{div}\ \delta\mathbb{V}_1 = -\frac{1}{\bar{\rho}_0}\delta\dot{\rho}_1 = -\left(\frac{\delta\dot{\rho}_1}{\rho_0}\right) = -\delta\dot{\bar{\rho}}_0 \tag{35}$$

$$\mathrm{div}\ \delta\mathbb{V} = -\frac{1}{\rho_0}\delta\dot{\rho} = -\left(\frac{\delta\dot{\rho}}{\rho_0}\right) = -\delta\dot{\bar{\rho}} \tag{36}$$

$$\mathrm{div}\ \delta\mathbb{V}^* = -\frac{1}{\rho_0}\delta\dot{\rho}^* = -\left(\frac{\delta\dot{\rho}^*}{\rho_0}\right) = -\delta\dot{\bar{\rho}}^* \tag{37}$$

where the superposed bar means the dimensionless form of the particle bulk density in terms of ρ_0. Eqs.(35)-(37) shows that $-\delta\dot{\bar{\rho}}_0$, $-\delta\dot{\bar{\rho}}$ and $-\delta\dot{\bar{\rho}}^*$ are equal to div $\delta\mathbb{V}_1$, div $\delta\mathbb{V}$ and div $\delta\mathbb{V}^*$ respectively. This fact simultaneously means that each of these changes in the particle bulk density satisfies a wave equation such as

$$\frac{\partial^2}{\partial t^2}(\delta\dot{\bar{\rho}}) = j\nu K^2\Delta(\delta\dot{\bar{\rho}}) \tag{38}$$

Eq.(38) can be easily modified to

$$\frac{\partial}{\partial t}\{\frac{\partial^2}{\partial t^2}(\delta\bar{\rho}) - j\nu K^2\Delta(\delta\bar{\rho})\} = 0 \quad (39)$$

The subject of our study is the transmission characteristics of the particle bulk density which is caused by a rupture in the powder layer. The rupture breakes out because of a plunger penetration in the layer which has had a uniform distribution of the particle bulk density initially. Threfore we can apply the following initial conditions to Eq.(39) to investigate it, at t=0, $\delta\rho(x,y,0) = 0$, $\delta\rho(x,y,0)_{,x}$ (or xx or t or tt) =0 and $\delta\rho(x,y,0)_{,y}$ (or yy or t or tt) =0.Then Eq.(38) reduces to

$$\frac{\partial^2}{\partial t^2}(\delta\bar{\rho}) = j\nu\Delta(\delta\bar{\rho}) \quad (40)$$

Each of $\delta\rho_1$, $\delta\rho$ and $\delta\rho$*satisfies Eq.(40),i.e. it propagates as a wave at a velocity $K\sqrt{\nu}$. Taking Eq.(32) into account, the following equation is obtained:

$$\delta\rho_1 = \rho_0\delta\bar{\rho} \times \delta\bar{\rho}^* \quad (41)$$

Eq.(41) states that the distribution of the particle bulk density $\delta\rho_1$ which is generated in a powder layer is proportional to the product of $\delta\rho$ and its conjugate $\delta\rho^*$.

It is clearly understood in the above derivation processes that basically the wave characteristics of $\delta\rho$ and $\delta\rho^*$ come from the dissipative energy,i.e. they relate to plastic deformation. Threfore a sudden change of a powder layer from a dynamic state to a static state has no influence upon it, i.e. the former distribution of the particle bulk density is maintained with no change in it.

CORRESPONDENCE BETWEEN THESE ANALYTICAL RESULTS AND EXPERIMENTS

As mentioned previously, if the transmission of the particle bulk density generated in the powder layer which is in a rupture state by external forces is conducted by the rotation velocity components of the particles, it was shown that the variance component of the particle velocity propagate as waves. Especially, as mentioned above, the former results mainly from the plastic deformation due to the dissipative energy, and as long as a sufficient external force acts upon the system the particle bulk density propagates in the powder layer. In the case, however, where the external force becomes smaller than a specific value or it is absent, the previous dynamic state in which the sufficient external force has been applied seems to be kept almost as it has been. These remarks are supported by the X-ray graph of the distribution of the particle density distribution around a plunger in our previous paper(10) shown in Fig. 1. Observation of the periodic nature, reflection, diffraction and interference phenomena gives experimental evidence for these analytical results. Fig. 2-6 compares the distribution of the particle bulk density around a single plunger and that around double plungers. It can be pointed out that an interference phenomenon similar to a wave motion is observed. The length scale of all the tests is given in Fig. 2.

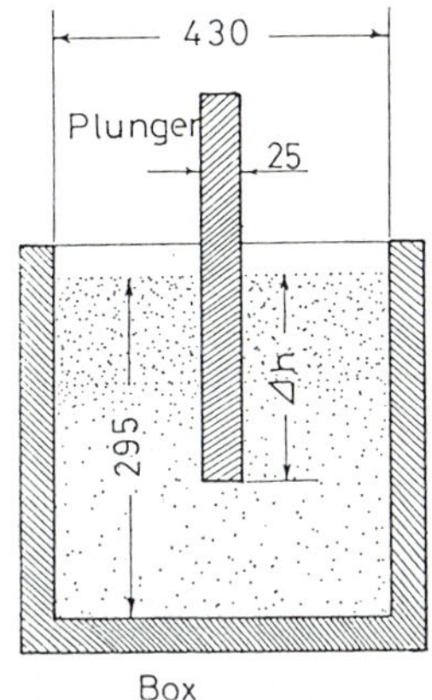

Fig. 1 Conceptional diagram of plunger penetrating experiment.

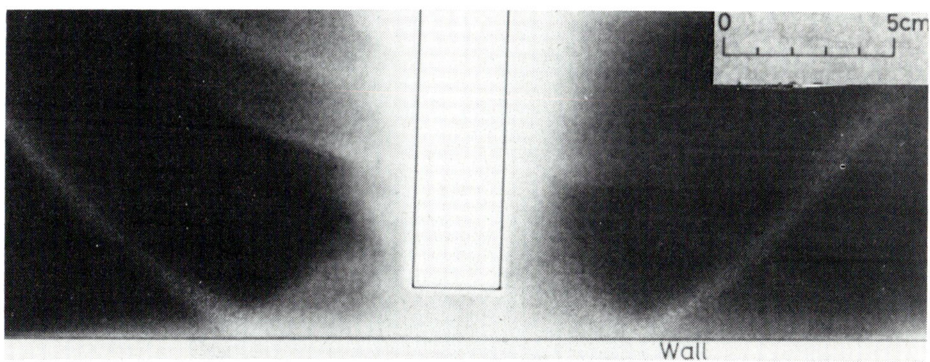

Fig. 2 Reflection, Δh = 15.3 cm.

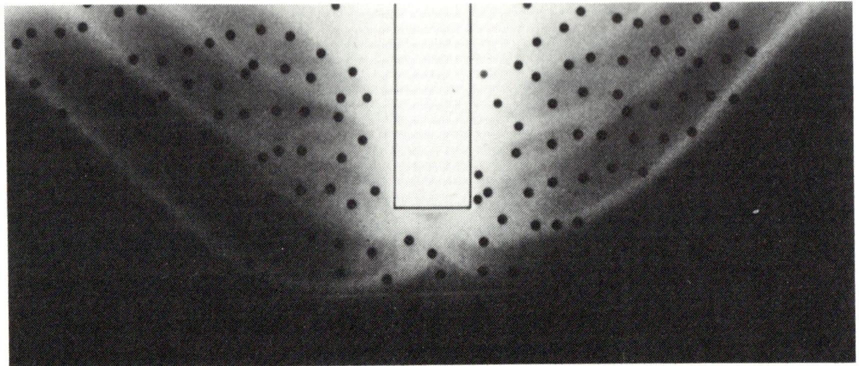

Fig. 3 Spatial periodicity, Δh = 7.5 cm.

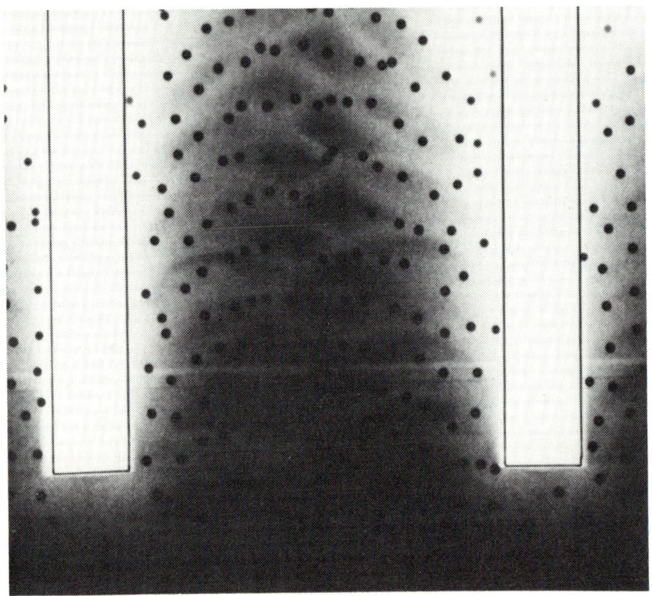

Fig. 4 Spatial periodicity and interference, Δh = 13.1 cm.

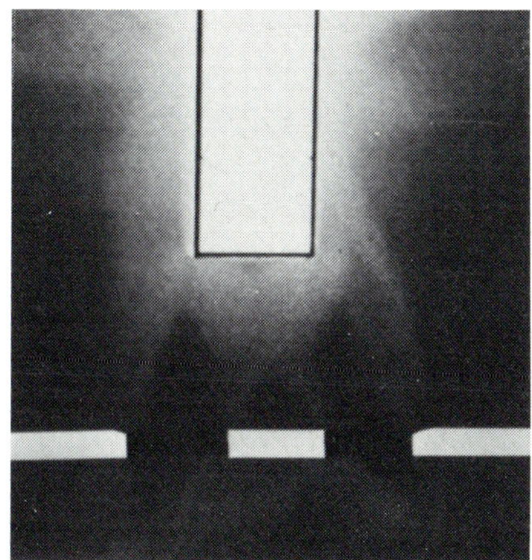

Fig. 5 Interference, Δh = 11.4 cm.

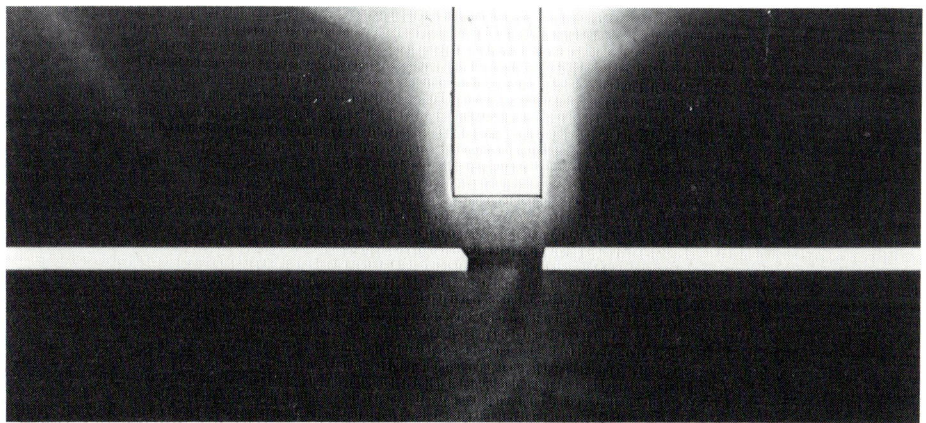

Fig. 6 Diffraction, Δh = 13.7 cm.

Furthermore the above result does not disagree with the result that the distribution of the particle bulk density of random particle system is possible to describe if we focus on the rotation component of particles, as one of the authors(5)(6) has previously pointed out.

On the other hand, the transmission of the particle bulk density is basically that of collisions among particles. Therfore the results of this report point out that acoustic emission from powder layer which are based on Eq.(40) is generated. This fact suggests that the spatial wave transmission characteristic of the particle buik density is directly contained in that of the acoustic emission as shown in Fig.7.

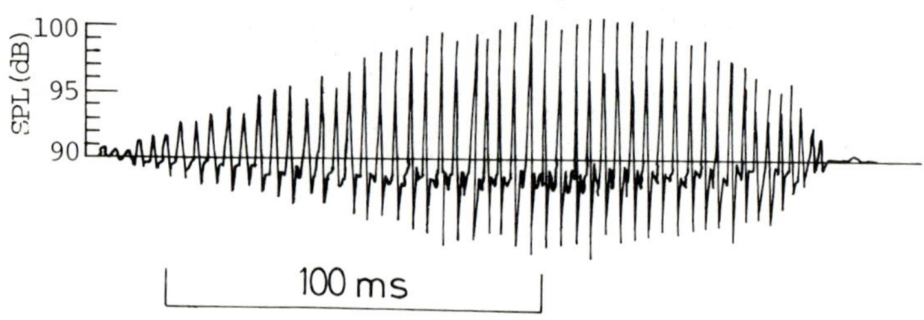

1)PRESSURE WAVE FORM OF ACOUSTIC EMISSION FROM POWDER LAYER

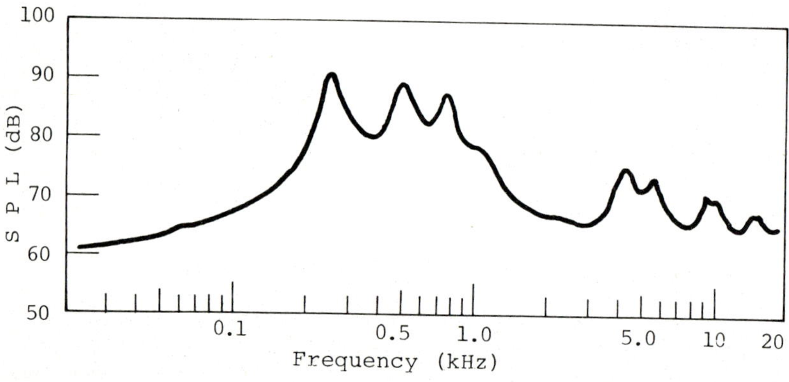

2)FREQUENCY SPECTRUM OF ACOUSTIC EMISSION

Fig.7, Example of acoustic emission generated in plunger experiment

CONCLUSION

In this report the transmission characteristics of the particle bulk density relating to plastic deformation of which is generated in a powder layer in a rupture state due to external forces were studied on the basis of the equation of motion. The results show that it satisfies the wave equation and propagates in the powder layer. These results were found to have no discrepancy with information that was obtained up to now. And the frequency characteristics of so-called acoustic emission which is generated by the rupture of a powder layer are closely related to those of the transmission of the particle bulk density mentioned above.

LITERATURE CITED

1 Kuno H.; J. of the Japan Soc. of powder and powder Metallurgy, 29, 82(1972)
2 Goodman M.A. and S.C. Cowin; J. Fluid Mech., 45, part 2, 321(1971)
3 Goodman M.A. and S.C. Cowin; Arch. Rational Mech. Anal., 44, 249(1972)
4 Kanatani K.; Int. Eng. Sci., 17. 419(1979)
5 Makino K.; J. of the Japan Soc. of Powder and Powder Metallurgy, 28,269(1981)
6 Makino K.; J. of the Japan Soc. of Powder and Powder Metallurgy, 28,276(1981)
7 Makino K., S. Itoh, J. Hidaka and T.P. Meloy; J. of Japan Soc. of Powder and Powder Metallurgy, to be submitted(1982)
8 Bransby P.L.; J. of Eng. for Ind., Trans. ASME, 17, 139(1973)
9 Cowin S.C.; Trans. of the Soc. of Rheology, 18, 247(1974)
10 Makino K., J. Hidaka, S. Miwa and T.P. Meloy; J. of the Japan Soc. of Powder and Powder Metallurgy, 29, to be published(1982)

Mechanics of Granular Materials: New Models and Constitutive Relations, edited by J.T. Jenkins and M. Satake, 1983
Elsevier Science Publishers B.V., Amsterdam — Printed in The Netherlands

GENERATION OF FINE DUST VIA COMPACTION OF GRANULAR MATERIALS*

S. L. SOO, S. CAI*, AND P. O. STERANKA, JR.

Department of Mechanical and Industrial Engineering, 1206 West Green Street
University of Illinois at Urbana-Champaign, Urbana, IL 61801

ABSTRACT

Crushing of granular materials by compaction has been studied with haydite in a piston-cylinder tester. Percent pulverized in seen to correlate well with 4/3 power of pressure suggested by applying Herz-Rayleigh relation to tetrahedral piling. The size distribution appears to be independent of pressure.

INTRODUCTION

The need for minimizing fugitive emission in the form of wind blown dust from storage piles of bulk materials such as coal and ores and for minimizing loss of material resources in millions of tons per year has been well documented (ref. 1). Various ways of reducing such a loss have been investigated (for instance, ref. 2). The present study addresses the method of stacking and, at the same time, compacting a pile using, for instance, a bulldozer. It is readily seen that in the course of stacking and compaction which reduces the porosity of the pile and hence reduced wind penetration (ref. 3), additional fine dust is generated. In searching for an optimum procedure, a preliminary study has been made to determine the relation between compaction and the generation of fines.

In performing this preliminary study, we have chosen a representative material, haydite, a consistent shale product which is kiln processed and crushed to nearly 10 mm size. This material is suitable for model testing of bulk materials because it is relatively homogeneous and is suitable for the study of compaction and generation of fines at contacting points in the laboratory scale.

---

*Visiting Scholar from Nanjing Institute of Technology, Nanjing, Peoples Republic of China.

AN IDEALIZED MODEL OF COMPACTION

Stresses inside a particulate aggregate such as soil have been treated extensively in soil mechanics (for instance, ref. 4 and 5) based on continuum theory. Compaction of a pile of bulk or granular material using a bulldozer actually corresponds to the cases of a strip load. However, these basic relations are not readily applied to compaction via crushing of granular materials.

An idealized model of an ore heap by Bagster (ref. 6) extending from an earlier study is based on a two-dimensional model assuming an array of circular rods. It is readily seen that a realistic approximation can be made using tetrahedral packing of spheres of diameter D as shown in Fig. 1 with gap 2δ to account for incidental looseness. In this case, the volume fraction of solid in the interior of a semi-infinite pile is given by

$$\phi = \pi[3(2^{1/2})]^{-1}\{1 - [\delta/(3D^{1/2})]\}^{-2}\ [1 - (2\delta/D)]^{-2} \qquad (1)$$

The force exerted on each of the top layers of spherical particles in the idealized system is seen to be

$$F_1 = P[(3^{1/2})/2]\ D^2[1 + (2\delta/D)]^2 \qquad (2)$$

where P is the force exerted per unit projected area of the top layer. This force, in turn, from each of the three adjacent top layer particles exerts forces P_2 at three points of an interior particle of

$$F_2 \sim (F_1/3\ \sin\theta)\ \{1 - [\delta/D(3^{1/2})]\}^{-2} = 0.408\ F_1\{1 - [\delta/D(3^{1/2})]\}^{-2} \qquad (3)$$

where θ = 54.74 degrees for exact tetrahedral piling $[\sin\theta = (2/3)^{1/2}]$. The increment of F_2 toward the bottom of the pile due to the weight of material above it is neglected.

For contacts of spherical particles, the radius a_1 of area of contact is given by Herz and Rayleigh (ref. 7).

$$a_1 = \{(3\pi/4)\ F_2\ (D/2)[(1 - \nu^2)/\pi E]\}^{1/3} \qquad (4)$$

for identical spheres where E is the modulus of elasticity and ν the Poisson ratio; $\nu \sim 0$ for coal and other ores. Simple tetrahedral piling is assumed. The distance of approach of the two adjacent spheres is approximately

$$\alpha = 1.23(4\ F_2^2/D\ E^2)^{1/3} \qquad (5)$$

and the maximum stress σ_c is given by

$$\sigma_c = 0.388[16\ F_2\ E^2/D^2]^{1/3} \tag{6}$$

For the case of a 20,000 kg bulldozer with 4 m^2 tread area on a coal pile approximated by 10 cm spheres in tetrahedral piling

$$P = 4.9 \times 10^4\ N/m^2,\ F_1 = 283\ N,\ F_2 = 115\ N,\ \sigma_c = 1.02 \times 10^8\ N/m^2.$$

For an average value of E for coal of 10^{10} N/m^2, nonsphericity readily gives an actual local compressive stress of the order of 10^{10} N/m^2 causing breakage. Any idealization of the configuration of a bulk material constituting a pile will be limited by this type of stress concentration.

The amount of fines generated by crushing can be estimated from the volume $\pi\ a_1^2\ \alpha/2$ which for fines of radius a gives $(3/8)\ a_1^2\ \alpha/a^3$ particles at each contact point. For a pile of volume V with $36\phi V/\pi D^3$ contact points, the mass of fine thus generated per unit volume is given by a fraction pulverized equal to

$$(1.23)\ 18(3/4)^{2/3}\ \bar{\rho}_p\ \phi(F_2/E\ D^2)^{4/3} = 3.486\ \bar{\rho}_p\ \phi\ [P/(\sin\theta)\ E]^{4/3} \tag{7}$$

We note that the fines generated by packing to $\phi = 0.70$ amounts to 0.38 g/(1000 kg) which is comparable in the order of magnitude measured by Cowherd (ref. 1).

Instead of a localized load on an infinite pile, an experimental study is facilitated by choosing a cylindrical container for the spherical granular material with load applied by a piston.

For the case of piston diameter nearly equal to the cylinder diameter, it is seen that all the interior spherical particles are each loaded at six points with force given by Eq. (3) while the boundary particles at the inner radius of the cylinder have a contact force of $F_2 \sin\theta$. This gives a radial pressure

$$P_r = 2\ F_2 \sin\theta/3^{1/2}\ D^2 = 3^{-1}\ P \tan\theta \tag{8}$$

from Eq. (2) for $\delta = 0$. The subsidence in this ideal case is given by the decrease in void according to $6\pi\ a_1^2\ \alpha/2(\pi/6)\ D^3 = 18\ a_1^2\ \alpha/D^3$

$$\Delta\phi = -18(1.23)(3/4)^{2/3}\ (F_2/E\ D^2)^{4/3} = -3.486(P/E\sin\theta)^{4/3} \tag{9}$$

The powder generated is assumed to fill in the gaps between the spheres.

In the case of loading over a localized area of a pile such as in the case of compaction with a bulldozer, the load transmitted including both the above forces and a shear force is distributed through the pile depending on the friction and cohesiveness of the material (ref. 5). The situation when applied to a cylindrical container with a loading piston of a smaller diameter than the cylinder is quite complicated. Because of the existence of shear stresses, the piston tends to penetrate into the pile with some of the granular material rising around it. The above forces exist only in a cone of apex angle $(\pi/2) - \theta$ beneath the piston. Outside of this region, a distribution of stresses exists depending on the boundary condition.

EXPERIMENTAL STUDY

The above example shows that compaction of an actual storage pile may generate a large amount of fines. In terms of percentage, however, the latter is not measurable in a laboratory experiment. Much larger compaction pressure than in the above example has to be used to produce a measurable percentage of pulverizing and compaction or subsidence.

Experimental Facility

To achieve some understanding of the generation of fine powder by the compaction of granular materials under externally exerted pressure, an experimental setup was designed and installed. The equipment as assembled is shown in Fig. 2.

A replaceable working piston (2) connected to the piston rod of a double acting high pressure hydraulic cylinder (1) and a 20.3 cm diameter test vessel for the specimen of granular materials (3) are the main working parts of the equipment which exert a variable load (pressure) on a given volume of test specimen. The high pressure oil pumped by a gear pump (7) was conducted to the top side or the bottom side or the piston of the hydraulic cylinder to control the direction of motion of the working piston with the aid of a four-way manually operated control valve (4). There are two adjustable relief valves in the system to change the working pressure according to the need of the experiment. One of them is a crude adjustment relief valve installed at the outlet of the oil pump. The other is a fine adjustment relief valve (5) which is installed on the pipe just before the inlet of a four-way hydraulic control valve.

Experimental Procedure

Three sizes of replaceable working pistons and five different values of oil pressure were used in our preliminary experiments to observe the effect of

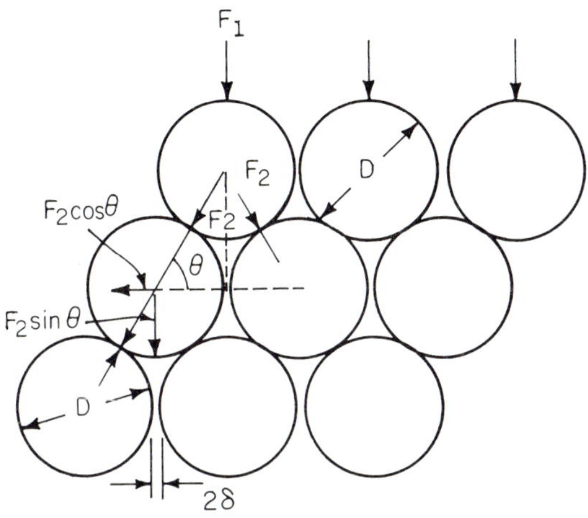

Figure 1 General Tetrahedral Piling of Spheres

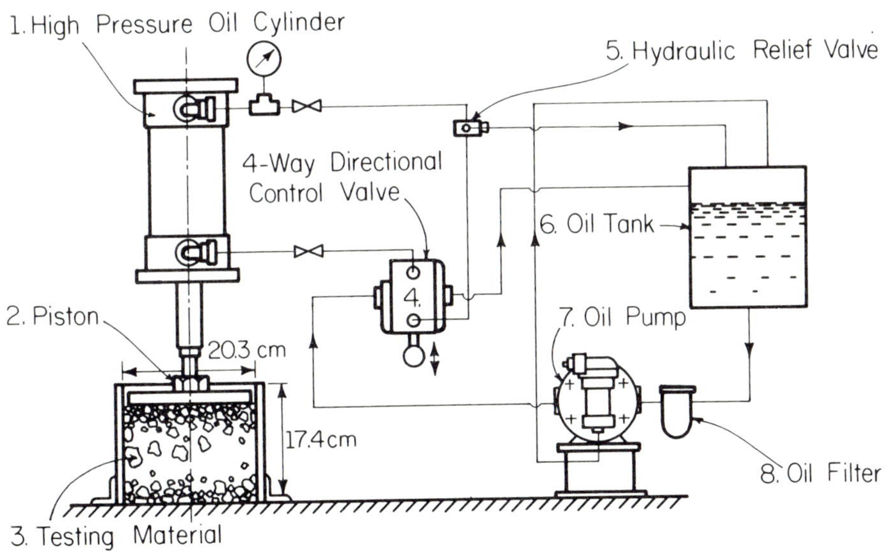

Figure 2 System of the Experimental Equipment

compaction on the particles at different distances from the bottom surface of the piston which exerts a specified pressure on the particles.

The test granules were divided into three layers, 2.54 cm thick each, by using thin plastic films for determining the percent pulverized at different distances from the piston. Experiments were conducted with the piston exerting pressure on sample material three times, each lasted one minute. The oil pressure, hence the respective pressure exerted on the particles, and the magnitudes of subsidence of the test specimen were measured in each experiment. The size distribution of the particles of different layers before and after experiment were measured by sieving.

The extent of breakage in terms of percent pulverized within each layer of specimens was determined by the following procedures:

1. Weigh the material within each layer.
2. Weigh all the particles passing each range of sieve below a 3.18 mm sieve: 425 μm, 30 μm, and 10 μm.
3. Compute the amount of particles under the projected area of the piston (Fig. 3).
4. The ratio of items (2) and (3) gives the fraction of breakage produced by the piston.

The extent of subsidence of the material was determined by measuring the change of the distance between the surface of the working piston to a fixed surface before and after each duration of pressurizing on the test specimen. This distance is divided by the total height of the column of granules in the vessel to give the percent subsidence.

Experimental Results

The experimental results are shown in Figs. 4, 5, and 6 and Table 1. Figure 4 shows the influence of piston diameter to the percent pulverized of test particles at different layers. In Fig. 4, the pressures exerted on the surface of specimen were 2.04 bar and 4.42 bar. Comparing these two sets of data, we note that for different pressure exerted on the granules, the trend of pulverizing is similar, i.e. with a reduced diameter of piston the percent of the granules pulverized increases. Yet the percent pulverized decreases as the depth from the bottom surface of working piston increases, more so for the small piston. The slope of these curves increases with increasing oil pressure. It is seen from the relation of percent pulverized to pressure exerted on particles for different diameters of piston in Fig. 5 that the percent pulverized increases rapidly when the pressure is raised. Table 1 presents the size distribution in the specimen after exerting pressure on it over predetermined durations. It appear that not only the size distribution is almost independent of the depth of layers but also the pressure exerted on the

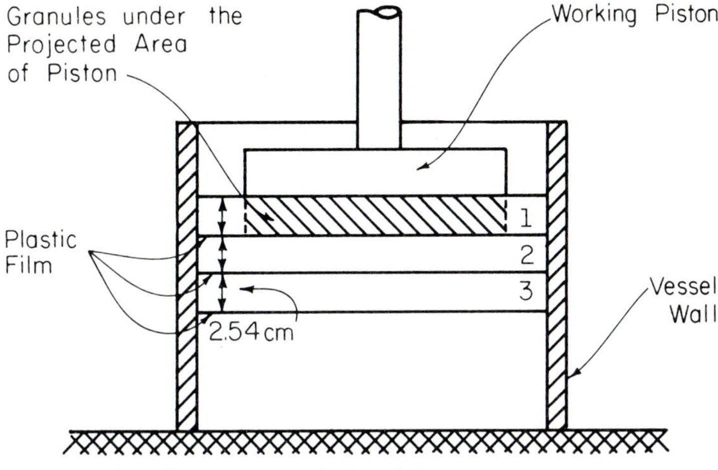

Figure 3 Scheme for Calculating the Percent Pulverized of Test Specimen

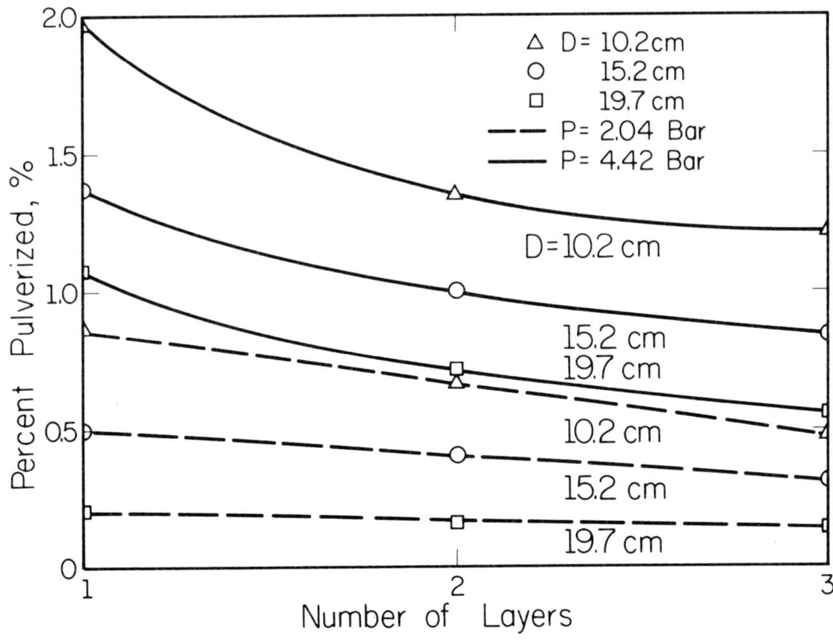

Figure 4 Influence of Piston Diameter on Percent Pulverized Particles at Different Layers

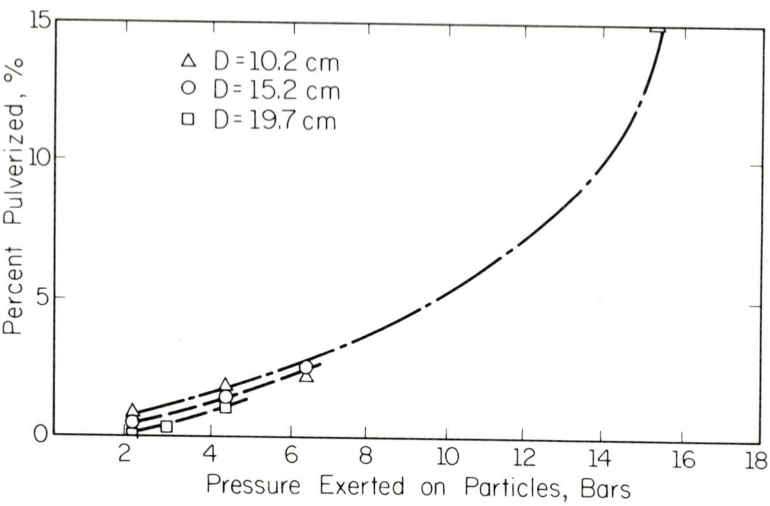

Figure 5 Relation of Percent Pulversized versus Pressure Exerted on Particles for Different Diameters of Pistons (top layer)

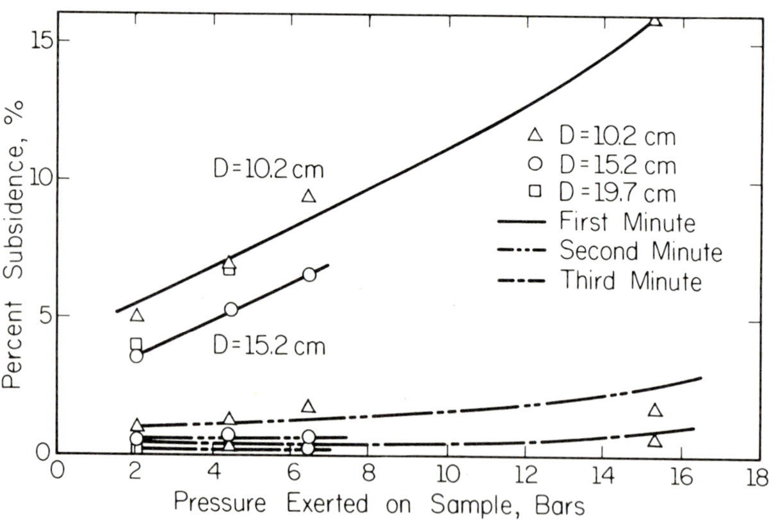

Figure 6 Relation between the Percent Subsidence and the Pressure Exerted on Sample and Repetition

TABLE I

Size distribution of particles (0 x 3.7 mm) after each application of pressure in percent

Piston Diameter, cm	10.2	15.2	19.7		
Top Layer (2.54 cm)					
425 μm-3.17 mm	73-86	74-83	74-81		
180 μm--425 mm	11-22	12-18	14-21		
< 180 μm	3-6	4-6	3-5		
Second Layer (2.54 cm)					
425 μm-3.17 mm	74-80	69-76	73-74		
180 μm-425 μm	13-21	21-26	20-22		
< 180 μm	4-7	7-8	5-7		
Third Layer (2.54 cm)					
425 μm-3.17 mm	67-78	65-78	74-76		
180 μm-425 μm	16-19	17-18	20-22		
<180 μm	6-8	5-6	3-5		
Average of All Tests				Average	
425 μm-3.17 mm				76	
180 μm-425 μm				19	
<180 μm				5	
Below 180 μm					(below 3.17 mm)
30 μm-180 μm				79.8	(3.99)
10 μm-30 μm				16.2	(0.81)
<10 μm				4.0	(0.20)

particles. Figure 6 shows the percent subsidence produced by the pressure exerted on specimen. It is seen that the first time of exerting pressure produced a large percent subsidence which increases with the pressure. Smaller subsidence was produced in the second and third times of application of pressure than in the first time. The percent subsidence was insignificant in the case of piston diameter nearly equal to the cylinder diameter.

DISCUSSION

Except at pressure P above 15 bars and for piston diameter much smaller than the cylinder, the percent pulverized of test specimens increases almost linearly with $P^{4/3}$ as suggested by Eq. (9). For piston diameter close to that of the cylinder, the percent pulverized in all three layers appears similar. However, for the case of piston diameter of half of the cylinder diameter, the percent pulverized decreases with depth as expected. Increased percent pulverized occurs for small piston diameters because of shear motion.

The pattern of breakage in terms of size distribution appears to be independent of the piston diameter and is related to the structure of the material which possesses certain breakage patterns.

The influence of pressure exerted on the specimen on the percent subsidence tends to be more significant at the first application than in subsequent applications because sharp corners tend to break off first. Reduced breakage occur when the sized particles are reused. The small piston causes greater subsidence than larger pistons by penetration into the specimen and shear motion.

The pressure used in the tests is far greater than that used in compaction of storage piles. The theoretical relation is valid for correlating compaction of actual storage piles. The trend in the experimental result suggests that the elastic model of spheres with localized fracture is applicable to the behavior of storage piles under compaction.

ACKNOWLEDGEMENTS

The joint support by the Industrial Environment Research Laboratory, Office of Energy, Materials, and Industry, U.S. EPA, Research Triangle Park, NC 27711, under Grant CR807558; and the American Iron and Steel Institute, Washington, DC 20036, under Project No. 78-394 is appreciated.

REFERENCES

1 C. Cowherd, Jr., R. Bohn, and T. Cascino, Jr., Iron and Steel Plant Open Source Fugitive Emission Evaluation, Report EOA-600/2-79-103, Midwest Research Institute, Kansas City, MO 64110, on Research Performed under Grant No. 68-02-2609, W.A. 3, Program Element No. IAB604, Industrial Environmental Research Laboratory, RTP, NC 27711., May 1979, p. 58.

2 S. L. Soo, J. C. Perez, and S. Rezakhany, Wind Velocity Distribution over Storage Piles and Use of Barriers, Proc., Symposium on Iron and Steel Pollution Technology for 1980, Office of Environmental Engineering and Technology, EPA-600/9-81-017 (F. A. Ayer, Compiler), Industrial Environmental Research Laboratory, RTP, NC 27711, 1981.

3 S. Cai, "A Method for Computing Blowing Velocity from a Porous Storage Pile In Wind," Master of Science Thesis, University of Illinois at Urbana-Champaign, December 1981.

4 P. L. Capper and W. F. Cassie, The Mechanics of Engineering Soils, E. and F. N. Sporn, Ltd., London, 6th ed., 1976, pp. 93-95 and 352-355.

5 C. R. Scott, Soil Mechanics and Foundations, Applied Science Publishers, Ltd., London, 2nd ed., 1974, pp. 126, 132, 287.

6 D. F. Bagster, J. Powder and Bulk Solids Technology, 2(1), (1978), 42-46.

7 S. Timoshenko, Theory of Elasticity, McGraw-Hill Book Co., New York, 1934, p. 339.

Mechanics of Granular Materials: New Models and Constitutive Relations, edited by
J.T. Jenkins and M. Satake, 1983
Elsevier Science Publishers B.V., Amsterdam — Printed in The Netherlands

MECHANICAL PROPERTIES OF IDEAL GRANULAR MATERIALS

Ken-ichi Kanatani

Department of Computer Science, Gunma University, Kiryu, Gunma 376 (Japan)

ABSTRACT

Theoretical relationships between macroscopic and microscopic mechanical properties of ideal granular materials composed of rigid spherical particles are derived by the use of variational principles and three dimensional tensor analysis. The stress both in the packing state and the flow state is determined, and macroscopic constitutive equations are derived. Some analyses are given for the flow on an inclined plane.

1. INTRODUCTION

Mechanical properties of granular materials such as powder, sand or grain differ according to the types of the material and the pattern of deformation. However, there are some properties common to all types of granular materials, reflecting the essential fact that they are composed of solid particles. Thus, we have to make distinction between fundamental characteristics and particular ones.

Here, aiming to know the fundamental characteristics of granular materials in general, we consider an idealized material composed of rigid spherical particles and derive mechanical laws governing the microscopic properties such as the distribution of the particle contacts, the contact force and the interparticle slips. They are derived by the use of a form of the variational principle, which demands, in general terms, that the work done by the interparticle interactions must be equal to the work described in terms of macroscopic field variables. Application of statistical treatment to them then yields macroscopic mechanical laws expressed in terms of the stress and the strain. (See also refs. 1, 2).

The stress in the flow regime is also determined in such a way that the energy dissipation described in terms of the stress is equal to that induced by interparticle friction and collisions. There, the particle rotation plays an important role. (See ref. 3) Thus, macroscopic constitutive equations are obtained from microscopic considerations, and they reveal some of the striking characteristics peculiar to granular materials in general.

2. THE STRESS AND THE CONTACT FORCES

Suppose the material is subject to a macroscopically uniform stress. The interparticle forces may vary from particle to particle (Fig. 1). Superpose all the contact forces in the assemblage on a hypothetical sphere, which we refer to as the *representative particle*, whose radius a is the average radius of the particles (Fig. 2). If the number of particles is sufficiently large, the contact force distribution on the representative particle is approximated by a continuous function of the contact direction n, the outward unit normal vector at the contact point. Let D(n)dn be the number of contact points in the differential solid angle dn divided by the number of the particles. Hence, D(n) is the *contact density*, and

$$N = \oint D(n)dn \tag{1}$$

is the *coordination number*, i.e., the average number of contacts per particle. Let $f_i(n)D(n)dn$ be the total force acting in the differential solid angle dn divided by the number of the particles. Hence, $f_i(n)D(n)$ is the *contact force density*, and $f_i(n)$ is the average contact force per single contact of the contact direction n. The equilibrium of force and torque implies

$$\oint f_i(n)D(n)dn = 0 , \qquad \oint f_{[i}(n)n_{j]}D(n)dn = 0 , \tag{2}$$

where [ ] designates the alternation of indices. Henceforth, we adopt the Cartesian tensor notation and the summation convention.

Consider a uniform linear deformation

$$x'_i = A_{ij}x_j , \tag{3}$$

according to which the point x_i moves to x'_i after the deformation. In terms of the *displacement* $u_i = x'_i - x_i$, eqn (3) is written as

$$u_i = F_{ij}x_j , \qquad F_{ij} \equiv A_{ij} - \delta_{ij} , \tag{4}$$

where δ_{ij} is the Kronecker delta. The *distortion tensor* F_{ij} is resolved into the symmetric part and the skew part, i.e.,

$$F_{ij} = e_{ij} + r_{ij} , \qquad e_{ij} \equiv F_{(ij)} , \qquad r_{ij} \equiv F_{[ij]} , \tag{5}$$

where ( ) designates the symmetrization of tensor indices. Here, e_{ij} is the *strain tensor*, and r_{ij} the *rotation tensor*.

If all the particles are rigid, any deformation that does not change the

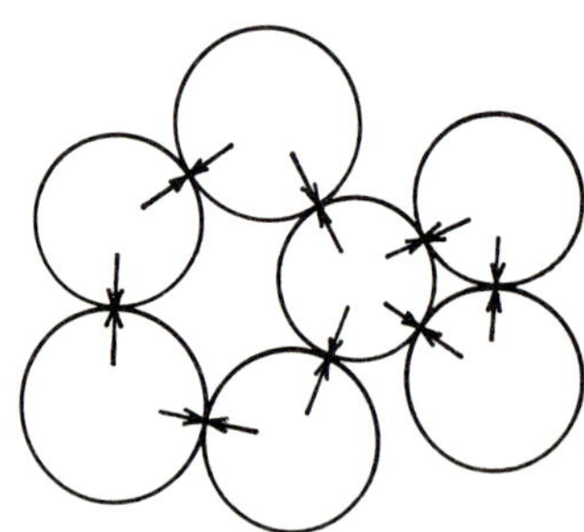

Fig. 1. Interparticle contact force distribution.

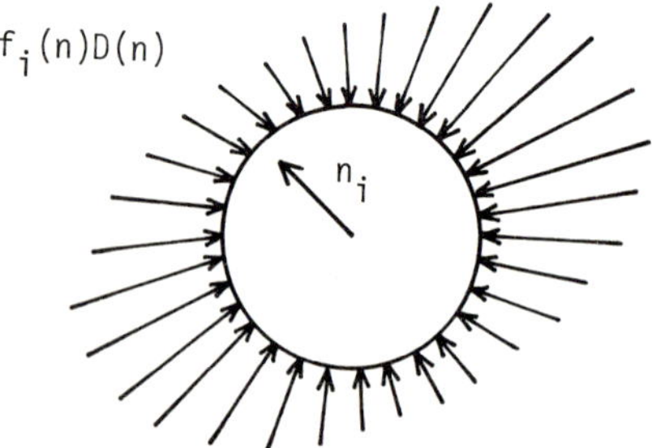

Fig. 2. The force distribution on the representative particle.

interparticle contact configuration is actually impossible except overall translations and rotations. However, we can hypothetically imagine this kind of deformation, i.e., *virtual deformation*. If we consider a virtual deformation described by eqn (3) or (4), it distorts the rigid representative particle into an ellipsoid (Fig. 3). The displacement $\zeta_i(n)$ at the contact point of the contact direction n is given by

$$\zeta_i(n) = aF_{ij}n_j \; . \tag{6}$$

Since the contact forces are assumed not to change during the virtual deformation, the *virtual work* done by the contact forces on the representative particle is

$$-\oint f_i\zeta_i D dn = -aF_{ij}\oint f_i n_j D dn = -ae_{ij}\oint f_{(i}n_{j)} D dn \; . \tag{7}$$

The last equality follows due to the torque balance $(2)_2$. If γ is the *solid volume fraction* of the assemblage, the number of particles in unit volume is $\gamma/(4/3)\pi a^3$. (Note that this relation *defines* the *average radius* to be exact.) Hence, the virtual work done in unit volume is given by

$$W = -\frac{3}{4}\frac{\gamma}{\pi a^2}\, e_{ij}\oint f_{(i}n_{j)} D dn \; . \tag{8}$$

On the other hand, the virtual work done per unit volume on the virtual strain e_{ij} by the stress σ_{ij} must be

$$W = \sigma_{ij}e_{ij} \; , \tag{9}$$

because the stress is assumed unchanged during the virtual deformation. Eqn (8) and eqn (9) must coincide for an arbitrary virtual deformation. Thus, we obtain the fundamental relation

$$\sigma_{ij} = -\frac{3}{4}\frac{\gamma}{\pi a^2}\oint f_{(i}n_{j)} D dn \; , \tag{10}$$

which relates the microscopic contact forces to the macroscopic stress.

3. THE CONTACT FORCE IN TERMS OF THE STRESS

The contact force density f_iD on the representative particle is regarded as a smooth function of the contact direction n when the average is taken over a sufficiently large number of randomly packed particles. Hence, it is expanded into a series of spherical harmonics. In our Cartesian tensor notation, the expansion takes the form

$$f_iD = A_i + B_{ij}n_j + C_{ijk}n_jn_k + \dots \; . \tag{11}$$

We retain only the first two terms, omitting the higher harmonics, and substitute this in the equilibrium conditions, eqns (2), using identities

$$\oint n_i dn = 0 \; , \qquad \oint n_i n_j dn = \frac{4}{3}\pi\delta_{ij} \; . \tag{12}$$

Then, we obtain $A_i = 0$ and $B_{[ij]} = 0$. Substitution of eqn (11) in the fundamental relation (10) yields $\sigma_{ij} = -(\gamma/a^2)B_{ij}$. Hence, we obtain

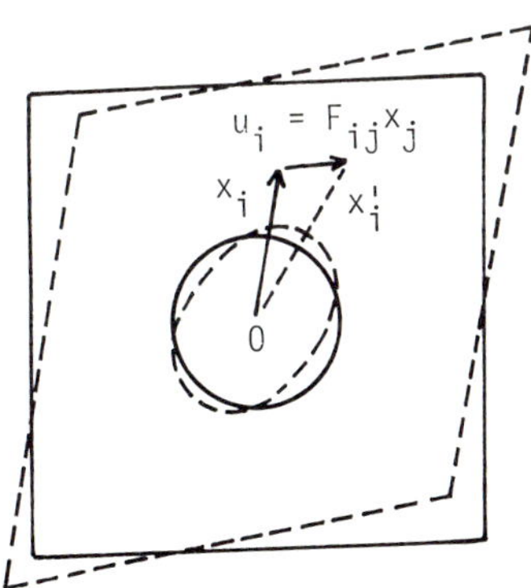

Fig. 3. Virtual deformation of a granular assemblage.

$$f_i(n)D(n) = -(a^2/\gamma)\sigma_{ij}n_j . \quad (13)$$

This result seems quite reasonable, if we take into account the fact that eqn (13) gives the force per unit solid angle of the representative particle. Namely, the force per unit area of the representative particle surface is $-(1/\gamma)\sigma_{ij}n_j$. If the material were a complete continuum, the force per unit area of a plane with unit normal n in the material would be $-\sigma_{ij}n_j$, whereas in this case it is divided by γ, the solid volume fraction, due to the existence of the voidage in the material. (We have adopted the convention that a compressive contact force is positive whereas the tensile stress is positive.)

There are experiments showing that the contact point density takes its maximum in the direction of maximum compression and its minimum in the direction of minimum compression (e.g., refs. 4, 5). This fact suggests that we may assume the following quadratic form

$$D(n) = -C(\gamma)\sigma_{ij}n_in_j . \quad (14)$$

Here, $C(\gamma)$ is a positive scalar coefficient which depends on the solid volume fraction γ. From eqn (1), the coordination number, which also depends on γ, is given by

$$N(\gamma) = -C(\gamma)\sigma_{ij}\oint n_in_j dn = -4\pi C(\gamma)p , \quad (15)$$

where $p \equiv -\sigma_{kk}/3$ is the *hydrostatic pressure*. Hence, $C(\gamma) = -N(\gamma)/4\pi p$ and consequently

$$D(\gamma) = -\frac{N(\gamma)}{4\pi p}\sigma_{ij}n_in_j , \quad (16)$$

which gives the contact point density per unit solid angle under a given stress. The empirical form of $N(\gamma)$ is discussed by a number of authors (e.g., ref. 6). Combination of eqn (13) and (16) yields

$$f_i(n) = \frac{4\pi a^2 p}{\gamma N(\gamma)}\sigma_{ij}n_j/\sigma_{kl}n_kn_l , \quad (17)$$

which expresses the average contact force per single contact in the direction n in terms of the solid volume fraction and the stress σ_{ij}.

4. PARTICLE SLIPS AND THE STRAIN

Now, we consider the relation between the interparticle slips and the resulting macroscopic strain. Let $\zeta_i(n)$ be the average displacement of the contacts in the direction n. The slip changes the contact direction from n_i to $n_i + \zeta_i/a$. If the macroscopic deformation is a smooth linear deformation and its distortion tensor is F_{ij}, then we can expect

$$\zeta_i/a = F_{ij}n_j \quad (18)$$

on the average. Multiplying this by n, and integrating it over the whole solid angle with respect to n, we have

$$\oint n_k\zeta_i/a dn = F_{ij}\oint n_kn_j dn = F_{ij}\times\frac{4}{3}\pi\delta_{jk} = \frac{4}{3}\pi F_{ik} . \quad (19)$$

Hence, we have

$$F_{ij} = \frac{3}{4\pi a}\oint\zeta_in_j dn . \quad (20)$$

Taking the symmetric part, we obtain

$$e_{ij} = \frac{3}{4\pi a}\oint \zeta_{(i}n_{j)}dn \ . \tag{21}$$

This expression gives the macroscopic strain when the number of contacts does not change. Since the displacement $\zeta_i(n)$ is assumed to be always perpendicular to the contact normal n, we can see that $e_{kk} = 0$, i.e., the volume does not change. If the solid volume fraction γ changes to $\gamma + \Delta\gamma$, (hence the coordination number $N(\gamma)$ changes accordingly,) the volumetric strain $-(\Delta\gamma/3\gamma)\delta_{ij}$ occurs. Thus, if we take into account the change of the number of contacts and the solid volume fraction, we have as the overall strain

$$e_{ij} = \frac{3}{4\pi a}\oint \zeta_{(i}n_{j)}dn - \frac{\Delta\gamma}{3\gamma}\delta_{ij} \ . \tag{22}$$

5. THE MACROSCOPIC FRACTURE CRITERION

The contact force $f_i(n)D(n)$ (per unit solid angle of the representative particle) is resolved into the normal compression $\nu = n_i f_i D$ and the tangential force $\tau_i = - f_i D + (-\nu)n_i$ (Fig. 4). Making use of eqn (13), we obtain

$$\nu(n) = -\frac{a^2}{\gamma}\sigma_{ij}n_i n_j \ , \qquad \tau_i(n) = \frac{a^2}{\gamma}(\sigma_{ij}n_j - \sigma_{jk}n_i n_j n_k) \ . \tag{23}$$

Microscopic *Coulomb's law* states that the slip at a contact may occur when the contact point becomes *critical*, i.e., when

$$\sqrt{\tau_i\tau_i} = \mu'\nu + a^2c' \tag{24}$$

is satisfied, where μ' is the static friction coefficient and c' is the cohesion force per unit area of the particle surface. The so called *Mohr-Coulomb criterion* states that the fracture begins whenever any one direction becomes critical, i.e., when

$$\max_n[\tau_i(n)\tau_i(n) - (\mu'\nu(n) + a^2c')^2] = 0 \tag{25}$$

is satisfied. However, even when one direction becomes critical, the overall fracture may not develop. If we assume that the overall fracture begins when all the directions become critical *on the average*, the criterion may be put

$$\oint[\tau_i(n)\tau_i(n) - (\mu'\nu(n) + a^2c')^2]dn = 0 \ . \tag{26}$$

Substituting eqns (23) in this, and making use of identities (12) and

$$\oint n_i n_j n_k dn = 0 \ , \qquad \oint n_i n_j n_k n_l dn = \frac{4}{15}\pi(\delta_{ij}\delta_{kl} + \delta_{ik}\delta_{jl} + \delta_{il}\delta_{jk}), \tag{27}$$

we finally obtain

$$\sqrt{\frac{1}{2}\tilde{\sigma}_{ij}\tilde{\sigma}_{ij}} = \sqrt{\frac{15\mu'^2}{2(3-2\mu'^2)}}\,p + \gamma\sqrt{\frac{15}{2(3-2\mu'^2)}}\,c', \tag{28}$$

where $p = -\sigma_{kk}/3$ is the hydrostatic pressure and

$$\tilde{\sigma}_{ij} = \sigma_{ij} - \frac{1}{3}\sigma_{kk}\delta_{ij} \tag{29}$$

is the *stress deviator*. Eqn (28) coincides with the so called *extended von Mises criterion*. If we define the *internal angle of friction* ϕ and the *cohesion force* c in such a way that eqn (28) coincides with the two dimensional Coulomb criterion in the case of plane strain (See refs. 7, 8, 9), we have

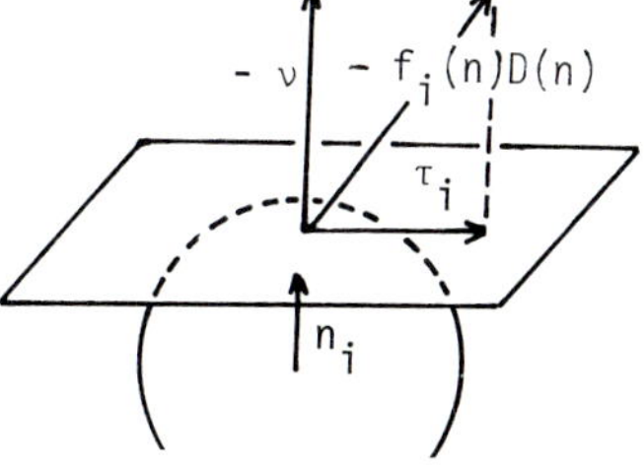

Fig. 4. Resolution of the contact force.

$$\tan\phi = \sqrt{\frac{15}{6 - 19\mu'^2}}\,\mu' , \qquad c = \gamma\sqrt{\frac{15}{6 - 19\mu'^2}}\,c' , \tag{30}$$

which express macroscopic parameters in terms of microscopic particle properties.

6. DISSIPATIVE STRESS IN THE FLOW STATE

Consider the flow state in which particles are moving with the average velocity v_i and rotating with the average rotational velocity ω_{ij}, which is a skew symmetric tensor (Fig. 5). (Here, by "average" we mean disregarding microscopic fluctuations and regarding v_i and ω_{ij} as smooth vector and tensor fields. The fluctuations are considered in the next section.) We put

$$D_{ij} = \partial_i v_j , \qquad \Omega_{ijk} = \partial_i \omega_{jk} , \tag{31}$$

the *velocity gradient* and the *rotation velocity gradient* respectively, where ∂_i designates $\partial/\partial x_i$. Consider particulatr two particles in contact in the flow (Fig. 6). We may put

$$v_j' = v_j + 2an_i D_{ij} , \qquad \omega_{jk}' = \omega_{jk} + 2an_i\Omega_{ijk} , \tag{32}$$

where the unit vector n designates the direction of contact. The relative velocity $\xi_i(n)$ at the contact point is given by

$$\xi_k(n) = 2a(n_j D_{jk} - n_j\omega_{jk} - n_i n_j n_k D_{ij} - an_i n_j \Omega_{ijk}) . \tag{33}$$

We regard the contact direction n as a random variable distributing uniformly over the whole solid angle and measure the amount of average relative velocity in the sense of the root-mean-square

$$\sqrt{\overline{\xi_i\xi_i}} \;\; (= [\,\frac{1}{4\pi}\oint \xi_i\xi_i dn]^{1/2}) . \tag{34}$$

Substituting eqn (33) in this and making use of identities (12) and (27), we finally obtain

$$\xi = \frac{2\sqrt{6}}{3}\,a\hat{\omega} , \tag{35}$$

where

$$\hat{\omega} = \sqrt{\frac{3}{10}E_{ij}E_{ij} + \frac{1}{2}R_{ij}R_{ij} + \frac{a^2}{10}(\Omega_{iik}\Omega_{jjk} + \Omega_{ijk}\Omega_{ijk} + \Omega_{ijk}\Omega_{jik})} . \tag{36}$$

Here, we have put

$$E_{ij} = \partial_{(i}v_{j)} , \qquad R_{ij} = \partial_{[i}v_{j]} - \omega_{ij} , \tag{37}$$

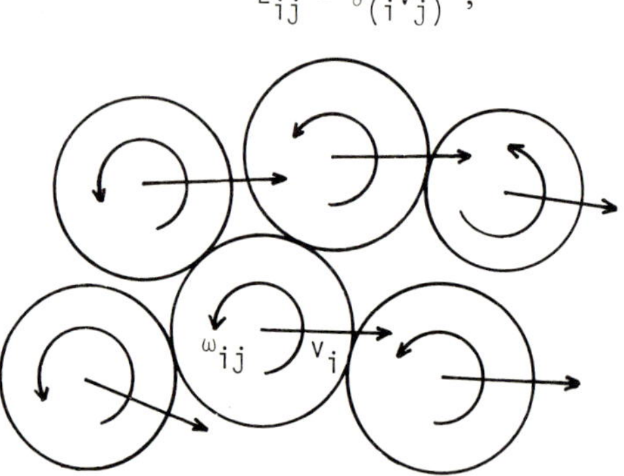

Fig. 5. Particles in the flow state.

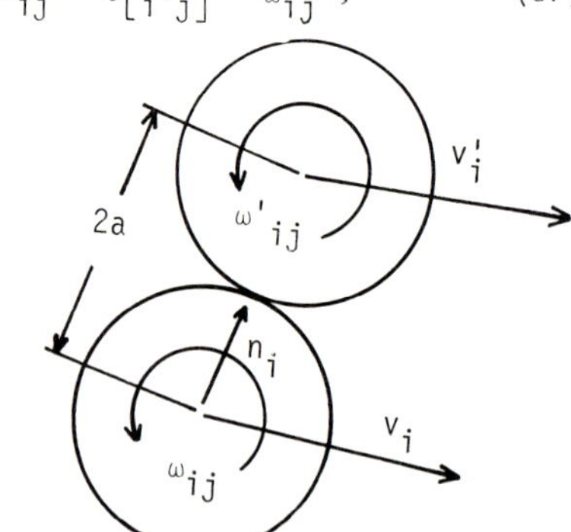

Fig. 6. Particular two particles in contact in the flow state.

the *strain rate* and the *relative rotation velocity* respectively. The quantity $\hat{\omega}$ may be termed the *equivalent amount of particle rotation*, because if $v_i = 0$ then $\hat{\omega} = \sqrt{\omega_{ij}\omega_{ij}/2}$.

According to eqn (17), the normal contact force per one contact point is $f_n \equiv f_i n_i = 4\pi a^2 p/\gamma N$. Thus, the energy dissipation rate per particle is approximated by $N\mu f_n \xi$, where μ is the kinetic friction coefficient of the particles. Then, energy dissipation rate per unit volume is given by

$$\phi = \sqrt{6}\mu p\hat{\omega} \; . \tag{38}$$

On the other hand, according to the *polar continuum theory* (e.g., refs. 3, 10, 11), we must have

$$\phi = \tilde{\sigma}_{ij}E_{ij} + \sigma_{[ij]}R_{ij} + \frac{1}{2}\mu_{ijk}\Omega_{ijk} \; , \tag{39}$$

where μ_{ijk} is the *couple stress*, i.e., the ij-component of the couple acting on unit area of the plane whose normal is in the j-direction. Noting that ϕ is a homogeneous form of degree 1 in E_{ij}, R_{ij} and μ_{ijk} and hence $\phi = E_{ij}\partial\phi/\partial E_{ij} + R_{ij}\partial\phi/\partial R_{ij} + \Omega_{ijk}\partial\phi/\partial\Omega_{ijk}$, we may put the equivalent macroscopic dissipative stresses to be

$$\tilde{\sigma}_{ij} = \frac{3\sqrt{6}\mu}{10}\frac{p}{\hat{\omega}}[\partial_{(i}v_{j)} - \frac{1}{3}\delta_{ij}\partial_k v_k] \; , \tag{40}$$

$$\sigma_{[ij]} = \frac{\sqrt{6}\mu}{2}\frac{p}{\hat{\omega}}[\partial_{[i}v_{j]} - \omega_{ij}] \; , \tag{41}$$

$$\mu_{ijk} = \frac{\sqrt{6}\mu a^2}{5}\frac{p}{\hat{\omega}}[\delta_{i[j}\partial_{|l}\omega_{l|k]} + \partial_i\omega_{jk} - \partial_{[k}\omega_{j]i}) \; , \tag{42}$$

where the hydrostatic pressure p is treated as an indeterminate variable determined by boundary conditions.

Let us consider the case where the particle size is sufficiently small compared with the size of the flow and hence the effects of the couple stress can be neglected. Then, the stress is not affected by the "magnitude" of the velocity gradient. Hence, no steady state may exist, or it may not be determined uniquely and the velocity profile is determined almost by initial conditions. The flow of this type is often referred to as the *rate-independent* flow (e.g., ref. 12). We can also see that the flow is purely plastic, i.e., the stress components are not mutually independent. In fact in the limit of $a \to 0$ they satisfy the following *extended von Mises criterion*:

$$\sqrt{\frac{1}{2}\tilde{\sigma}_{ij}\tilde{\sigma}_{ij}} = \frac{3\sqrt{10}}{10}\mu p \; . \tag{43}$$

If we apply eqns (40) - (42) to the flow on an inclined plane, we can conclude that the flow exists only when the inclination angle θ of the plane exceeds the *angle of repose* θ^* given by

$$\tan\theta^* = \frac{3\sqrt{10}}{10}\mu \; . \tag{44}$$

7. THE PRESSURE DUE TO PARTICLE COLLISIONS

In a rapid flow, the particles repeat collisions with neighboring particles. We now consider the pressure generated by the collisions in the flow in the form of an *equation of state* similar to that in the statistical mechanics. Imagine that a particle is repeating collisions with a spherical wall surrounding the particle (Fig. 8). The reciprocal of the number of paritcles in unit volume is $V=m/\rho$, where m is the average mass of a particle and ρ the bulk density of the assemblage. Hence, V is the volume of the space equally assigned to each particle. The radius of a sphere of volume V is $r = (3V/4\pi)^{1/3}$. We put the radius of the cell to be r and put r_0 to be the radius when the bulk density is $\rho = \rho_0$, the value at the packing state, i.e., the minimum possible value of ρ. We also put v' to be the average fluctuation velocity in the flow.

According to our model, the collisions take place $v'/2(r - r_0)$ times per unit time. The momentum given to the wall is 2mv' for each collision. Hence, the pressure acting on the wall is $p = mv'^2/4\pi r^2(r - r_0) = \rho\rho_0 v'^2/3(\rho_0 - \rho^{1/3}\rho_0{}^{2/3}) \simeq \rho\rho_0 v'^2/(\rho_0 - \rho)$. (Note $\rho_0 - \rho^{1/3}\rho_0{}^{2/3} = (\rho_0 - \rho)/3 + (\rho_0 - \rho)^2/9\rho + \ldots$.) Next, we must determine the value of v'. We assume that the particle fluctuation is given in terms of $\hat{\omega}$, the amount of interparticle friction. If a particle is rotating with the equivalent rotation velocity $\hat{\omega}$, the rotational kinetic energy is $(1/5)ma^2\hat{\omega}^2$. (The moment of inertia around the center is $(1/10)ma^2$.) On the other hand, the kinetic energy of fluctuation is $(1/2)mv'^2$. We now postulate that the latter energy is proportional to the former energy, putting $(1/2)mv'^2 = T(1/5)ma^2\hat{\omega}^2$, where T is a nondimensional constant. Thus, we get $v' = (\sqrt{10}/5)\sqrt{T}a\hat{\omega}$. Substituting this in the previous expression for p, we obtain the *equation of state*:

$$p = \frac{2}{5} Ta^2 \frac{\rho_0\rho}{\rho_0 - \rho} \hat{\omega} . \tag{45}$$

The energy dissipation rate in unit volume is determined by the same argument in the previous section to be

$$\Phi = \sqrt{6}\mu \frac{a}{r} p\hat{\omega} . \tag{46}$$

Following the previous argument, we obtain the dissipative stresses as follows:

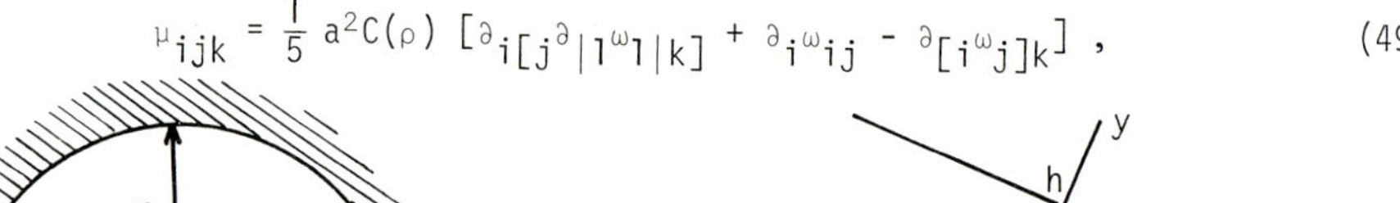

$$\tilde{\sigma}_{ij} = \frac{5}{10} C(\rho) [\partial_{(i}v_{j)} - \frac{1}{3} \delta_{ij}\partial_k v_k] , \tag{47}$$

$$\sigma_{[ij]} = \frac{1}{2} C(\sigma) [\partial_{[i}v_{j]} - \omega_{ij}] , \tag{48}$$

$$\mu_{ijk} = \frac{1}{5} a^2 C(\rho) [\partial_{i[j}\partial_{|l}\omega_{l|k]} + \partial_i\omega_{ij} - \partial_{[i}\omega_{j]k}] , \tag{49}$$

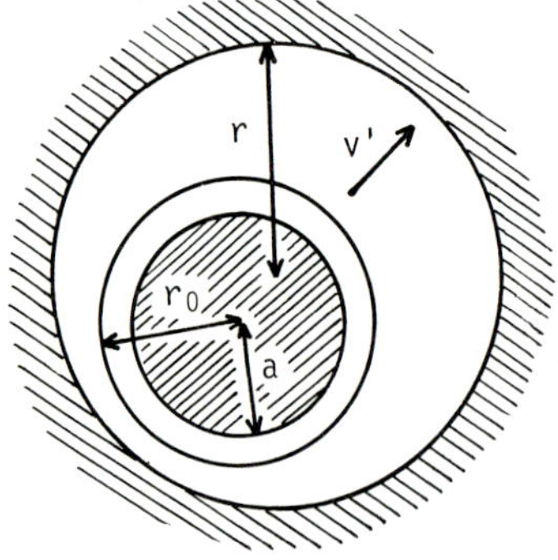

Fig. 7. A particle colliding with a spherical wall.

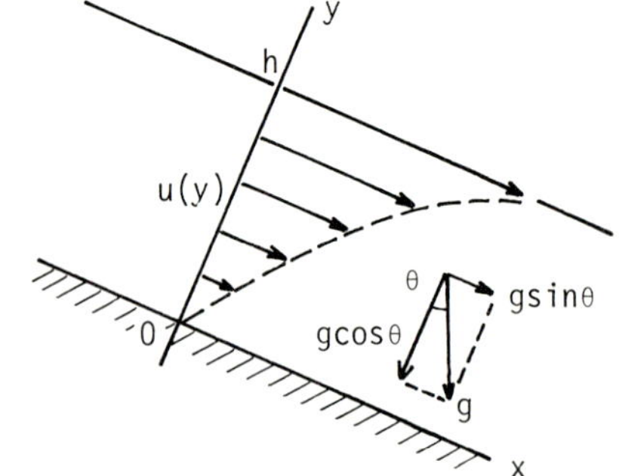

Fig. 8. The gravity flow on an inclined plane.

where

$$C(\rho) \equiv \frac{2\sqrt{6}}{5} T\mu \frac{a^3}{r} \frac{\rho_0 \rho}{\rho_0 - \rho} . \tag{50}$$

Here, the density ρ is treated as an independent variable determined by the equation of continuity and boundary conditions.

8. THE GRAVITY FLOW ON AN INCLINED PLANE

Consider the inclined gravity flow shown in Fig. 8. We consider the case $a/h \ll 1$, where h is the depth of the flow. In this case we can neglect the effect of the couple stress and put $\Omega_{ijk} = 0$ and $R_{ij} = 0$. Then the relevant equations are

$$p = \frac{\sqrt{6}}{40\mu} \frac{r}{a} C(\rho)(\frac{du}{dy})^2 , \qquad \sigma_{yx} = \frac{3\sqrt{15}}{200} C(\rho)(\frac{du}{dy})^2 , \tag{51}$$

where u is the x-component of the velocity as is illustrated in Fig. 8. Both the shear stress and the pressure are proportional to the *square* of the velocity gradient $(du/dy)^2$ as was experimentally confirmed by Bagnold (ref. 13) and others. If the flow is a simple shear flow, we can integrate the equation of motion to give the density profile $\rho(y)$ in the following implicit equation:

$$y/h = 1 + c(1 - \log\frac{c\rho}{\rho_0 - \rho} - \frac{\rho_0}{\rho_0 - \rho}) , \qquad c \equiv \frac{3Ta}{50gh_0\cos\theta} (\frac{du}{dy})^2 . \tag{52}$$

Here h_0 is the depth of the slab when all the particles are at rest. It is seen that the increase of the shearing leads to the increase of particle collisions, which in turn cause dilatation of the flow (Fig. 9). Using this form of $\rho(y)$, we can obtain the acceleration profile $\partial u/\partial t$. If there is no slip at the bottom

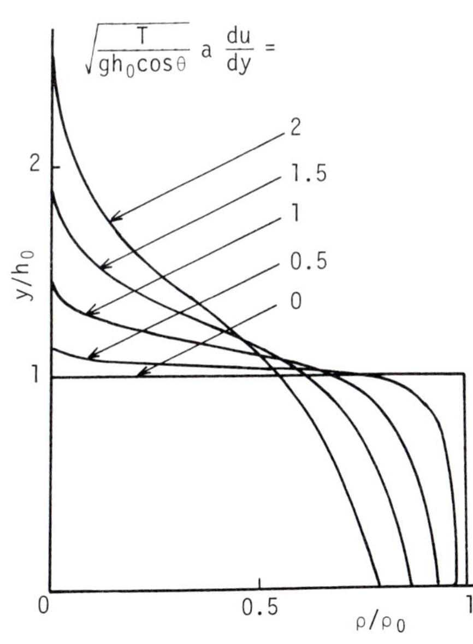

Fig. 9. The density profile of the flow.

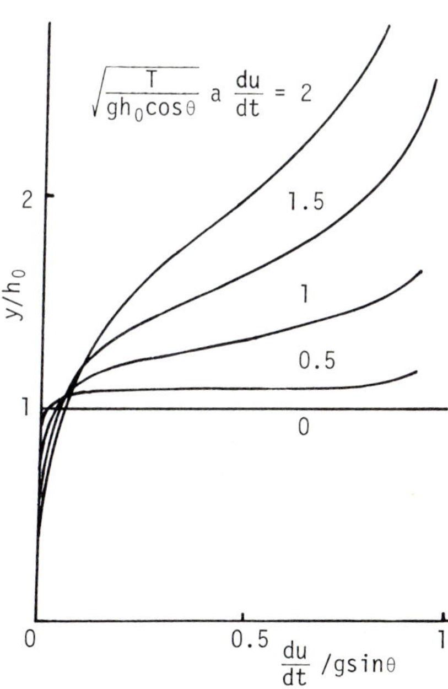

Fig. 10. The acceleration profile of the flow.

y = 0, we get

$$\frac{du}{dt} = g \sin\theta \left(1 - \left(\frac{\rho(y)}{\rho(0)}\right)^{1/3} \frac{4\rho_0 - \rho(y)}{4\rho_0 - \rho(0)}\right) \tag{53}$$

(Fig. 10). We can observe that the acceleration is especially large in the upper layer where a large amount of dilatiation occurs.

9. CONCLUDING REMARKS

The first half of this paper is an analysis of packed rigid particles. This is an important subject, and recently considerable attention is paid to the contact point distribution, or the *fabric* (e.g., refs. 14, 15).

The second half is for the flow where interparticle friction is dominant. Although the effect of particle collisions is incorporated, we have assumed that the material is fairly dense and used the expansion of ρ around the packed density ρ_0. Thus, our theory deals with the *friction-dominant* flow with *infinitesimal* collisions. If we are to consider the *collision-dominant* regime, we have to develop a full set of equations describing particle collisions. Since the proposal of the present theory in its original form in 1979 (ref. 3), a considerable progress has been made on this line (e.g., refs. 16, 17, 18 and papers in this volume).

On the other hand, the flow state is also viewed from the standpoint of the plasticity theory, and plastic dilatant deformation is described consistently by a new version of the associated flow rule (refs. 7, 8, 9). The packing state of particles is characterized by the entropy in the statistical sense (ref. 2). A quantity like the thermodynamic entropy is also introduced in the flow state (ref. 3).

REFERENCES

1 K. Kanatani, Powder Technol., 28 (1981) 167-172.
2 K. Kanatani, Powder Technol., 30 (1981) 217-223.
3 K. Kanatani, Int. J. Engng Sci., 17 (1979) 419-432.
4 M. Oda, Soils and Foundations, 12-2 (1972) 1-18.
5 M. Oda and J. Konish, Soils and Foundations, 14-4 (1974) 25-53.
6 M. Oda, Soils and Foundations, 17-2 (1977) 29-42.
7 K. Kanatani, Int. J. Engng Sci., 20 (1982) 1-13.
8 K. Kanatani, Int. J. Engng Sci., 20 (1982) 879-884.
9 K. Kanatani, Deformation and Failure of Granular Materials, Proc. IUTAM Conf. Delft (Eds. P. A. Fermeer and H. J. Luger), A. A. Balkema, Rotterdam, 1982, pp. 119-127.
10 N. Oshima, RAAG Memoirs, 1 (1955) 563-572.
11 A.C. Eringen (Ed.), Continuum Physics IV - Polar and Nonlocal Field Theories, Academic Press, New York, 1976.
12 S.B. Savage, in this volume.
13 R.A. Bagnold, Proc. Roy. Soc. London, A295 (1966) 219-232.
14 M. Satake, in this volume.
15 M.M. Mehrabadi and S. Nemat-Nasser, in this volume.
16 S. Ogawa, A. Umemura and N. Oshima, ZAMP 31 (1980) 483-493.
17 H. Shen and N.L. Ackermann, Trans. ASCE EM, Oct. (1982).
18 J.T. Jenkins and S.B. Savage, J. Fluid Mech. submitted.

Mechanics of Granular Materials: New Models and Constitutive Relations, edited by
J.T. Jenkins and M. Satake, 1983
Elsevier Science Publishers B.V., Amsterdam — Printed in The Netherlands

KINEMATICALLY DETERMINED AXIALLY SYMMETRIC DEFORMATIONS OF GRANULAR MATERIALS

A. J. M. SPENCER

Department of Theoretical Mechanics, University of Nottingham, Nottingham, England

ABSTRACT

We consider axially symmetric deformations of incompressible granular materials which deform according to the Coulomb-Mohr yield condition and the 'double-shearing' theory formulated by Mandel [1] and Spencer [2,3]. For the case in which the hoop stress σ_θ is strictly greater or strictly less than the other two principal stress components, the equations for the velocity field uncouple from those for the stress field, and are independent of the angle of internal friction. Accordingly, velocity solutions derived by Lippmann [7,8] for material which satisfies Tresca's yield condition are equally valid for this ideal granular material. The stress associated with these velocity solutions is also determined.

1 INTRODUCTION

The 'double-shearing' theory for plane deformations of incompressible ideal granular materials was formulated by Mandel [1] and Spencer [2]. Recently Spencer [3] has extended this theory to three-dimensional deformations, and in particular has formulated the equations for axially symmetric deformations. Previously, Cox, Eason and Hopkins [4] had formulated a theory of axially symmetric stress and deformation of granular materials which was based on Shield's [5] generalisation of the Coulomb-Mohr yield condition and the flow rule associated with this yield condition. The theories of Cox, Eason and Hopkins [4] and of Spencer [3] are in agreement as regards the equations which govern the stress field, but propose different equations for the velocity field. Both theories reduce to Shield's [6] theory for axially symmetric deformations of ideal rigid-plastic materials which satisfy Tresca's yield condition when the angle of internal friction ϕ is zero.

For axially symmetric stress and velocity fields, various possibilities arise depending on the relative magnitudes of the hoop stress σ_θ and the principal stress components in the meridional planes. Some of these cases lead to only a limited range of solutions. Hitherto interest has centred on the 'Haar-von Karman' regimes, in which σ_θ is equal to one of the other principal stress components, and some solutions for these regimes are described in [3]. However, for frictionless materials which satisfy Tresca's yield condition, Lippmann [7,8] has pointed out that interest also attached to the case in which σ_θ is, strictly, the greatest or least principal stress, and has obtained two non-trivial exact solutions for these regimes. In this paper we exploit the observation made in [3] that, in the 'double-shearing' theory for these regimes, the velocity equations are independent of ϕ. It follows that Lippmann's solutions are equally applicable to frictional granular materials, although the associated stress does depend on ϕ. In Sections 4 and 5 we re-state Lippmann's solutions and determine the stress associated with them.

2 AXIALLY SYMMETRIC STRESS STATES

The stress and deformation are referred to a system of cylindrical polar coordinates r,θ,z. We consider axially symmetrical stress states, and denote the non-zero stress components by σ_r, σ_θ, σ_z and σ_{rz} (tensile stresses are taken to be positive). Then σ_θ is one of the principal stress components and the θ-direction is a principal axis of stress. The other two principal components of stress are denoted by σ_1 and σ_2 and are ordered so that $\sigma_1 \geqslant \sigma_2$. The principal axes of stress corresponding to σ_1 and σ_2 lie in the (r,z) planes. We denote by ψ the angle which the axis corresponding to σ_1 makes with the radial direction. Then

$$\begin{aligned} \sigma_r &= \tfrac{1}{2}(\sigma_1+\sigma_2) + \tfrac{1}{2}(\sigma_1-\sigma_2)\cos 2\psi , \\ \sigma_z &= \tfrac{1}{2}(\sigma_1+\sigma_2) - \tfrac{1}{2}(\sigma_1-\sigma_2)\cos 2\psi , \\ \sigma_{rz} &= \tfrac{1}{2}(\sigma_1-\sigma_2)\sin 2\psi . \end{aligned} \qquad (2.1)$$

Shield [5] formulated a three-dimensional generalisation of the Coulomb-Mohr yield condition which is analogous to Tresca's yield condition for frictionless plastic materials. If the principal stress components are denoted by σ_I, σ_{II}, σ_{III}, with $\sigma_I \geqslant \sigma_{II} \geqslant \sigma_{III}$, then Shield's condition takes the form

$$\sigma_I - \sigma_{III} = 2c\cos\phi - (\sigma_I+\sigma_{III})\sin\phi , \qquad (2.2)$$

where c is the cohesion and ϕ is the angle of internal friction. Cox, Eason and Hopkins [4] applied Shield's theory to axially symmetric stress states. Various possibilities arise depending on the magnitude of σ_θ relative to the magnitudes of σ_1 and σ_2. The possible cases are illustrated in Fig. 1, which shows a section of the yield surface in $\sigma_1,\sigma_2,\sigma_\theta$ space (which is a hexagonal pyramid) by a plane σ_θ = constant. Accessible stress states lie in or on the hexagon ABCDEF, with deformation being possible only for stress states which lie on the hexagon. The various possibilities are described by Cox, Eason and Hopkins [4] and by Spencer [3], where further details may be found. Analysis shows that the stress states which correspond to the vertices B and E and the side AF of the hexagon (since $\sigma_1 \geqslant \sigma_2$, it is sufficient to consider BAFE) give rise to only a limited range of solutions. In the past attention has concentrated on the regimes which

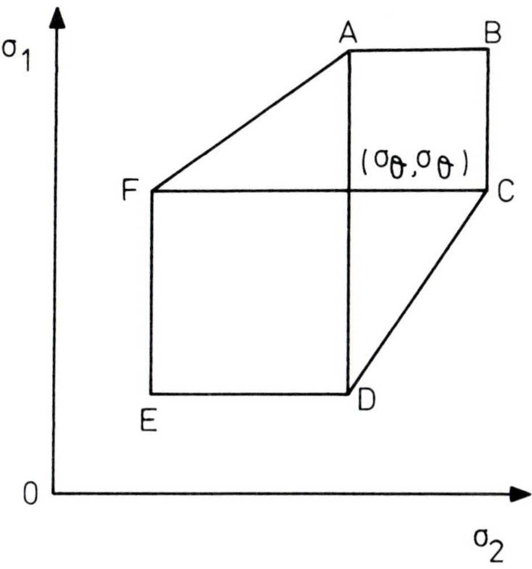

Fig. 1 Yield hexagon for axially symmetric stress states

correspond to the vertices A and F (which are called the 'Haar-von Karman' regimes). In this paper we consider some solutions for the plastic regimes associated with the sides AB and EF of the hexagon. In these regimes σ_θ is either the greatest (regime EF) or the least (regime AB) of the principal stress components, and (2.2) becomes

$$\text{Regime AB:} \quad \sigma_1(1+\sin\phi) = 2c\cos\phi + \sigma_\theta(1-\sin\phi) . \tag{2.3}$$

$$\text{Regime EF:} \quad \sigma_\theta(1+\sin\phi) = 2c\cos\phi + \sigma_2(1-\sin\phi) . \tag{2.4}$$

The equations of motion are

$$\frac{\partial\sigma_r}{\partial r} + \frac{\partial\sigma_{rz}}{\partial z} + \frac{\sigma_r - \sigma_\theta}{r} + \rho b_r = \rho f_r , \tag{2.5}$$

$$\frac{\partial\sigma_{rz}}{\partial r} + \frac{\partial\sigma_z}{\partial z} + \frac{\sigma_{rz}}{r} + \rho b_z = \rho f_z , \tag{2.6}$$

where ρ is density, b_r and b_z are components of body force, per unit mass, in the r and z directions, and f_r and f_z are components of acceleration in the r and z directions.

Equations (2.3) or (2.4), with (2.5) and (2.6), are three equations for the four non-zero stress components. Thus, even for quasi-static deformations, it is not possible to solve for the stress without reference to the deformation.

3 KINEMATICALLY DETERMINED AXIALLY SYMMETRIC DEFORMATIONS

A general three-dimensional theory for flow of granular materials whose mechanical behaviour is governed by the yield condition (2.2) is described in Spencer [3]. This theory employs the 'double-shearing' model, in which deformation is assumed to occur by shearing on the surfaces on which the shear stress τ takes the critical value $c + \sigma \tan\phi$, where σ is the normal pressure on the surface. Also in Spencer [3], this theory was specialised to formulate the velocity equations for axially symmetric deformations in the various cases which correspond to the edges and vertices of the hexagon ABCDEF in Fig. 1. We only consider here the regimes which correspond to the edges EF and AB.

Components of velocity in the radial and axial directions are denoted by v_r and v_z respectively. The material is assumed to be incompressible, so that

$$\frac{\partial v_r}{\partial r} + \frac{v_r}{r} + \frac{\partial v_z}{\partial z} = 0 . \tag{3.1}$$

We denote by $\underset{\sim}{e}_r$, $\underset{\sim}{e}_\theta$, $\underset{\sim}{e}_z$ the unit vectors in the r, θ, z directions respectively. Then, for axially symmetric deformations, the rate of deformation tensor $\underset{\sim}{d}$ may be expressed in components referred to r,θ,z coordinates as

$$\underset{\sim}{d} = d_{rr}\underset{\sim}{e}_r \otimes \underset{\sim}{e}_r + d_{rz}(\underset{\sim}{e}_r \otimes \underset{\sim}{e}_z + \underset{\sim}{e}_z \otimes \underset{\sim}{e}_r) + d_{zz}\underset{\sim}{e}_z \otimes \underset{\sim}{e}_z + d_{\theta\theta}\underset{\sim}{e}_\theta \otimes \underset{\sim}{e}_\theta , \tag{3.2}$$

where

$$d_{rr} = \frac{\partial v_r}{\partial r}, \qquad d_{\theta\theta} = \frac{v_r}{r}, \qquad d_{zz} = \frac{\partial v_z}{\partial z}, \qquad d_{rz} = \tfrac{1}{2}\left(\frac{\partial v_r}{\partial z} + \frac{\partial v_z}{\partial r}\right) . \tag{3.3}$$

Alternatively, $\underset{\sim}{d}$ may be referred to base vectors $\underset{\sim}{e}_1$, $\underset{\sim}{e}_2$, $\underset{\sim}{e}_\theta$, where $\underset{\sim}{e}_1$ and $\underset{\sim}{e}_2$ are unit vectors in the directions of the principal axes of stress in the (r,z) planes, so that

$$\begin{pmatrix} \underset{\sim}{e}_1 \\ \underset{\sim}{e}_2 \end{pmatrix} = \begin{pmatrix} \cos\psi & \sin\psi \\ -\sin\psi & \cos\psi \end{pmatrix} \begin{pmatrix} \underset{\sim}{e}_r \\ \underset{\sim}{e}_z \end{pmatrix} , \qquad \begin{pmatrix} \underset{\sim}{e}_r \\ \underset{\sim}{e}_z \end{pmatrix} = \begin{pmatrix} \cos\psi & -\sin\psi \\ \sin\psi & \cos\psi \end{pmatrix} \begin{pmatrix} \underset{\sim}{e}_1 \\ \underset{\sim}{e}_2 \end{pmatrix} . \tag{3.4}$$

In terms of $\underset{\sim}{e}_1$, $\underset{\sim}{e}_2$ and $\underset{\sim}{e}_\theta$, $\underset{\sim}{d}$ can be expressed as

$$\underset{\sim}{d} = d_{11}\underset{\sim}{e}_1 \otimes \underset{\sim}{e}_1 + d_{12}(\underset{\sim}{e}_1 \otimes \underset{\sim}{e}_2 + \underset{\sim}{e}_2 \otimes \underset{\sim}{e}_1) + d_{22}\underset{\sim}{e}_2 \otimes \underset{\sim}{e}_2 + d_{\theta\theta}\underset{\sim}{e}_\theta \otimes \underset{\sim}{e}_\theta , \tag{3.5}$$

where

$$\begin{pmatrix} d_{11} & d_{12} \\ d_{12} & d_{22} \end{pmatrix} = \begin{pmatrix} \cos\psi & \sin\psi \\ -\sin\psi & \cos\psi \end{pmatrix} \begin{pmatrix} d_{rr} & d_{rz} \\ d_{rz} & d_{zz} \end{pmatrix} \begin{pmatrix} \cos\psi & -\sin\psi \\ \sin\psi & \cos\psi \end{pmatrix} , \tag{3.6}$$

$$\begin{pmatrix} d_{rr} & d_{rz} \\ d_{rz} & d_{zz} \end{pmatrix} = \begin{pmatrix} \cos\psi & -\sin\psi \\ \sin\psi & \cos\psi \end{pmatrix} \begin{pmatrix} d_{11} & d_{12} \\ d_{12} & d_{22} \end{pmatrix} \begin{pmatrix} \cos\psi & \sin\psi \\ -\sin\psi & \cos\psi \end{pmatrix} . \tag{3.7}$$

For the regimes EF and AB, the strict inequality $\sigma_I > \sigma_{II} > \sigma_{III}$ holds. In these circumstances, the theory proposed in Spencer [3] postulates that deformation occurs by shearing in the planes of the principal axes which correspond to σ_I and σ_{III}. For definiteness consider the regime AB, in which $\sigma_I = \sigma_1$ and $\sigma_{III} = \sigma_\theta$. Then deformation occurs by shear in the planes of $\underset{\sim}{e}_1$ and $\underset{\sim}{e}_\theta$, and so $d_{12} = 0$ and $d_{22} = 0$. Hence, from (3.3) and (3.6),

$$\sin 2\psi \left(\frac{\partial v_r}{\partial r} + \frac{\partial v_z}{\partial z}\right) - \cos 2\psi \left(\frac{\partial v_r}{\partial z} + \frac{\partial v_z}{\partial r}\right) = 0 , \tag{3.8}$$

$$\left(\frac{\partial v_r}{\partial r} + \frac{\partial v_z}{\partial z}\right) - \cos 2\psi \left(\frac{\partial v_r}{\partial r} - \frac{\partial v_z}{\partial z}\right) - \sin 2\psi \left(\frac{\partial v_r}{\partial z} + \frac{\partial v_z}{\partial r}\right) = 0 . \tag{3.9}$$

Equation (3.8) shows that for these regimes the principal axes of the stress and rate of deformation tensors coincide, even though this coaxiality does not hold in general in the 'double-shearing' theory.

Equations (3.1), (3.8) and (3.9) may be replaced by the simpler equivalent system

$$\begin{aligned} &\frac{\partial v_r}{\partial r} + \cos^2 \psi \frac{v_r}{r} = 0 , \\ &\frac{\partial v_z}{\partial z} + \sin^2 \psi \frac{v_r}{r} = 0 , \\ &\frac{\partial v_r}{\partial z} + \frac{\partial v_z}{\partial r} + \sin 2\psi \frac{v_r}{r} = 0 . \end{aligned} \tag{3.10}$$

By eliminating ψ from (3.8) and (3.9), it follows that

$$\left(\frac{\partial v_r}{\partial r} + \frac{\partial v_z}{\partial z}\right)^2 = \left(\frac{\partial v_r}{\partial r} - \frac{\partial v_z}{\partial z}\right)^2 + \left(\frac{\partial v_r}{\partial z} + \frac{\partial v_z}{\partial r}\right)^2 . \tag{3.11}$$

The problem is kinematically determined in the sense that (3.1) and (3.11) are two equations for the velocity components v_r and v_z, which do not involve the stress. It appears, however, to be preferable to regard (3.10) as a system of equations for v_r, v_z and ψ. Given a solution of these equations, the corresponding stress is determined by (2.1), (2.3), (2.5) and (2.6).

The analysis for the regime EF leads to similar results, except that ψ is replaced by $\psi+\frac{1}{2}\pi$.

Cox, Eason and Hopkins [4] formulated velocity equations for the regimes AB and EF (among others) on the assumption that the material deforms according to the flow rule associated with the Coulomb-Mohr yield condition. In their theory (3.1) is replaced by the equation

$$\frac{\partial v_r}{\partial r} + \frac{1 \mp \sin\phi}{1 \pm \sin\phi} \frac{v_r}{r} + \frac{\partial v_z}{\partial z} = 0 , \tag{3.12}$$

where the upper and lower signs are selected according to whether v_r is positive or negative.

An important feature of equations (3.10) is that they do not involve the angle of internal friction ϕ, and so the deformation is independent of ϕ (however, ϕ enters into the determination of the stress through (2.3) or (2.4)). In particular, (3.10) are identical to the equations which occur, in this regime, in the analysis of deformation of a frictionless material, as in metal plasticity. These equations were formulated, in the metal plasticity context, by Shield [6]. Cox, Eason and Hopkins [4] showed that they form a hyperbolic system. Lippmann [7,8] analysed the equations in detail, derived the characteristic relations and outlined numerical methods of solution. Lippmann's analysis may be carried over in its entirety to the present case, the only change necessary being to employ (2.3) or (2.4) in place of the corresponding equations with $\phi = 0$. Lippmann [7,8] also derived two exact solutions of (3.10); we consider these and their corresponding stress fields in the next two sections.

4 EXTENSION, RADIAL CONTRACTION AND TELESCOPIC SHEAR OF A TUBE

By considering solutions in which v_r and ψ depend only on r, Lippmann [8] obtained the following solution of (3.10):

$$v_r = -A\left(\frac{d^2+r^2}{r}\right) ,$$

$$v_z = 2Az - 2\sqrt{2}Ad\left[\ell n\left\{\frac{d-(d^2-r^2)^{\frac{1}{2}}}{r}\right\} + \frac{(d^2-r^2)^{\frac{1}{2}}}{d}\right] + e , \tag{4.1}$$

$$\cos\psi = \left(\frac{d^2-r^2}{d^2+r^2}\right)^{\frac{1}{2}} , \qquad \sin\psi = -\left(\frac{2r^2}{d^2+r^2}\right)^{\frac{1}{2}} ,$$

where A, d and e are constants, and for regime AB, A is positive (a similar solution holds for regime EF with A negative and ψ replaced by $\psi+\frac{1}{2}\pi$). The

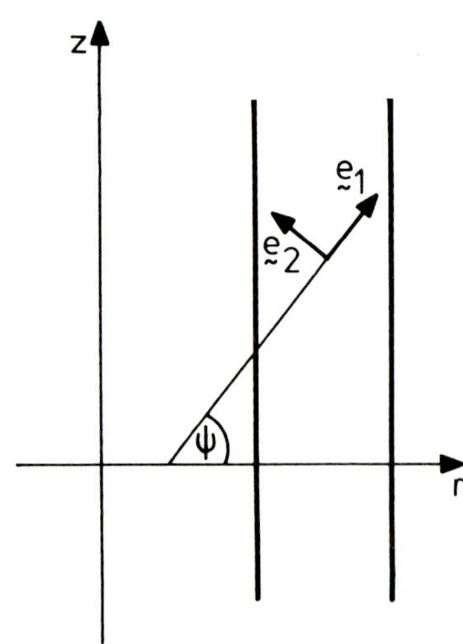

Fig. 2 Deformation of a circular cylinder

deformation (4.1) represents an axial extension and radial contraction, accompanied by a telescopic shear, of the circular cylindrical tube which is illustrated in Fig. 2.

For the associated stress field we assume for simplicity that $b_r = f_r = f_z = 0$, and that b_z is constant. The extension to the case in which $b_r - f_r$ and $b_z - f_z$ are functions of r is straightforward. Following Lippmann [8] we seek solutions in which the stress depends only on r. Then (2.6) gives

$$\sigma_{rz} = \frac{B}{r} - \tfrac{1}{2}\rho r b_z , \tag{4.2}$$

where B is constant.

From (2.1) it follows that

$$\sigma_r = \sigma_1 - \sigma_{rz}\tan\psi , \qquad \sigma_z = \sigma_1 - \sigma_{rz}\cot\psi , \tag{4.3}$$

and from (2.3) that

$$\sigma_\theta = \sigma_1 \frac{1+\sin\phi}{1-\sin\phi} - \frac{2c\cos\phi}{1-\sin\phi} . \tag{4.4}$$

Hence the equilibrium equation (2.5) can be expressed in the form

$$\frac{d\sigma_1}{dr} - \frac{2\sigma_1\sin\phi}{r(1-\sin\phi)} = \frac{1}{r}\frac{d}{dr}(r\sigma_{rz}\tan\psi) - \frac{2c\cos\phi}{r(1-\sin\phi)} .$$

The solution of this equation is

$$\sigma_1 = c\cot\phi + Dr^n + \sigma_{rz}\tan\psi + (n+1)r^n \int r^{-n-1}\sigma_{rz}\tan\psi\, dr , \tag{4.5}$$

where D is constant and

$$n = 2\sin\phi/(1-\sin\phi) . \tag{4.6}$$

Then σ_r, σ_z and σ_θ are given by (4.3) and (4.4). The admissible values of the constants B and D are restricted by the requirement that $\sigma_2 > \sigma_\theta$ throughout the deforming region.

5 FLOW BETWEEN CONCENTRIC SPHERICAL SURFACES

The other exact solution of (3.10) described by Lippmann [7,8] is

$$v_r = -Bz(r^2+z^2)^{\frac{1}{2}}/r , \qquad v_z = B(r^2+z^2)^{\frac{1}{2}} , \tag{5.1}$$

$$\sin\psi = r/(r^2+z^2)^{\frac{1}{2}} , \qquad \cos\psi = -z/(r^2+z^2)^{\frac{1}{2}} , \tag{5.2}$$

where B is a constant which is positive for regime AB. A similar solution holds for regime EF with B negative and ψ replaced by $\psi+\frac{1}{2}\pi$. We consider the case $B > 0$. The solution is more conveniently referred to spherical polar coordinates R,Θ,θ, where

$$r = R\sin\Theta , \qquad z = R\cos\Theta . \tag{5.3}$$

If v_R and v_Θ denote velocity components in the spherical polar coordinates, then (5.1) and (5.2) may be expressed as

$$v_R = 0\,, \qquad v_\Theta = -BR/\sin\Theta\,, \qquad \psi = \pi - \Theta\,. \tag{5.4}$$

The solution is illustrated in Fig. 3.

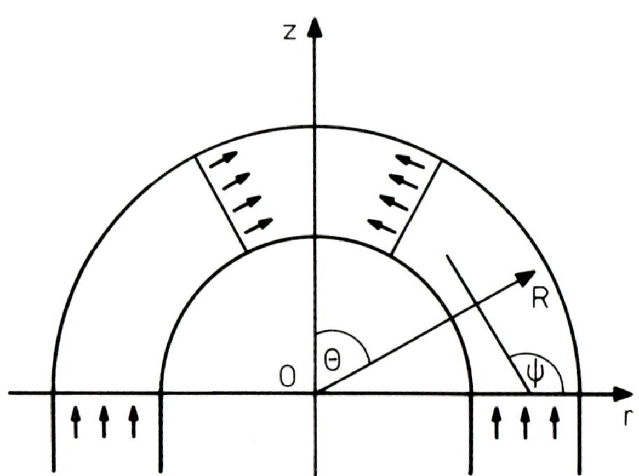

Fig. 3 Flow between concentric spherical surfaces

The velocity is everywhere directed tangentially to the spherical surfaces R = constant, along the meridional curves and towards the North poles. The solution therefore represents a possible converging flow between two concentric spherical surfaces. In metal plasticity, Lippmann employed this solution in an analysis of the tube-nosing operation. For granular materials, this application is perhaps artificial, but the solution is nevertheless of interest as one of a very small number of known exact solutions.

From $(5.4)_3$, the σ_1 principal stress direction is tangential to the meridional curves R = constant, and so the R and Θ directions are the principal stress directions in the meridional planes. If σ_R and σ_Θ denote the corresponding principal stress components, then

$$\sigma_1 = \sigma_\Theta\,, \qquad \sigma_2 = \sigma_R\,, \tag{5.5}$$

and (2.3) becomes

$$\sigma_\Theta(1+\sin\phi) = 2c\cos\phi + \sigma_\theta(1-\sin\phi)\,. \tag{5.6}$$

If body and inertia forces are neglected, the equilibrium equations reduce to

$$\frac{\partial\sigma_R}{\partial R} + \frac{2\sigma_R - \sigma_\Theta - \sigma_\theta}{R} = 0\,, \qquad \frac{\partial\sigma_\Theta}{\partial\Theta} + (\sigma_\Theta - \sigma_\theta)\cot\Theta = 0\,. \tag{5.7}$$

The general solution of (5.6) and (5.7) is

$$\sigma_R = c\cot\phi + \frac{2g(R)}{R^2\cos^2\phi}\sin^n\Theta + \frac{1}{R^2}h(\Theta) ,$$

$$\sigma_\Theta = c\cot\phi + \frac{g'(R)}{R(1+\sin\phi)}\sin^n\Theta , \tag{5.8}$$

$$\sigma_\theta = c\cot\phi + \frac{g'(R)}{R(1-\sin\phi)}\sin^n\Theta ,$$

where the constant n is again defined by (4.6) and g(R), h(Θ) are arbitrary functions of their arguments which must be such that $\sigma_\Theta > \sigma_R > \sigma_\theta$. The functions g(R) and h(Θ) are determined by suitable boundary conditions. If, for example, as in Lippmann [7,8], we specify that

$$\sigma_\Theta = 0 \quad \text{when} \quad \Theta = \Theta_0 , \qquad \text{and} \qquad \sigma_R = -P \quad \text{when} \quad R = R_0 , \tag{5.9}$$

then (5.8) become

$$\sigma_R = -\frac{R_0^2}{R^2}P - \left(\frac{R_0^2}{R^2} - 1\right)\left(1 - \frac{1}{1-\sin\phi}\,\frac{\sin^n\Theta}{\sin^n\Theta_0}\right)c\cot\phi ,$$

$$\sigma_\Theta = \left(1 - \frac{\sin^n\Theta}{\sin^n\Theta_0}\right)c\cot\phi , \qquad \sigma_\theta = \left(1 - \frac{1+\sin\phi}{1-\sin\phi}\,\frac{\sin^n\Theta}{\sin^n\Theta_0}\right)c\cot\phi , \tag{5.10}$$

where the admissible range of values of P is restricted by the requirement that σ_R is the intermediate principal stress throughout the deforming region.

ACKNOWLEDGEMENT

The author thanks the University of Queensland for a Visiting Professorship in the Department of Mathematics, during the tenure of which the work described in this paper was carried out.

REFERENCES

1 J. Mandel, Sur les lignes de glissement et le calcul des déplacements dans la déformation plastique, C. r. Acad. Sci. Paris, 225 (1947), 1272-1273.
2 A. J. M. Spencer, A theory of the kinematics of ideal soils under plane strain conditions, J. Mech. Phys. Solids, 12 (1964), 337-351.
3 A. J. M. Spencer, Deformation of ideal granular materials, in H. G. Hopkins and M. J. Sewell (Eds.), Mechanics of Solids: the Rodney Hill Anniversary Volume, Pergamon Press, Oxford, 1982, 607-652.
4 A. D. Cox, G. Eason and H. G. Hopkins, Axially symmetric plastic deformations in soils, Phil. Trans. R. Soc. London, A254 (1961), 1-45.
5 R. T. Shield, On Coulomb's law of failure in soils, J. Mech. Phys. Solids, 4 (1955), 10-16.
6 R. T. Shield, On the plastic flow of metals under conditions of axial symmetry, Proc. R. Soc. London, A233 (1955), 267-287.
7 H. Lippmann, Principal line theory of axially-symmetric plastic deformation, J. Mech. Phys. Solids, 10 (1962), 111-122.
8 H. Lippmann, Statics and dynamics of axially-symmetric plastic flow, J. Mech. Phys. Solids, 13 (1965), 29-39.

Mechanics of Granular Materials: New Models and Constitutive Relations, edited by
J.T. Jenkins and M. Satake, 1983
Elsevier Science Publishers B.V., Amsterdam — Printed in The Netherlands

CONSEQUENCES OF A THEORY FOR FLOWING GRANULAR MATERIALS

S. L. Passman
Computational Physics and Mechanics Division II
Sandia National Laboratories
Albuquerque, N. M. 87185
USA

ABSTRACT

The fact that most geological materials are not simple materials leads to possible nonexistence of universal motions. Thus there are difficulties in interpreting the results of experiments. The difficulties are alleviated by detailed analysis of the boundary value problems associated with the experiments. This program is being carried out for flowing granular materials. I discuss the results obtained to date.

MOTIONS OF SIMPLE AND NONSIMPLE MATERIALS

Many ordinary engineering materials are completely characterized in pure mechanics by a single constitutive equation. The simple material, for example, is defined as one in which the stress is a functional of the local deformation history. Elastic and viscoelastic materials and many commonly studied fluids are either special or asymptotic cases of such materials (ref. 1). It is a straightforward task to show that, for materials of this type, homogeneous motions are dynamically possible, that is, they satisfy the equations of motion. Furthermore, they are universal. A universal motion for a given class of materials is one which is dynamically possible, and may be produced by surface tractions alone for every material in that class. The significance of such motions and, in particular of homogeneous motions for simple materials, is that they are useful for materials testing in the sense that measurememt of tractions and deformations on the exterior of the body suffices to entirely characterize the constitutive equation of the body. Most machines designed for materials testing, such as triaxial devices and viscometers, take this fact for granted.

When one looks at geological materials such as soils or rocks, it is immediately obvious that not all such materials are simple. This is apparent for granular materials in that, in addition to the gross deformation, the porosity is a significant kinematical variable. For rocks, a similar statement holds, except that the concept of damage may replace the concept of porosity. I confine most of my subsequent discussion to granular materials or materials with a granular component.

With respect to theory, each case must be examined individually. Generally, however, there is no a priori reason to expect that a homogeneous motion will be a universal solution for a class of nonsimple materials. With respect to experiment, it is often difficult or impossible to produce homogeneous motions in granular materials even in experimental apparatus similar or even identical to those in which homogeneous motions are trivially produced in some other engineering materials. This fact is tacitly recognized in most modern books on soil mechanics (refs. 2,3) and is pointed out explicitly by Cowin (ref. 4).

It then follows that some material tests on granular materials are not well characterized by measurements of tractions and deformations on the exterior of the tested body. Thus, measurement of these surface effects only does not directly determine the constitutive equations of the material. The surface effects are a consequence not just of the constitutive equation, but also the equations of motion and the interaction of the body with the testing machine. A rather spectacular comparison, for an elastic body--so that homogeneous motions are universal--of an actual constitutive equation to the apparent constitutive equations measured in various types of testing machines, is given by James (ref. 5).

Since, then, conventional testing methods may not be significant for granular materials, the question remains as to what can be done. Squarely blocking the way is a major unsolved problem in mechanics: _Given a physical material, what constitutive equation or class of constitutive equations should we assign to it?_[a] Until this problem is solved, we are pretty much stuck with the traditional technique--roughly, trial and error. For this technique, one guesses a putative constitutive equation, then solves a number of boundary value problems, in closed form if possible. Subsequently, one does experiments corresponding to the boundary value problems. Agreement of experiment with theory, of course, does _not_ "verify" or "validate" the theory, for there could exist situations, as yet unexplored, in which the same theory does not match with experiment. However, as the number of experiments which agree with the theory increases, so does one's confidence in the theory. Naturally, disagreement of the theory with a single experiment decreases one's confidence in that theory, as a model for the particular material tested, to nil. This painful process has proved quite successful for the cases which are now considered "classical," _e. g._, linear viscosity and linear elasticity. The process will, of course, be even more painful for nonsimple granular materials because of the difficulty of dealing with them theoretically and because matching experimental and theoretical results may require observation of deformation and porosity

---

[a]The idea persists in some quarters that kinetic/statistical theories are appropriate for solving this problem. That could be the case. However, in the cases of which I am aware in which this has been tried, the answers have been assumed, not deduced.

fields throughout the whole body in an experiment, rather than just on a surface. However, the process has been initiated. Observations of this type are being carried out by Savage (ref. 6). I report on the theoretical aspects to date, with respect to a particular constitutive model, in the balance of this paper.

A most interesting and appealing theory for flowing granular materials was presented a number of years ago by Goodman and Cowin (refs. 7,8). This is not a simple material, does not generally have homogeneous motions as universal solutions and, roughly speaking, models packing but not orientation fabric in the sense explained by Oda (ref. 9). In response to comparisons with various experiments and various theoretical studies the model has undergone minor modifications. However, its basic structure remains intact, and sufficient theoretical work has been done on it so that its physical appropriateness can now be tentatively examined.

THEORETICAL RESULTS FOR DRY GRANULAR MATERIALS

Nunziato, Passman and Thomas (ref. 10)[a] and Passman, Nunziato, Bailey and Thomas (ref. 11)[b] have solved the field equations for a slight modifications of the theory of Goodman and Cowin (refs. 7,8) obtaining velocity, solid volume fraction, and pressure fields for the following boundary value problems:

1. Simple shearing,
2. Flow in a vertical channel under the influence of gravity,
3. Flow down an inclined plate under the influence of gravity.

The constitutive model is much like that of a linearly viscous fluid, except that it allows for spatial variation of volume fraction. Nonetheless, the velocity profiles found are quite different from those of a linearly viscous fluid subject to the same geometric boundary conditions and body forces. Generally speaking, the material exhibits very high shear rates near solid boundaries, and prefers to flow nearly as a rigid body, that is, at very low shear rates, elsewhere. The high shear rates are accompanied by low solid volume fractions and _vice versa_.[c] For the cases of simple shearing and flow in a vertical channel, normal forces are exerted on the walls as a result

[a,b]There is an internal sign error in the numerical solution of two of these problems which renders some of the graphs given in these two papers qualitatively incorrect. A list of corrections is forthcoming.

[c]The term "dilatancy" is sometimes used to describe this phenomenon. It is also used to describe various other phenomena.

of the flow.[a] In pure fluid mechanics, normal forces of this type are not exerted by linearly viscous fluids, but are exerted by almost all nonlinearly viscous fluids. Other granular material models have been proposed (refs. 12, 13) which exhibit normal forces.

Of a fundamentally different nature from the shearing flows discussed above are elongational flows, as exemplified by uniaxial and triaxial creep tests. It is not clear in the context of fluid rheology that, given a fluid with a constitutive equation which is known to accurately describe that fluid in shearing flows, whether the same constitutive equation suffices to describe the fluid in elongational flows. That may well also be case for the theory of Goodman and Cowin with respect to granular materials. However, the matter is worth pursuing. Passman and Trucano (ref. 14) have worked out the case of one-dimensional creep in detail. Here, in some cases, homogeneous motions can exist. If inertial effects are neglected, the solid volume fraction is governed by a nonlinear ordinary differential equation. Closed form solution of this equation involves analytical difficulties which can be proven to be insuperable. However, it may be analyzed qualitatively by standard techniques from the theory of ordinary differential equations. The equations are sufficiently rich so that a number of different phenomena may occur, depending on the sign and magnitude of the load compared to various material constants and the initial state of the body. In particular, the body may exhibit asymptotically stable and unstable equilibrium states. Thus, four types of behavior are immediate, depending on the loading and material parameters and initial state: the body, if started in an unstable equilibrium configuration, will stay in that configuration; otherwise, it may evolve toward a stable configuration; it may evolve toward a maximal packing; or it may fail. In each case, the body evolves toward a constant deformation rate ("strain rate"). The last three cases are stable in senses which may be made precise. However, that is not all. There are conditions under which the solution to the equation for the volume fraction <u>ceases to exist</u> after a finite time. This does not mean that the theory is invalid. Only the homogeneous solution ceases to exist after a finite time. Thus some sort of localization of deformation must occur. Such phenomena are known in other contexts and have been observed in soils. Study of the relevant nonlinear partial differential equation has just begun.

---

[a]The term "dilatancy" is sometimes used to describe this phenomenon.

THEORETICAL RESULTS FOR WET GRANULAR MATERIALS

Generally, constructing theoretically sound mixture theories is a complicated task, even for ordinary continua (ref. 15). In the case where additional kinematical variables, such as the volume fraction, must be taken into account, account, the task is even more difficult. There are a number of papers (refs. 16, 17, 18) in which the appropriate mixture theory is developed from the theory of Goodman and Cowin. Of the theories of "mixtures of structured continua", this particular theory is probably the best developed.

Boundary value problems for nontrivial mixture theories always present difficulties because, even in their linear forms, they are so complicated. Thus, closed form solutions are usually out of the question. Not only that, but the large number of material parameters makes it impractical to do numerical parameter studies for the full physical ranges of all of the pertinent dimensionless variables, as was done in the problems discussed here for dry granular materials. Thus, assignments must be made for the constitutive constants whose values are known, and order-of-magnitude estimates must be made for the others so that meaningful parameter studies may be carried out.

A number of boundary value problems have been looked at in this context, including some for reacting continua (ref. 19). Of particular interest are two shearing flow problems: flow in a pipe (ref. 20), and simple shearing (ref. 21). Of these only preliminary results exist for the first. The second has been looked at in some detail for a sand-water mixture. For this mixture, since the theory places no restrictions on concentrations, everything from flow in a packed bed to flow of a dilute suspension may be handled by varying the material constants without changing the forms of the equations. In brief, we find that the solid component tends to flow as described in the solution to the problem for shearing of a dry granular material while the fluid prefers a linear velocity profile. Increasing the drag coefficient tends to modify each velocity profile towards the other while decreasing it sufficiently gives essentially independent flows. Imposing no relative motion between the constituents, as is sometimes done in more elementary theories, corresponds to a drag coefficient so large that it cannot be rationalized physically.

ACKNOWLEDGEMENTS

The work reported here was done in cooperation with P. B. Bailey, J. W. Nunziato, K. W. Reed and T. G. Trucano. It was supported by the United States Department of Energy under contract DE-AC04-76DP00789 to Sandia National Laboratories.

REFERENCES

1 C. Truesdell and W. Noll, The Non-Linear Field Theories of Mechanics, Handbuch der Physik, III/3, Springer-Verlag, Berlin, 1965.

2 T. W. Lambe and R. V. Whitman, Soil Mechanics, SI Version, John Wiley, New York, 1979.

3 S. C. Cowin and M. Satake (Eds.), Proceedings of the U. S.-Japan Seminar on Continuum Mechanical and Statistical Approaches in the Mechanics of Granular Materials, Gakujutsu Bunken Fukyu-Kai, Tokyo, 1978.

4 S. C. Cowin, Microstructural Continuum Models for Granular Materials, (ref. 3 above), 162-170.

5 R. D. James, Co-Existent Phases in the One-Dimensional Static Theory of Elastic Bars. Arch. Rat. Mech. Anal., 72 (1979), 99-140.

6 S. B. Savage, Gravity Flow of Cohesionless Granular Materials in Chutes and Channels, J. Fluid Mech., 92 (1979), 53-96.

7 M. A. Goodman and S. C. Cowin, Two Problems in the Gravity Flow of Granular Materials, J. Fluid Mech., 45 (1971), 321-339.

8 M. A. Goodman and S. C. Cowin, A Continuum Theory for Granular Materials, Arch. Rat. Mech. Anal., 44 (1972), 249-266.

9 M. Oda, Significance of Fabric in Granular Materials, (ref. 3 above), 7-26.

10 J. W. Nunziato, S. L. Passman and J. P. Thomas, Jr., Gravitational Flows of Granular Materials with Incompressible Grains, J. Rheology, 24 (1980), 395-420.

11 S. L. Passman, J. W. Nunziato, P. B. Bailey and J. P. Thomas, Jr., Shearing Flows of Granular Materials, J. Engr. Mech. Div., ASCE, 106 (1980), 773-783.

12 D. F. McTigue, A Nonlinear Constitutive Model for Granular Materials: Applications to Gravity Flow, J. Appl. Mech., 49 (1982), 291-296.

13 S. B. Savage and D. J. Jeffrey, The Stress Tensor in a Granular Flow, J. Fluid Mech., 110 (1981), 255-272.

14 S. L. Passman and T. G. Trucano, A Theory Describing Creep, Fracture and Instability in Rocks, to appear.

15 C. Truesdell, Rational Thermodynamics, McGraw-Hill, New York, 1969.

16 S. L. Passman, Balance Laws for Mixtures of Granular Media, 169-185 in J. F. Hutton, J. R. A. Pearson and K. Walters (Eds.) Advances in Theoretical Rheology, Applied Science Publishers, Barking, England, 1975.

17 S. L. Passman, Mixtures of Granular Materials, Intl. J. Engr. Sci., 15, 1977, 117-129.

18 J. W. Nunziato and E. K. Walsh, On Ideal Multiphase Mixtures With Chemical Reactions and Diffusion, Arch. Rat. Mech. Anal., 73, 1980, 285-311.

19 M. E. Kipp, J. W. Nunziato, R. E. Setchell and E. K. Walsh, Hot Spot Initiation of Heterogeneous Explosives, Proceedings of the 7th Symposium (International) on Detonation, to appear.

20 S. L. Passman, J. W. Nunziato and P. B. Bailey, Flow of a Suspension in a Vertical Circular Tube. Abstract only appears in J. Rheology, 26 (1982), 85.

21 S. L. Passman, J. W. Nunziato, P. B. Bailey and K. W. Reed, Shearing Flow of a Suspension, to appear.

Mechanics of Granular Materials: New Models and Constitutive Relations, edited by
J.T. Jenkins and M. Satake, 1983
Elsevier Science Publishers B.V., Amsterdam — Printed in The Netherlands

GRANULAR FLOWS DOWN ROUGH INCLINES - REVIEW AND EXTENSION

Stuart B. Savage
Department of Civil Engineering and Applied Mechanics
McGill University, Montreal H3A 2K6, Canada

ABSTRACT

Previous continuum and microstructural models that have been applied to predict the flows of dry, cohesionless granular materials down inclined surfaces are reviewed and discussed in the light of experimental information on such flows. A new model which adds a rate-independent stress contribution to the dynamic stresses given by the theory of Jenkins and Savage is proposed and applied to the chute flow problem. Comparisons of predicted velocity distributions for various bed inclinations show reasonable agreement with the experimental measurements of Ishida and Shirai. The theory involves no empirical fitting of constitutive coefficients or empirical information other than the values of the usual material properties.

INTRODUCTION

The free-surface flow of particulate solids down an incline is a fascinating example of granular flow that not only has a number of engineering applications but also takes place in several types of natural phenomena. Inclined chutes are used in materials handling engineering for the transportation of bulk solids and for mineral and powder processing. Rock falls, debris flows, snow avalanches, and sub-aqueous grain flows are instances of granular flows down inclines that occur in the context of geophysics. Furthermore, for the theoretician concerned with constitutive modelling, the inclined chute flow can serve as a viscometric flow. In this role the chute flow is particularly discriminating since small changes in the constitutive modelling often result in significant changes in overall flow field properties. Despite its apparent simplicity, the chute flow, in fact, embodies a surprising complexity. And, in my view, an entirely satisfactory theoretical solution, which at least displays the primary features of the observed flow behaviour, has not yet been proposed.

In the present paper, I shall regard the chute flow in the main as a test flow and use it to examine the accuracy and/or completeness of various constitutive theories. The paper begins with a brief outline of the essential constitutive behaviour for rapid granular flows as obtained from some recent annular shear cell experiments. Previous experimental observations and theoretical analyses of chute flows are briefly but critically reviewed, the intention being to pinpoint important aspects of each theoretical formulation that are either inappropriate or unnecessary, or that might on the other hand, be absent from the constitutive theory. A new and simple proposal for the constitutive equations is made and applied to predict the granular flow down a rough incline. The predicted velocity profiles are compared with the experiments of Ishida and Shirai (ref. 1).

CONSTITUTIVE BEHAVIOUR FROM SHEAR CELL TESTS

At the last U.S.-Japan Seminar I reported on some preliminary stress-shear rate tests (ref. 2) performed with an annular shear cell apparatus. Since that time a considerable amount of experimental data has been obtained with this device, primarily by one of my Ph.D. students M. Sayed (ref. 3,4,5). These more extensive and more accurate tests have clarified some of the anomalous behaviour observed in the preliminary tests (ref. 2). Tests have been carried out with a variety of dry granular materials. To study particle density and diameter dependence effects, monosized spherical particles of glass ballotini and polystyrene were used with the nominal diameter varied between tests. Binary mixtures of particles of different sizes were used to investigate size distribution effects. To study the effects of particle angularity, experiments were performed using crushed walnut shells.

By rotating the lower half of the annular shear cell with the upper half fixed an apparent shear rate $\bar{u}/H$ was developed in the granular material contained within, where $\bar{u}$ is the linear velocity of the lower disk at the mid-annulus radius and H is the height of granular material contained between the roughened shear walls of the apparatus. The apparatus can determine normal stress and shear stress as functions of shear-rate and solids fraction ν.

Although differences occurred in these stress-strain-rate tests from material to material, the various sets of data all showed the same general trends. Some typical results are shown in Figure 1 for a bimodal mixture of polystyrene beads (30% by weight of 0.55 mm spheres and 70% of 1.65 mm spheres). The plot shows nondimensional normal stress $\tau_{22}/\rho_p g\sigma$ versus nondimensional apparent shear rate $(\sigma/g)^{\frac{1}{2}}\ \bar{u}/H$ for constant values of solids fraction ν, where ρ_p is the individual particle mass density, σ is the mean particle diameter and g is the gravitational acceleration. Compressive stresses are taken as positive. At lower concentrations, the nondimensional stress versus shear-rate plotted on log-log paper has a slope of 2, i.e. the stress varies as the square of shear-rate, consistent with the predictions of Bagnold (ref. 6) in his <u>grain-inertia</u> flow regime. There are strong increases in stresses with relatively slight increases in solids concentration. At higher concentrations the curves tend to flatten out, with a slope of less than 2, as the shear-rate is decreased. A very similar behaviour was observed for the shear stresses.

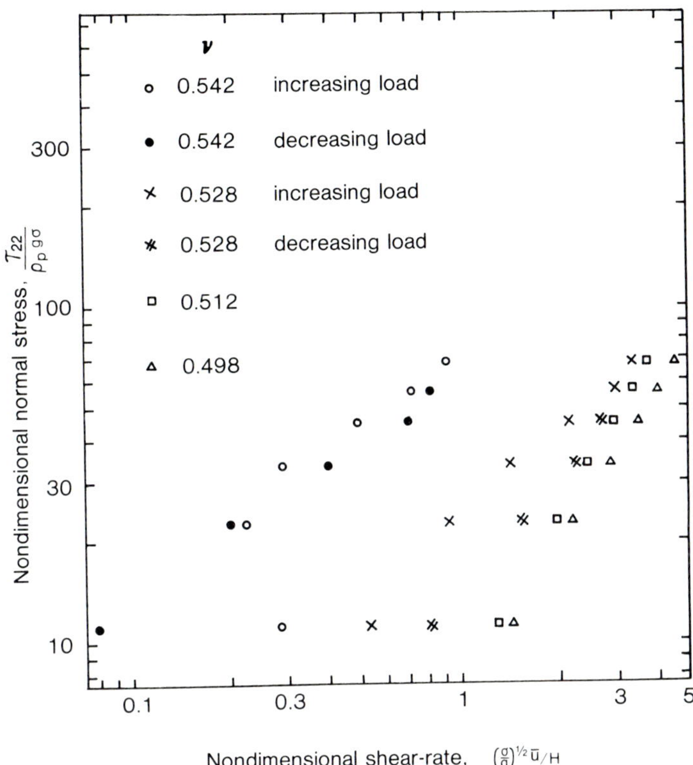

Fig. 1. Shear cell data for bimodal mixture of polystyrene beads (30% by weight of 0.55 mm spheres and 70% 1.65 mm spheres).

It was found that by making a least-square fit to each set of data at a particular value of ν, it is possible to represent the total measured normal stress (or shear stress) at a shear-rate du_1/dx_2 as the sum of two parts,

$$\tau_{total} = \underbrace{\tau(\nu)_{Coulomb}}_{\text{Rate-independent dry friction part}} + \underbrace{\mu(\nu)\,\rho_p\sigma^2\left(\frac{du_1}{dx_2}\right)^2}_{\text{Rate-dependent viscous part}} \qquad (1)$$

We may regard the 'dynamic' or 'viscous' contribution to be a result of momentum transfer during collisions between particles roughly as described by Bagnold (ref. 6). The rate-independent part is due to dry Coulomb friction and particle overriding during enduring contacts. The stress contribution $\tau_{Coulomb}$ and the coefficient μ are monotonic increasing functions of ν.

At high concentrations and low shear-rates the rate-independent term is dominant, whereas at low concentration and high shear rates this term is negligible and the collisional transfer term prevails.

A cautionary note. In the least-square curve fits for these particular tests, we chose to represent $\tau_{Coulomb}$ by regarding it to be a function of ν. It is important to recognize that $\tau_{Coulomb}$ may not be a unique function of ν. The $\tau_{Coulomb}$ dependence upon ν may only be a parametric one for these particular tests in the sense that ν is not really the relevant variable and variations with ν are only one reflection of changes in some other more significant variable.

From the test data we can determine the ratio of shear stress to normal stress τ_{12}/τ_{22}, which we define as tan ϕ_D, where ϕ_D is the 'dynamic friction angle'. Figure 2 shows typical results for tan ϕ_D plotted versus the non-dimensional apparent shear-rate for various concentrations. Note that the dynamic friction angle ϕ_D is not greatly different from the quasi-static internal friction angle ϕ. For a given ν, the dynamic friction angle is only a weak function of shear-rate. There is, however, a small but noticeable decrease in ϕ_D with increasing ν. Perhaps more to the point is the observation that ϕ_D increases as the rate-independent term becomes a more significant part of the total stress.

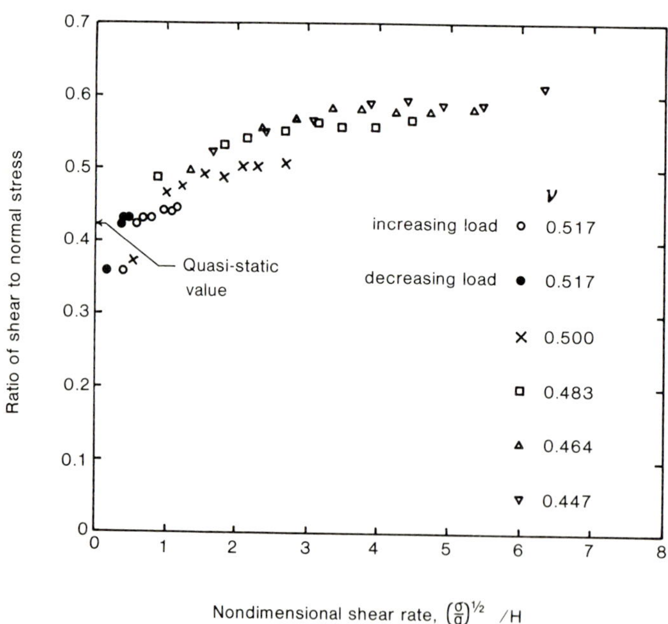

Fig. 2. Dynamic friction angle for 1.32 mm polystyrene beads.

INCLINED CHUTE FLOW EXPERIMENTS

Experiments on granular flows down inclines have been discussed recently by Savage, Nedderman, Tuzun and Houlsby (ref. 7) as a part of a more general review of flows of granular materials. Here we mention only those experiments in which the velocity profiles over the depth were either measured directly or inferred from other measurements.

First we note that for the steady, two-dimensional, fully developed, free-surface granular flow down a rough plane inclined at an angle ζ as shown in Figure 3, the normal and shear stresses on a plane parallel to the bed are given by the linear momentum equations as

$$\tau_{22} = g \cos \rho \int_0^{x_2} \rho dx_2 \tag{2}$$

$$\tau_{12} = g \sin \rho \int_0^{x_2} \rho dx_2 \tag{3}$$

where ρ is the bulk density $\rho = \nu\rho_p$. From equations (2) and (3) we see that

$$\frac{\tau_{12}}{\tau_{22}} = \tan \zeta = \text{const.} \tag{4}$$

throughout the depth. For such a flow to exist, the constitutive behaviour of the material, and in particular the dynamic friction angle ϕ_D, must be consistent with (4) such that $\phi_D = \zeta$ throughout the depth.

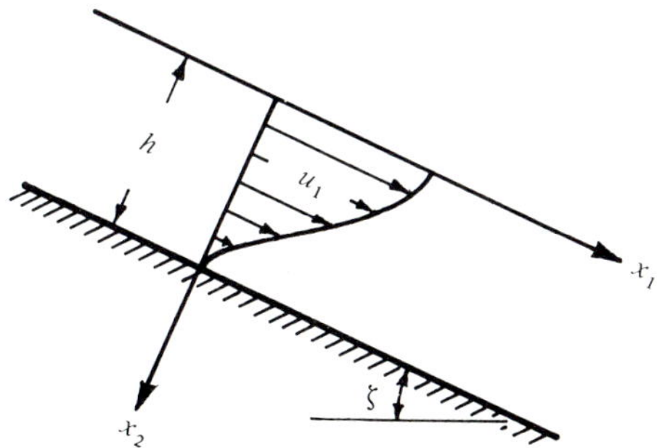

Fig. 3. Two-dimensional gravity flow of granular material down an inclined plane.

Velocity profiles for such flows have been experimentally determined by Augenstein and Hogg (ref. 8), Bailard (ref. 9), Savage (ref. 10), Ishida and Shirai (ref. 1) and Ishida, Hatano and Shirai (ref. 11). These experiments indicated that steady fully developed, constant depth flows down inclines having a roughness at least of the size of the flowing particles were possible only over a small range of bed inclination angles between the angle of repose ϕ_r and some upper limit ϕ_M, i.e. for flow $\phi_r<\zeta<\phi_M$. For $\zeta>\phi_M$ the flow will accelerate down the chute.

Augenstein and Hogg (ref. 8) and Bailard (ref. 9) determined the velocity profiles from the trajectories of particles leaving the end of the chute. In both of these tests the velocity profiles were found for the most part to be blunt or at least of a convex shape. The bulk density profiles were low near the bed and the upper surface and reached a maximum around mid-depth.

On the other hand, Savage (ref. 10) and Ishida and Shirai (ref. 1), who both measured the velocities directly with fibre-optic probes, found the velocity profiles in general to be concave in shape. At the higher slopes, the profiles of Ishida and Shirai tended to become triangular in shape. From visual observations through the glass side wall in Savage's (ref. 10) tests, it appeared that the bulk density increased monotonically with depth.

In experiments of granular chute flows that were fluidized by an upward air flow through the channel bed, Ishida, et al. (ref. 11) were able to discern 5 distinct types of flow patterns depending upon bed slope and fluidizing air flow velocity. With zero fluidizing air flow they observed three types of granular flows:

1. Immature Sliding Flow. This occurs when the bed slope ζ is close to the angle of repose of the material. There can be a stationary layer on top of the bed, thicker on the upstream than on the downstream side. An effective bed surface forms within the material itself such that the inclination of the zero velocity surface is greater than the inclination of the channel itself. The velocities near the bed are quite low and the profile has a concave shape.

2. Sliding Flow. With a small increase in bed inclination ζ, the velocities increase, the dead-flow region disappears, and all the particles are in motion. The velocity profile fills out and tends towards a triangular shape. The free surface is fairly distinct, saltation of particles there may occur but individual particle jumps are weak and infrequent.

3. Splashing Flow. Further increase in bed slope makes the velocity profile more blunt with a higher shear-rate near the bed than near the upper surface. Saltation is vigorous and the top of the flow consists of a low density cloud of particles rather than a distinct free surface. The motions of the saltating

particles and those in the upper layers are affected by air drag which acts to restrain the particle velocities.

It seems plausible that the velocity profiles observed by Augenstein and Hogg (ref. 8) and Bailard (ref. 9) correspond to the splashing type of flow, whereas those observed by Savage (ref. 10) and Ishida and Shirai (ref. 1) correspond to the sliding and immature sliding type.

We shall be concerned in what follows with flows of the sliding and immature sliding type where it seems permissible to neglect air drag and interstitial fluid effects. For later reference, Figure 4 shows velocity profiles for the flow of glass beads (diameters ranging between 0.35 and 0.50 mm) down a sandpaper roughened inclined chute and is taken from the paper by Ishida and Shirai (ref. 1).

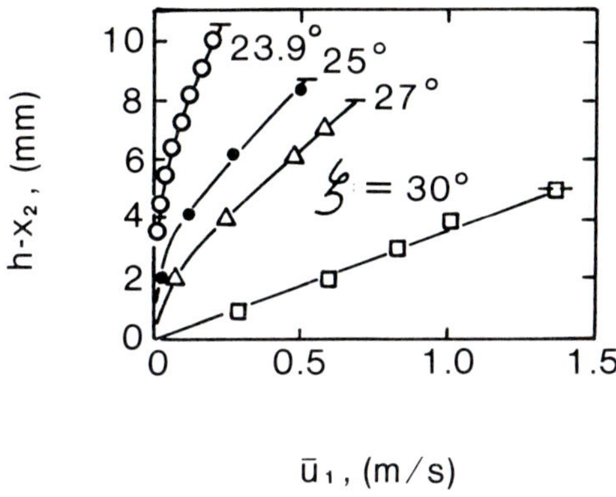

Fig. 4. Velocity profiles for flow of glass beads down inclined channel (after Ishida & Shirai (ref. 1)).

CONSTITUTIVE THEORIES APPLIED TO THE INCLINED CHUTE FLOW PROBLEM

Table 1 summarizes the main characteristics of a number of constitutive theories for granular flow. I have included only analytical models that have specifically been applied to predict granular flows down inclines. Thus, for example, the interesting theory of Shen and Ackermann (ref. 12) does not appear, nor do the computer modelling studies of the two dimensional flows of disk-like particles by Walton (ref. 13) and Campbell (ref. 14). In what follows I shall merely note what appear to me to be weak points in each of these theories related to the determination of the stresses.

TYPE OF THEORY	INVESTIGATORS	STRESS CONTRIBUTIONS		FLUCTUATION ENERGY EQUATION
		Rate-Independent Part	Rate-Dependent Part	
CONTINUUM (Goodman-Cowin type)	Goodman & Cowin (ref. 15)	Mohr-Coulomb-like pressure dependent flow or yield stress. Stresses primarily a function of grad ν, secondary dependence upon ν.	Newtonian, linear dependence upon strain-rate.	NONE
	Savage & Cowin (ref. 10)		Nonlinear, square dependence upon strain-rate, experimentally based ν dependence.	
	Nunziato, Passman, & Thomas (ref. 16)		Linear dependence upon strain-rate experimentally based ν dependence.	
CONTINUUM (using input from microstructural theories & experiments)	McTigue (ref. 17)	Mohr-Coulomb-like stresses primarily dependent upon ν.	Fully nonlinear form suggested by microstructural theory based upon collisional momentum transfer and by shear cell experiments.	NONE
	Sayed & Savage (ref. 18)			
MICROSTRUCTURAL	Bagnold (ref. 6)	NONE	Based on simple collisional momentum transfer auguments; partial neglect of fluctuations.	NONE
	Ogawa, Umemura & Oshima (ref. 20)		Based on collisional momentum calculations for inelastic, rough particles having no rotary inertia. Coupled linear momentum and fluctuation energy equations used for stress determinations.	General form proposed for translational fluctuations; flux terms later neglected
	Savage (ref. 5)		Based upon Jenkins & Savage (ref. 21) theory which considers collisional momentum transfer for smooth inelastic particles. Stress determination coupled to fluctuation energy equation.	Full translational fluctuation energy equation used.

TABLE 1 Summary of the Characteristics of Some Theories Applied to the Calculation of Granular Flows Down Inclined Chutes

Continuum Theories

The first group of theoretical models are Continuum Theories of the Goodman-Cowin (ref. 15) type. The stress tensor is expressed as the sum of two parts; a rate-independent part and a rate-dependent part. The rate-independent part has Mohr-Coulomb-like characteristics in that slip or shear occurs on a plane when there is some specified relationship between the normal and shear stresses acting on the plane. The problem is that the stresses are related to the gradients of solids fraction ν. There seems to be little physical justification for such a proposal. It gives the physically unlikely result that the stresses are necessarily zero when ν is uniform. One can think of other simple boundary value problems for which the theory cannot provide the stresses that must exist and are necessary for equilibrium (ref. 10).

Goodman and Cowin (ref. 15) used a Newtonian form for the rate-dependent part of the stress tensor in which there was a linear dependence upon strain-rate. Nunziato, Passman and Thomas (ref. 16) proposed a similar linear dependence upon strain-rate but used a non-linear, experimentally based ν-dependence. In the forms of the stress tensors proposed by Goodman and Cowin (ref. 15) and Nunziato, et al. (ref. 16), there were no limits on the ratio of shear to normal stresses acting on the shear surfaces. Thus such theories predict that fully developed steady chute flows can occur for <u>any</u> bed inclination angle, whereas they are observed for only a small <u>range</u> of inclination angles. To avoid such difficulties Savage and Cowin (ref. 10) used a rate-dependent stress tensor of the Reiner-Rivlin type in which both normal and shear stresses depended upon the square of the strain-rate. The various constitutive coefficients were chosen so as to limit the ratio of shear to normal stresses acting on the shear surfaces and thus to be capable of simulating the experimentally observed chute flows.

Although the predicted velocity profiles and overall behaviour were qualitatively similar to those observed by Savage (ref. 10), this agreement may have been little more than fortuitous. The doubts about the grad ν dependence of the rate-independent part of the stress tensor that were raised but brushed aside in Savage (ref. 10), are in my present view more serious.

McTigue (ref. 17) and Sayed and Savage (ref. 18) have proposed two very similar continuum theories which avoided the grad ν dependence in the rate-independent part of the stress tensor. They used fully nonlinear forms for the rate-dependent part that were chosen on the basis of experimental data and information obtained from microstructural theories which consider collisional interactions between particles. In the rate-independent part of the stress tensor, a unique relationship between the mean normal stress p and the solids fraction ν was proposed. This kind of $p(\nu)$ relationship is reminiscent of the critical voids ratio concept used in soil mechanics (Roscoe, Schofield and Wroth (ref. 19)). However, the stress levels typically encountered in the context of soil mechanics are far higher than those experienced in granular chute flows. At high stress levels under static conditions, the variations in ν can result from deformations of individual grains which comprise the bulk material. At the low stress levels of typical granular chute flows the individual grains are essentially rigid and the existence of a unique $p(\nu)$ relationship is debatable. Nonetheless, with suitable choices for the constitutive coefficients, these theories can also predict the main qualitative features of experimental chute flows.

Microstructural Theories

Bagnold, in his grain-inertia analysis (ref. 6), derived the first rudimentary form of a microstructural theory for a simple shear flow. By considering the momentum transfer arising from glancing collisions as the particles of one layer overtook those of an adjacent layer, he determined that both shear and normal stresses were proportional to the square of the shear-rate. To solve for the velocity profile developed during steady flow down a rough inclined chute he assumed a uniform solids concentration over the depth. His analysis predicted a simple velocity profile for one bed inclination angle.

As a result of collisions arising in a shear flow, the particles acquire random fluctuating spin and translational fluctuating velocities in addition to their mean translational velocities. The particles are inelastic and frictional, and hence energy is dissipated with each collision. The more complete microstructural theories take specific account of these random fluctuations. To obtain the stresses, Ogawa, Umemura and Oshima (ref. 20) considered the collisional momentum transfer for inelastic, rough particles having negligible rotary inertia. A general form for the translational fluctuation energy was proposed, but the flux terms were neglected in later analyses. The stresses were determined by solving the coupled linear momentum and fluctuation energy equations. Ogawa, et al. (ref. 20) applied this theory to the inclined chute flow problem. This analysis was incomplete in the sense that all the flow field variables were not explicitly determined. They assumed a linear bulk density profile and determined the mean and fluctuating velocity profiles.

Jenkins and Savage (ref. 21) developed a theory for smooth, inelastic spherical particles that was in much the same spirit as the theory of Ogawa, et al. but they treated the collision dynamics and the statistics in a more sophisticated and more accurate manner. The full translational fluctuation energy equation was written down and the collisional flux and dissipation terms were explicitly derived. Again stress and flow fields were determined by the solution of the coupled conservation of mass, linear momentum and fluctuation energy equations. Savage (ref. 5) applied this theory to determine the velocity, solids fraction and fluctuation velocity profiles, developed during steady chute flow. The theory predicted a single triangular velocity profile at one bed inclination angle. For conditions corresponding to the experiments of Ishida and Shirai (ref. 1), it was found that the predicted velocities were similar in magnitude to the experimental measurements for a 30° bed slope (see Figure 4).

Concluding Remarks

The continuum theories described above did not explicitly involve the particle velocity fluctuations. The rate-dependent part of the stress tensor was a specified function of solids fraction ν and strain-rate. In order to obtain solutions for the solids concentration ν and velocity profiles for the chute flows at various bed inclination angles, these theories used a rate-independent part of the stress tensor which was some unique function of ν or its gradients. The realism of such assumptions is questionable. The values of the constitutive coefficients were chosen so as to make the theory give reasonable agreement with the observed chute flow behaviour. The level of agreement achieved does not necessarily demonstrate the general validity of the assumed constitutive behaviour; the procedure followed may in fact be a rather elaborate form of curve fitting, perhaps appropriate for the particular flow studied but lacking in general applicability.

The microstructural models considered collisional fluxes which occurred during binary collisions and the more sophisticated ones took specific account of the velocity fluctuations and the associated collisional energy dissipation. The stresses were found to depend upon ν, the strain-rate and the magnitude of the velocity fluctuations. The stresses for a given flow situation were determined as part of the solution to the coupled conservation of mass, momentum and energy equations. The theoretical models were based upon the assumption of essentially instantaneous binary collisions. Even with such a simplifying assumption the theories can involve considerable complexity. It is far from obvious how one might, within a similar framework, consider collisions between more than two particles and the case of enduring contacts which give rise to the rate-independent kinds of stresses. For this reason the stress tensor obtained from these microstructural theories has only a rate-dependent part. In the chute flow problem, these theories yield only a single velocity profile for one bed inclination angle.

The microstructural models which make use of fluctuation energy equations embody much of the relevant physics for rapid shear flows and they are promising candidates for further developments. However, in order to reproduce the experimentally observed results for chute flows it appears essential that a rate-independent contribution to the stress tensor be included in the theoretical model. In view of the formidable difficulties that are faced in trying to generalize the individual particle interactions to produce such rate-independent effects, it is tempting to seek a much cruder but far simpler way of achieving the desired result. A first attempt at such an approach is described in the next section.

A SIMPLE CONSTITUTIVE PROPOSAL

We begin by recalling the typical theoretical treatment of problems involving granular materials in a state of limiting static equilibrium (see Sokolovski (ref. 22), Spencer (ref. 23)). The material is often assumed to be incompressible and the stress fields are obtained by solving the equations of static equilibrium in conjunction with some Mohr-Coulomb type of yield criterion which relates the normal and shear stress components to one another. A relationship between the mean stress and solids fraction is relatively unimportant and is, in fact, unnecessary to obtain stress solutions. For example, the stresses at some point in a contained mass of static granular material are primarily functions of the surcharge and the overburden pressure and depend far less upon the variations in bulk density above that point.

This suggests a very simple way to incorporate rate-independent stresses into the dynamic flow problems. First, we suppose that the microstructural theories can be used to determine the rate-dependent stress contribution. With increasing ν, the assumptions used in these theories become less valid; however, the dynamic stresses constitute a smaller fraction of the total stress, so the errors involved may not be too great. The fluctuation energy equation determines the magnitude of the velocity fluctuations and hence the 'dynamic' stresses. The linear momentum equations are satisfied in part by total stresses which we assume to be the sum of this rate-independent part and an additional amount which we regard as the rate-independent stress contribution. We take these quasi-static stresses to satisfy a Mohr-Coulomb type yield criterion and a flow rule but we unhook these stresses from any explicit kind of ν-dependence. The rate-independent stresses we have introduced are thus analogous to the 'indeterminate' static pressure which occurs in incompressible fluid mechanics problems.

In physical terms, we might visualize the formation, at the higher concentrations, of clusters or columns of particles which agglomerate and break up in intermittent fashion. Such columns of particles in contact could form force networks which span the full width of the shear region giving rise to rate-independent stresses. In this case there would be not only microscopic fine scale variations in density, velocity, etc. on the order of a particle diameter but also coarse scale temporal and spatial variations on the order of many collision times and several particle diameters. In the very simple model proposed above, involving a simple superposition of dynamic and rate-independent parts of the stress tensor, we regard the stresses to be averaged values, smoothed or smeared out not only on the fine scale but over the coarse scale fluctuations as well.

At the very highest concentrations and lowest shear rates, almost all the particles would be in continuous contact, the inertia forces associated with collisions and particle overriding would be negligible and the stresses are essentially rate-independent.

ANALYSIS OF FLOW DOWN A ROUGH INCLINED PLANE

Governing Equations

The usual equations for the conservation of mass and linear momentum are

$$\frac{d\rho}{dt} = -\rho\nabla\cdot\underline{u} \tag{5}$$

$$\rho\frac{d\underline{u}}{dt} = \rho\underline{g} - \nabla\cdot\underline{p} \tag{6}$$

where $\rho = \nu\rho_p$ is the bulk mass density, $\underline{u} = \langle\underline{c}\rangle$ is the bulk velocity equal to the ensemble average of the individual particle velocities $\underline{c}$, $\underline{g}$ is the gravitational acceleration and $\underline{p}$ is the stress tensor. We assume the stress tensor to be composed of the sum of quasi-static or rate-independent part $\underline{p}_s$ and a collisional momentum flux part $\underline{p}_c$; we neglect any kinetic contribution to the stress tensor coming from diffusion or translation of particles from one shear layer to another, thus

$$\underline{p} = \underline{p}_s + \underline{p}_c \tag{7}$$

The quasi-static stress $\underline{p}_s$ is assumed to be of the form

$$\underline{p}_s = p_o\underline{I} - p_o \sin\phi \frac{\underline{D}}{(\frac{1}{2}\mathrm{tr}\underline{D}\underline{D})^{\frac{1}{2}}} \tag{8}$$

where p_o is the mean quasi-static normal stress, ϕ is the quasi-static friction angle which might in general be a function of ν and $\underline{D}$ is the rate of deformation tensor

$$\underline{D} = \tfrac{1}{2}(u_{i,j} + u_{j,i}) \tag{9}$$

Equation (8) has the form of a Mohr-Coulomb yield criterion and stipulates a coincidence between principal axes of quasi-static stress and strain-rate.

The use of a more general double-shearing model (see Spencer (ref. 23)) instead of (8) might be more correct, but for the present purposes the differences in results are not likely to be significant and we assume equation (8) primarily because of its simplicity.

For the collisional part of the stress tensor we make use of the microstructural theory of Jenkins and Savage (ref. 21) (see Savage (ref. 5) for further discussion and justification for this theory). They obtained the following expressions for the collisional stress $\underset{-}{p}_c$, the collisional flux of translational fluctuation energy $\underset{-}{q}_c$ and the rate of collisional energy dissipation γ,

$$\underset{-}{p}_c = [\frac{\kappa}{\sigma} (\pi T)^{\frac{1}{2}} - \frac{\kappa}{5} (2+\alpha) \text{ tr } \underset{-}{D}] \underset{-}{I} - \frac{2\kappa}{5} (2+\alpha) \underset{-}{D} \tag{10}$$

$$\underset{-}{q}_c = - \kappa \nabla T \tag{11}$$

$$\gamma = \frac{6(1-e)\kappa}{\sigma^2} [T + (\frac{\pi}{4} - \frac{\alpha}{3}) \sigma (\frac{T}{\pi})^{\frac{1}{2}} \text{ tr } \underset{-}{D}] \tag{12}$$

$$\text{where } \kappa = 2\nu^2 (1+e)\ g_o\ \rho_p \sigma\ (T/\pi)^{\frac{1}{2}} \tag{13}$$

g_o is the radial distribution function, σ is the uniform spherical particle diameter, e is the particles' coefficient of restitution, $3T/2 = \langle C^2 \rangle /2$ is the fluctuation specific kinetic energy, $\underset{-}{C} = \underset{-}{c} - \underset{-}{u}$ and α is a numerical coefficient we can take to be unity (Savage (ref. 5)). We take the radial distribution function $g_o(\nu)$ to be that proposed by Carnahan and Starling (ref.24) on the basis of numerical simulations

$$g_o(\nu) = \frac{1}{(1-\nu)} + \frac{3\nu}{2(1-\nu)^2} + \frac{\nu^2}{2(1-\nu)^3} \tag{14}$$

Neglecting kinetic terms, the 'granular temperature' T is governed by the fluctuation energy equation

$$\frac{3}{2} \rho \frac{dT}{dt} = - \underset{-}{p}_c : \nabla \underset{-}{u} - \nabla \cdot \underset{-}{q}_c - \gamma \tag{15}$$

Analysis

Consider the steady fully developed, two-dimensional free-surface flow of cohesionless granular material down a rough plane inclined at an angle ζ to the horizontal as shown in Figure 3. For such a flow, the only variations in flow properties are in the x_2-direction, normal to the inclined plane. Integrating the linear momentum equations (6) yields

$$p_{22} = g \cos \zeta \int_0^{x_2} \rho \, dx_2 \tag{16}$$

$$p_{21} = p_{12} = g \sin \zeta \int_0^{x_2} \rho \, dx_2 \tag{17}$$

and $$\frac{p_{12}}{p_{22}} = \tan \zeta = \text{const.} \tag{18}$$

We make the approximation that

$$\int_0^{x_2} \rho \, dx_2 \simeq \bar{\rho} x_2 = \rho_p \bar{\nu} x_2 \tag{19}$$

where $\bar{\rho}$ and $\bar{\nu}$ are average values of ρ and ν over the depth h. With $\alpha = 1$, the constitutive equation (7) to (13) for this flow reduce to

$$p_{22} = p_{c_{22}} + p_{s_{22}} = \frac{\kappa}{\sigma} (\pi T)^{\frac{1}{2}} + p_o \simeq g \cos \zeta \, \rho_p \, \bar{\nu} \, x_2 \tag{20}$$

$$p_{12} = p_{21} = p_{c_{21}} + p_{s_{21}} = - \frac{3\kappa}{5} u_{1,2} - p_o \sin \phi \left(\frac{u_{1,2}}{|u_{1,2}|}\right) \simeq g \sin \zeta \, \rho_p \, \bar{\nu} x_2 \tag{21}$$

$$\frac{p_{12}}{p_{22}} = \tan \zeta = \frac{- \frac{3\kappa}{5} u_{1,2} - p_o \sin \phi \, \text{sgn} \, (u_{1,2})}{\frac{\kappa}{\sigma} (\pi T)^{\frac{1}{2}} + p_o} = k \tag{22}$$

$$\gamma = 6(1-e) \, \kappa T/\sigma^2 \tag{23}$$

$$\underline{q}_c = - \kappa \frac{dT}{dx_2} \underline{e}_2 \tag{24}$$

and the quasi-static friction angle ϕ is assumed constant.

The fluctuation energy equation (13) reduces to

$$\frac{3}{2} \rho \frac{dT}{dt} = 0 = - p_{c_{12}} u_{1,2} + (\kappa T_{,2})_{,2} - \gamma \tag{25}$$

which represents a balance between shear work, the energy flux gradient and rate of dissipation.

Equation (22) <u>can</u> be satisfied by taking the ratio of mean quasi-static to mean dynamic normal stress, r, to be a constant at every depth. (This will give a particular class of solutions.) After taking the sign of $u_{1,2}$ to be negative for the flow under consideration, we may write

$$r = \frac{\sigma p_o}{\kappa (\pi T)^{\frac{1}{2}}} = \frac{\Gamma}{\left[1 - \frac{\sin \phi}{\tan \zeta}\right]} \tag{26}$$

where $$\Gamma = - \frac{3}{5} \frac{\sigma u_{1,2}}{k(\pi T)^{\frac{1}{2}}} - 1 = \text{const.} \tag{27}$$

From equations (21), (26) and (27) we find

$$- \kappa u_{1,2} = B x_2 \tag{28}$$

where $B = \frac{5}{3} g \sin \zeta \rho_p \bar{\nu}/A$ (29)

and

$$A = 1 - \frac{\Gamma}{(\Gamma+1)k [1 - \frac{\sin \phi}{\tan \zeta}]} \quad (30)$$

From equation (27)

$$- u_{1,2} = C\, T^{\frac{1}{2}}/\sigma \quad (31)$$

where $C = \frac{5}{3} \pi^{\frac{1}{2}} k [\Gamma+1]$ (32)

Substituting (31) into (28) yields

$$\kappa\, T^{\frac{1}{2}} = B\sigma x_2/C \quad (33)$$

From equations (21) and (28)

$$p_{c_{12}} = \frac{3}{5} B x_2 \quad (34)$$

Using equations (23), (24), (31) and (34) in the fluctuation energy equation (25) we obtain

$$\frac{d}{dx_2} [x_2 \frac{df}{dx_2}] + \frac{\delta x_2 f}{\sigma^2} = 0 \quad (35)$$

where $f = T^{\frac{1}{2}}$

$$\delta = \frac{5\pi k^2}{6} [\Gamma+1]^2 - 3(1-e) \quad (37)$$

$$= \frac{9}{10} R^2 - 3(1-e) \quad (38)$$

and R is a parameter defined previously by Savage and Jeffrey (ref. 25) as the ratio of the characteristic mean shear velocity to the rms fluctuation velocity

$$R = \sigma |du_1/dx_2| / \langle C^2 \rangle^{\frac{1}{2}} \quad (39)$$

It is interesting to note that equation (35) is of exactly the same form as that obtained previously by Savage (ref. 5) using an analysis which neglected the rate-independent stresses and thus δ for that analysis corresponds to the case of $\Gamma = r = 0$.

Equation (35) has the form of Bessel's equation of zeroth order and is subject to a number of possible boundary conditions. For the splashing type of flows described earlier, the definition of the upper 'surface' is somewhat ambiguous because of the strong saltation of particles. As a result of the air drag there would be some shear stress imposed at the upper surface and some flux of fluctuation energy upwards at this 'surface'. We deal here with the sliding and immature sliding flows where the boundary conditions are less ambiguous. For these cases we can take the stresses and the fluxes of fluctuation energy at the upper surface to be zero. At the bed $x_2 = h$, we specify the energy flux. We shall describe the general case elsewhere and here we consider only the case of zero energy flux into the bed. It turns out that this corresponds to $\delta = 0$. Thus from (37) and (38)

$$R = \frac{5}{3} \left(\frac{\pi}{3}\right)^{\frac{1}{2}} k \, [\Gamma+1] \tag{40}$$

$$= \left[\frac{10}{3} (1-e)\right]^{\frac{1}{2}} \tag{41}$$

From equations (26) and (40) we may solve for the ratio r

$$r = \left[\frac{3}{5} \left(\frac{3}{\pi}\right)^{\frac{1}{2}} R/\tan \zeta - 1\right] \left[1 - \frac{\sin \phi}{\tan \zeta}\right]^{-1} \tag{42}$$

Applying (20) at the bed and using (14) and (26) we find the granular temperature at the bed is

$$T_b = \frac{\bar{\nu} \, g \, h \cos}{2(1+r) \, \nu_b^2 (1+e) \, g_o(\nu_b)} \tag{43}$$

where the subscript b refers to values at the bed $x_2 = h$.

For the case of $\delta = 0$, the granular temperature

$$T = T_b = \text{const.} \tag{44}$$

throughout the depth. From equations (31), (32) and (26) with T = const. we obtain

$$\frac{du_1}{dx_2} = - \frac{5}{3\sigma} (\pi T)^{\frac{1}{2}} \left[1+r\left(1- \frac{\sin \phi}{\tan \zeta}\right)\right] \tan \zeta = - R(3T)^{\frac{1}{2}}/\sigma \tag{45}$$

Assuming the no-slip condition at the bed, we obtain the velocity profile

$$u_1 = \frac{R}{\sigma} (3T_b)^{\frac{1}{2}} (h-x_2) \tag{46}$$

From (33) and (13) we can determine an expression capable of yielding $\nu(x_2)$ thus

$$\frac{\nu^2 g_o(\nu)}{\nu_b^2 \, g_o(\nu_b)} = \frac{x_2}{h} \tag{47}$$

After defining

$$\bar{\nu} = \frac{1}{h} \int_0^h \nu \, dx_2 \tag{48}$$

and specifying a value of ν_b and the physical properties of the granular material, equations (46), (44) and (47) give the desired velocity, granular temperature and solids fraction profiles.

These results are for the case of zero fluctuation energy flux at the upper surface and at the bed. Although we shall not deal with more general boundary conditions here, we note that in the solutions for the case in which energy is put into the flow through the bed, the granular temperature is higher near the bed, the material flows more readily (at lower inclinations) and the velocity profile changes from a linear one to a blunt or convex shape. If energy is absorbed at the bed, the granular temperatures are decreased there, material flows less readily and the velocity profiles take on a concave shape.

COMPARISON WITH EXPERIMENTAL RESULTS OF ISHIDA AND SHIRAI

It is interesting to make comparisons of the theoretical velocity profiles just obtained with those measured by Ishida and Shirai for the flow of glass beads (between 0.35 and 0.50 mm) down an inclined chute which had its bed roughened with sandpaper (see Figure 4). To make this comparison we choose reasonable values of e, ϕ, ν_b and h of say

$e = 0.8$
$\phi = 24^o$
$\nu_b = 0.5$
$h = 6$ mm

and take $\sigma = (0.35+0.5)/2 = 0.425$ mm.

From equation (41) we obtain $R = \sqrt{2/3}$. The ratio of mean quasi-static to mean dynamic normal stress, r is determined from equation (42) and plotted in Figure 5 as a function of bed inclination ζ. At the maximum value of $\zeta = 25.6^o$ for steady flow, r = 0 and with decreasing ζ the value of r rapidly increases, tending to ∞ at $\zeta = 22.1^o$ when the flow comes to rest. From equation (47) and (14) we obtain the solids concentration profile which is shown in Figure 6. The value of $\bar{\nu}/\nu_b$ can be obtained numerically from the ν profile. The granular temperatures are uniform with depth, i.e. $T = T_b$ and the variations with ζ are shown in Figure 6. The triangular velocity profiles obtained from equation (46) are also shown on Figure 6 for different ζ.

Comparing the velocity profiles from Figure 4 and 6 we see that the theory predicts velocities of approximately the correct magnitude. Because of the inclusion of the rate-independent stresses, steady flows are predicted for a range of bed slopes from 22.1^o to 25.6^o. While the predicted upper limit is somewhat lower than the experimental ζ of 30° for the triangular profile shown in Figure 4, it should be noted that the analysis is for an infinitely wide channel, whereas the experiments were performed in a channel of finite width. Because of friction on the side walls, a larger ζ would be required to develop a flow similar to what might occur if the side walls were perfectly smooth or infinitely far apart. Furthermore, various simplifying assumptions

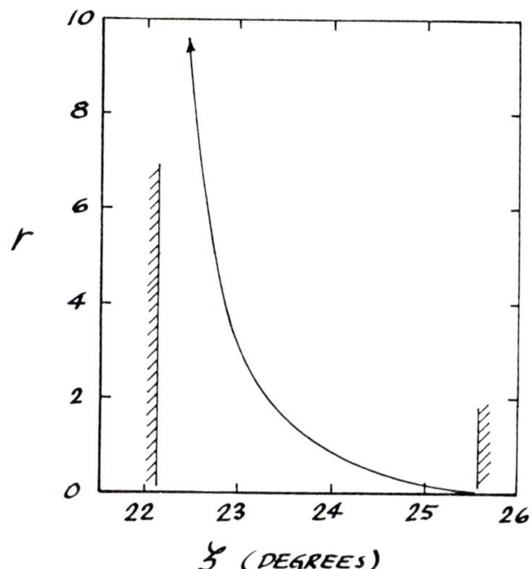

Fig. 5. Ratio of mean quasi-static to mean dynamic normal stress, r, as a function of bed inclination angle ζ. Steady flows possible for $22.1^{\circ} < \zeta < 25.6^{\circ}$.

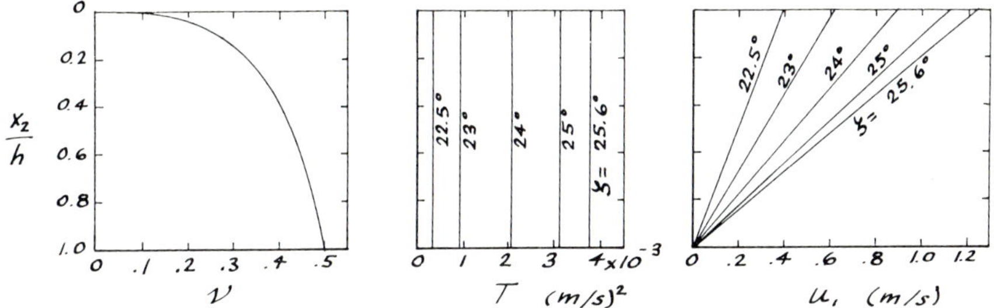

Fig. 6. Theoretical solids fraction, granular temperature and velocity profiles for flow down rough chute inclined at various angles.

were made in the modelling of the dynamic stresses, which no doubt affect the 'dynamic' friction angle for the flowing material.

With decreasing bed slope, the experimental velocities decrease and the profile becomes more concave, with an approximately linear upper portion above a nearly dead-flow zone close to the bed. All the predicted velocity profiles which assumed no energy flux into the bed are triangular in shape. The difference may be due to an inappropriate boundary condition near the bed. A dead-flow zone near the bed could dissipate velocity fluctuations in the neighbouring flow and in effect act as an energy absorbing bed. It should be noted that calculations with the boundary condition $dT/dx_2 < 0$ at the bed give velocity profiles that are concave in shape.

CONCLUSIONS

In this paper we have been concerned with both theoretical and experimental aspects of the flow of dry cohesionless granular materials down rough inclined chutes. Experimental information on constitutive behaviour and observations of flow field properties during chute flow have been briefly reviewed. Previous continuum and microstructural theoretical models have been reviewed in the light of the available experimental information. A simple ad hoc proposal to add rate-independent stress terms to the dynamic stresses obtained by the microstructural theory of Jenkins and Savage (ref. 21) was put forth. This theory was applied to the solution of the chute flow problem and the predicted velocity profiles were compared with the experimental measurements of Ishida and Shirai (ref. 1). The theoretical velocities are similar in magnitude to the measured ones. Steady fully developed profiles are predicted for a range of bed inclination angles somewhat larger than the angle of repose of the material. The theory involves no empirical fitting of constitutive coefficients or any empirical constants other than the easily determined material properties such as individual particle mass density and diameter, coefficient of restitution, and quasi-static internal friction angle. The success achieved suggests that further developments of this theory would be worth pursuing. Studies to include the effects of particle surface friction and rotary inertia in the dynamics are now being undertaken by J.T. Jenkins, C. Lun and myself.

ACKNOWLEDGEMENT

This work was supported by the Natural Sciences and Engineering Research Council of Canada.

REFERENCES

1 M. Ishida, T. Shirai, Velocity Distributions in the Flow of Solid Particles in an Inclined Open Channel, J. Chem. Eng. of Japan, 12 (1979) 46-50.

2 S.B. Savage, Experiments on Shear Flows of Cohesionless Granular Materials, In S.C. Cowin and M. Satake (Eds), Proc. U.S.-Japan Seminar on Continuum Mechanical and Statistical Approaches in the Mechanics of Granular Materials, (1978), pp. 241-254.

3 M. Sayed, Theoretical and Experimental Studies of the Flow of Cohesionless Granular Materials, Ph.D. Thesis, McGill Univ., Montreal, 1981, 157 pp.

4 S.B. Savage and M. Sayed, Experiments on Dry Cohesionless Materials in an Annular Shear Cell at High Strain Rates. Presented at EUROMECH 133 - Statics and Dynamics of Granular Materials, Oxford Univ. (1980).

5 S.B. Savage, Granular Flows at High Shear Rates, in R.E. Meyer (Ed.) Theory of Dispersed Multiphase Flow, Academic Press, New York 1982.

6 R.A. Bagnold, Experiments on a Gravity Free Dispersion of Large Solid Spheres in a Newtonian Fluid Under Shear, Proc. R. Soc. London, Ser. A225 (1954) 49-63.

7 S.B. Savage, R.M. Nedderman, U. Tuzun and G.T. Houlsby, Flow of Granular Materials - III, Rapid Shear Flows, Chem. Eng. Sci., 37 (1982).

8 D.A. Augenstein and R. Hogg, An Experimental Study of the Flow of Dry Powders Over Inclined Surfaces, Powder Techn., 19 (1978) 205-15.

9 J. Bailard, An Experimental Study of Granular-Fluid Flow, Ph.D. Thesis, Univ. of Calif., San Diego, 1978, 172 pp.

10 S.B. Savage, Gravity Flow of Cohesionless Granular Materials in Chutes and Channels, J. Fluid Mech., 92 (1979) 53-96.

11 M. Ishida, H. Hatano, and T. Shirai, The Flow of Solid Particles in an Aerated Inclined Channel, Powder Techn., 27 (1980) 7-12.

12 H. Shen, N.L. Ackermann, Constitutive Relationships for Fluid-Solid Mixtures, Submitted to Proc. ASCE, J. Eng. Mech., (1982).

13 O. Walton, Particle Dynamics Modelling of Geological Materials, Lawrence Livermore Laboratory Rept. UCRL-52915, Univ. of Calif., (1980).

14 C.E. Campbell, Shear Flows of Granular Materials, Ph.D. Thesis, Calif. Inst. of Techn., Pasadena, 1982.

15 M.A. Goodman and S.C. Cowin, Two Problems in Gravity Flow of Granular Materials, J. Fluid. Mech., 45 (1971) 321-39.

16 J.W. Nunziato, S.L. Passman, and J.P. Thomas, Gravitational Flows of Granular Materials with Incompressible Grains, J. Rheol., 24 (1980) 395-420.

17 D.F. McTigue, A Nonlinear Continuum Model for Flowing Granular Materials, Ph.D. Thesis, Stanford Univ., Stanford, (1979) 165 pp.

18 M. Sayed and S.B. Savage, Rapid Gravity Flow of Cohesionless Granular Materials Down Inclined Chutes, Submitted to J. Appl. Math. Phys. (ZAMP) (1982).

19 K.H. Roscoe, A.N. Schofield, and C.P. Wroth, On the Yielding of Soils, Geotech., 8 (1958) 22-53.

20 S. Ogawa, A. Umemura, and N. Oshima, On the Equations of Fully Fluidized Granular Materials, J. Appl. Math. Phys. (ZAMP), 31 (1980) 483-93.

21 J.T. Jenkins, and S.B. Savage, A Theory for the Rapid Flow of Identical, Smooth, Nearly Elastic Spherical Particles, submitted to J. Fluid. Mech. (1982).

22 V.V. Sokolovski, Statics of Granular Media, Pergamon, Oxford, 1965.

23 A.J.M. Spencer, Deformation of an Ideal Granular Material. In H.G. Hopkins and M.J. Sewell (Eds.), Mechanics of Solids, Rodney Hill 60th Anniv. Volume, Pergamon, Oxford 1981.

24 N.F. Carnahan, and K.E. Starling, Equations of State for Non-Attracting Rigid Spheres, J. Chem. Phys., 51 (1969) 635-36.

25 S.B. Savage and D.J. Jeffrey, The Stress Tensor in a Granular Flow at High Shear Rates, J. Fluid Mech., 110 (1981) 255-72.

Mechanics of Granular Materials: New Models and Constitutive Relations, edited by J.T. Jenkins and M. Satake, 1983
Elsevier Science Publishers B.V., Amsterdam — Printed in The Netherlands

RAPID PLANE FLOW OF GRANULAR MATERIALS DOWN A CHUTE

Kolumban HUTTER and Thomas SCHEIWILLER

Laboratory of Hydraulics, Hydrology and Glaciology, ETH Zurich (Switzerland)

ABSTRACT

The continuum model, describing rapid flow of granular materials, which was recently deduced by Jenkins & Savage (8) from considerations of statistical mechanics, is taken up here to formulate and solve numerically gravity driven flow down a chute. In analysing this problem a sliding condition of the basal boundary is conjectured. The solutions of the nonlinear boundary value problem depend on the coefficient of restitution, the ratio of the particle diameter to the thickness of the shear layer, a parameter characterizing the shear, two sliding coefficients and the inclination angle of the chute. Velocity profiles show the qualitative features observed by Savage (12) but density profiles differ from those reported by others. It is shown that, on occasion, solutions may critically depend on boundary conditions. This points at the need for a deduction of boundary conditions from principles which are as rigorous as those leading to the field equations.

INTRODUCTION

Rapid deformations of dry, relatively dense granular materials have in recent years found increasing attraction of scientists in many fields of industrial and geophysical applications. Yet, until recently no systematic theoretical model was available in which both, the structure of the model and the phenomenological equations would have been deduced with equal rigor from a common fundament. We only mention here the formulations of Googman & Cowin (2,3), Savage (12), Kanatani (9), Jenkins & Cowin (6) and Jenkins & Savage (7). All these models are defective to a certain degree, if only in their rigor how the constitutive relations are introduced.

A first attempt to deduce a continuum model for the rapid flow of granular materials from statistical mechanics is due to Savage & Jeffrey (13), but it is incomplete as only the stress tensor was analysed. A complete statistical model for hard, smooth identical spherical balls was developed later by Jenkins & Savage (8). It includes the derivation of balance laws for mass, momentum and collisional fluctuation energy including deduction of constitutive relations for the stress tensor, the flux of fluctuation energy and its annihilation. Physically, Jenkins and Savage's model is simple: only three parameters enter the field equations, namely the particle diameter, the coefficient of restitution in binary collisions and a coefficient characterizing the anisotropy of the particle distribution function at collision whose value is known.

Explicit demonstration whether this model is capable of producing the correct velocity and density distributions in typical boundary value problems has so far not been given to our knowledge. However, rapid flows of granular materials down a chute have experimentally been studied, see Savage (12). Velocity profiles deviate substantially from those for Newtonian fluids as they may have a point of inflexion. Furthermore, the limited available observational data seem to indicate that the density of the bulk material is largest in a layer midway between the bottom and the free upper surface; for a review of experiments see Savage (12).

Boundary conditions that must be formulated in a typical boundary value problem are not yet well understood. At a rigid wall, depending on its roughness, sliding may be assumed. In fluid mechanics of polymers (see e.g. Pearson & Petrie, (11), Uhland (16, 17)) and other non-classical fluid applications (Hutter (4,5)) the sliding law consists of a (nonlinear) relation between shear traction and sliding velocity (usually a power law). This law will also be adopted here, but must be complemented by another relationship involving the fluctuation energy. We propose a relation between the fluctuation energy and the sliding velocity and study its significance in a typical boundary value problem, namely the rapid flow of a granular material down a chute. Boundary conditions at a free surface are better understood; they require the vanishing of the fluxes of momentum and fluctuation energy perpendicular to the surface.

Analytical solutions of the complicated nonlinear two-point-boundary value problem being inaccessible, we perform a numerical integration. To this end, the governing equations are non-dimensionalized. Thereby the physically important scales become explicitly apparent and the significance of the various different scales can be studied by performing the integration for a wide range for the non-dimensional numbers which govern the problem. We investigate in detail the variations of the coefficient of restitution, of the inclination angle and of the sliding conditions. Results indicate that velocity profiles with an inflexion point emerge when the density is non-uniformly distributed over depth. For a uniform density profile the velocity profile is blunt and flow conditions are near Newtonian. Which of these flow features are attained depends on the parameters which govern the problem. Both can be obtained by either varying the parameters arising in the field equations and keeping those of the boundary conditions fixed, or vice versa. The statistical model leading to field equations should therefore, be complemented by an equally sound deduction of boundary conditions.

THE BOUNDARY VALUE PROBLEM

Consider steady plane shear flow down a flat plate with angle of inclination α^* (see Fig. 1). Let (x,y) be Cartesian coordinates in the direction of flow and perpendicular to it and let $u(y)$ and $\Theta(y)$ be streamwise velocity and fluctuation energy per unit mass, respectively, which in this steady gravity flow are merely

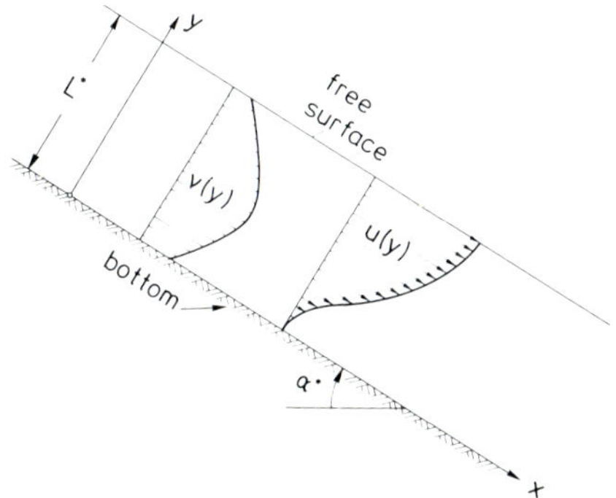

Fig. 1. Definition of steady flow geometry: Gravity driven flow of a layer with thickness L^* and inclination angle α^*. Typical velocity profile $u(y)$ and density distribution as surmised by the literature (see Scheiwiller & Hutter (14)).

functions of the coordinate y. For this situation, using the field equations of Jenkins and Savage (8), it is easy to see that balance of momentum and fluctuation energy reduce to the equations (see Scheiwiller & Hutter (14)).

$$\frac{2}{5}(2+\alpha)(\kappa k)' + \rho g \sin\alpha^* = 0,$$

$$\frac{\sqrt{\pi}}{\sigma}(\kappa\sqrt{\Theta})' + \rho g \cos\alpha^* = 0, \tag{1}$$

$$(\kappa\Theta')' - \frac{6(1-e)}{\sigma^2}\kappa\Theta + \frac{4}{5}(2+\alpha)\kappa k^2 = 0,$$

in which $(\ )' = d(\ )/dy$ and

$$k \equiv \frac{1}{2}u', \quad \kappa = \frac{2\hat{\rho}\sigma(1+e)}{\sqrt{\pi}}\phi_0(\nu)\sqrt{\Theta}, \quad \rho = \nu\hat{\rho}. \tag{2}$$

Here k is the shear rate, κ a conductivity, ρ the bulk density of the material under consideration, g the gravity acceleration, ν the volume distribution function and $\hat{\rho}$ the density of granuls. Furthermore, $\phi_0(\nu)$ is related to the isotropic statistical correlation function $g_0(\nu)$ at contact, which according to Carnahan and Starling (1) is given by

$$g_0(\nu) = \frac{1}{1-\nu} + \frac{3\nu}{2(1-\nu)^2} + \frac{\nu^2}{2(1-\nu)^3},$$

$$\phi_0(\nu) = \nu^2 g_0(\nu), \tag{3}$$

and is a good approximation for $\nu \lesssim 0.5$. The remaining parameters in equations

(1), σ, e and α are the diameter of the spherical particles, the coefficient of restitution ($0 \ll e \leq 1$) and a parameter describing the anisotropy of the distribution function. Generally $\alpha = \alpha(\nu)$ but we shall assume $\alpha = 1$ as suggested by Jenkins & Savage (8).

Boundary conditions, to which equations (1) are subject are a viscous sliding law at the bottom,

$$u = f(\tau^2)\tau\ , \qquad \Theta = g(u^2)u^2, \qquad \text{at } y = 0, \tag{4}$$

with $\tau = \frac{2}{5}(2 + \alpha)\,\kappa\, k$ and f and g depending on surface roughness, and no flux of momentum and fluctuation energy perpendicular to the free surface,

$$k = 0, \quad \Theta = 0, \quad \Theta' = 0, \qquad \text{at } y = L^*. \tag{5}$$

As far as sliding is concerned, we shall restrict ourselves to the power laws

$$f(x^2) \equiv \mathcal{C}[x^2]^{(n-1)/2}, \qquad g(x^2) \equiv \mathcal{D}[x^2]^{(m-1)/2}, \tag{6}$$

with sliding coefficients $\mathcal{C}$ and $\mathcal{D}$ and power law coefficients n and m. The limit $\mathcal{C} \to 0$ and $\mathcal{D}$ bounded corresponds to the no slip condition. The first two of conditions (5) are the expressions of no shear traction and no surface pressure, the third expresses the fact that there is no flux of fluctuation energy.

Equations (1), (4) and (5) with the specifications (6) form a nonlinear two-point-boundary-value problem for the variables ν, u and Θ. The construction of its solutions is the aim of this article. To this end, the equations must be non-dimensionalized. The boundary value problem arises then in a form in which the physically important scales appear as dimensionless parameters, see Scheiwiller & Hutter (15). These parameters are the coefficient of restitution ϵ, the angle of inclination α^* of the chute and

$$\mathbb{C} = \frac{\mathcal{C}\, D^{*n}}{U^*}\ , \quad \mathbb{D} = \frac{\mathcal{D}}{2\,\varepsilon^2}\, U^{*m-1}\ , \quad \mathbb{A} = \frac{\sigma}{L^*}\ , \quad \mathbb{R} = \frac{g\,\nu^*\, L^{*3}}{(U^*\,\sigma)^2} = \frac{g\,\nu^*\, L^*}{U^{*2}}\,\mathbb{A}^{-2}\ , \tag{7}$$

in which

$$D^{*2} = \frac{4(1+e)^2\,\hat{\rho}^2(2+\alpha)^2}{25\pi\,(1-e)}\qquad \left(\frac{U^*\,\sigma}{L^*}\right)^4, \qquad \varepsilon^2 = \frac{1}{6(1-e)}\,\mathbb{A}^2.$$

ν^*, U^*, L^* are characteristic values of the volume distribution function, a typical longitudinal velocity and the thickness of the sheared layer. There are, thus eight parameters, which can be varied, namely e, α^*, $\mathbb{A}$, $\mathbb{R}$, $\mathbb{C}$, $\mathbb{D}$, m and n. $\mathbb{A}$ is an aspect ratio and generally small, $\mathbb{R}$ is the product of $\mathbb{A}^{-2}$ with an inverse Froude number and measures the gravity induced shear and $\mathbb{C}$ and $\mathbb{D}$ are dimensionless sliding ciefficients. In order that sliding is significant these must differ from zero. In subsequent calculation we shall limit attention to linear sliding $n = 1$, $m = 0$.

Realistic maximal and minimal values for the scalings are, perhaps, $\nu^* = (0.1, 0.74)$, $L^* = (0.5, 5)$m, $U^* = (1, 100)$m/s, $\sigma = (10^{-3}, 10^{-1})$m so that, roughly, $10^{-3} < \mathbb{A} < 10^{-1}$, $5\text{x}10^2 < \mathbb{R} < 10^4$ and $10^{-6} < (1\text{-}e) < 10^{-4}$. We shall also choose $10^0 < \alpha^* < 90^0$, $10^{-3} < \mathbb{C}\ 10^{-1}$, $10^0 \leq \mathbb{D} \leq 10^4$.

DISCUSSION OF RESULTS

Our primary interest is in profiles of the volume distribution function, of the velocity and the fluctuation energy and in orders of magnitudes of the surface velocity and maximal values of the fluctuation energy as functions of the above mentioned parameters. In particular, we shall emphasize the significance of boundary conditions. The analysis is performed mostly for a chute with $\alpha^* = 45^0$, and a layer of the granular material of 0.5 m thickness. Also only linear sliding ($n = 1$, $m = 0$) will be considered (see also Scheiwiller & Hutter (15)).

Variation of the coefficient of restitution

For $\mathbb{A} = 2\text{x}10^{-2}$, $\mathbb{R} = 925$, $\mathbb{C} = 10^{-3}$, $\mathbb{D} = 10$, $\alpha^* = 45^0$, the profiles of the volume distribution, velocity and fluctuation energy are shown in Fig. 2, from

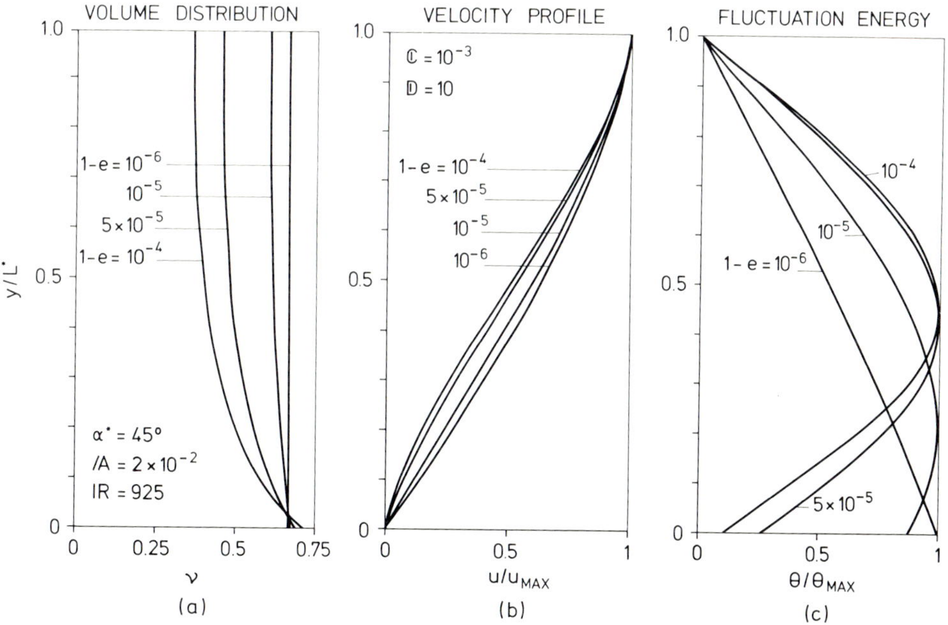

Fig. 2. Profiles of volume distribution (a), velocity (b) and fluctuation energy (c) as functions of the coefficient of restitution, e. Calculations were performed for a chute with $L^* = 0.5$ m, $\alpha^* = 45^0$ and $\mathbb{A} = 2\text{x}10^{-2}$, $\mathbb{R} = 925$, $\mathbb{C} = 10^{-3}$, $\mathbb{D} = 10$.

which the following facts are evident:

(i) The density distribution function is a monotonic function of depth; density is largest at the bottom and smallest at the free surface. This is contrary to the qualitative results displayed in Fig. 1 and reported by Savage (13) to be obtained observationally and also disagrees with the findings of Nunziato et al. (10), which are based on a different theoretical model.

(ii) With decreasing values of e, that is increasing plasticity of the collisions the non-uniformity of the volume distribution function increases. For $(1-e) \leq 10^{-6}$ uniformity is attained.

(iii) The velocity profile is blunt (close to a parabola) when $(1-e) \leq 10^{-6}$, but has an inflexion point for values $(1-e) > 10^{-5}$. This latter fact is in agreement with the observations reported by Savage (13).

(iv) When $(1-e) \leq 10^{-6}$ the fluctuation energy is monotonic; it increases from top to bottom. With increasing plasticity of the collisions this monotonicity breaks down; the maximum of the fluctuation energy moves into the layer, and values at the boundaries are now smaller.

The figures suggest that the flow is Newtonian-like when the density profile is nearly constant or the fluctuation energy linearly distributed over depth, which arises when $(1-e) \lesssim 10^{-5}$. Further calculations corroborate this behavior.

Surface velocities and maximal values of the fluctuation energy for a chute with $\alpha^* = 45^0$, $L^* = 0.5\,m$ are shown in Fig. 3. Notice that changes in u_{Max} and Θ_{Max} are appreciable for $(1-e) > 10^{-5}$, but negligible when $(1-e) \lesssim 10^{-6}$. This brings about very clearly the significance of collosional inelasticity. For e very near unity collisions are elastic and flow behavior apparently Newtonian-like. To obtain a velocity profile with inflexion point a certain inelasticity of the collisions is necessary. However, it need not be large and $(1-e) = O(10^{-4})$ suffices to obtain velocity profiles which qualitatively agree with those observed by Savage (13). It should also be recognized that, depending on the values of $\mathbb{D}$, u_{Max} may be very large, while $\Theta_{Max}^{1/2}$ is one to two orders of magnitude smaller compared to it. The strong dependence of u_{Max} and Θ_{Max} on the frictional coefficient $\mathbb{D}$ also points at the importance of a clear understanding of boundary conditions. This will further be emphasized below. Finally, it should be stressed that for values $(1-e) > 10^{-4}$ no convergence of the numerical integration scheme was achieved. This requires further scrutiny because Fig. 3 suggests this to be an interesting range of the coefficient of restitution.

Variation of the inclination angle

By varying the inclination angle profiles for ν/ν_{Max}, u/u_{Max} and Θ/Θ_{Max} change

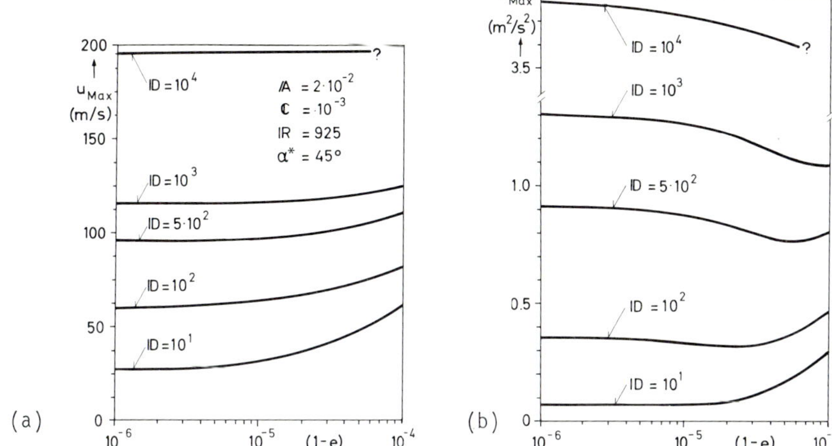

Fig. 3. Surface velocity (a) and maximal values of the fluctuation energy (b) as functions of coefficient of restitution, e, and for various values of the friction parameter ID, valid for a chute with $L^* = 0.5$ m, $\alpha^* = 45^o$, /A = $2x10^{-2}$, IR = 925 and C = 10^{-3}. Question marks indicate domains where numerical integration procedure did not converge.

very little, but absolute values are affected by varying α^*; ν_{Max} decreases while u_{Max} increases with α^*. The relation $\Theta_{Max}(\alpha^*)$ is non-monotonic. For further details see Scheiwiller & Hutter (15).

Variation of the particle size

The parameters governing the particle size is /A = σ/L^*, representative values of which are smaller than 0.2, approximately. Calculations indicate that /A has a dominant effect on the profile forms of ν, u and Θ. Generally, the smaller /A, the less Newtonian-like is the flow and the more will be the velocity profile resemble the shape with an inflexion point as observed by Savage (12). Furthermore, since ν decreases with /A it is not surprising that both, u_{Max} and Θ_{Max} grow with growing /A. For a detailed discussion see Scheiwiller & Hutter (15).

VARIATION OF THE SLIDING CONDITIONS

Because deduction of boundary conditions at a rigid, rough wall is largely conjectural, a sensitivity analysis of the solutions of the boundary value problem to the postulated sliding conditions is particularly instructive. Fig. 4 shows the three transverse profiles of density, velocity and fluctuation energy using a linear basal sliding law according to equations (6), varying C and keeping ID fixed. Fig. 5 summarizes the same result when ID is varied but C is

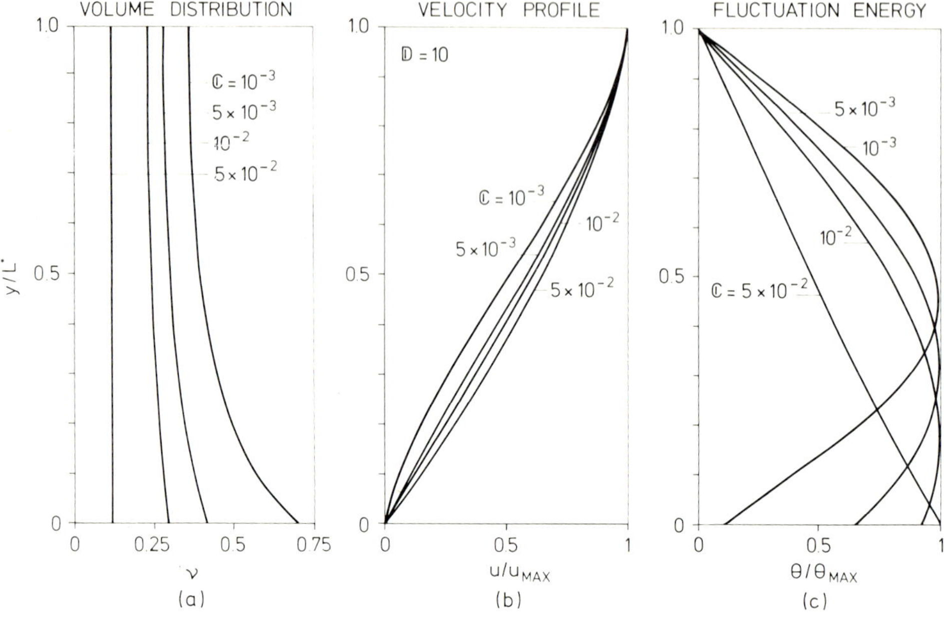

Fig. 4. Profiles of the volume distribution function (a), velocity (b) and fluctuation energy (c) as functions of the sliding parameter $\mathbb{C}$, keeping $(1-e) = 10^{-4}$, $\mathbb{D} = 10$, $\mathbb{A} = 2x10^{-2}$, $\mathbb{R} = 925$, $\alpha^* = 45^0$ fixed. A linear sliding law with n = 1, m = 0 was used.

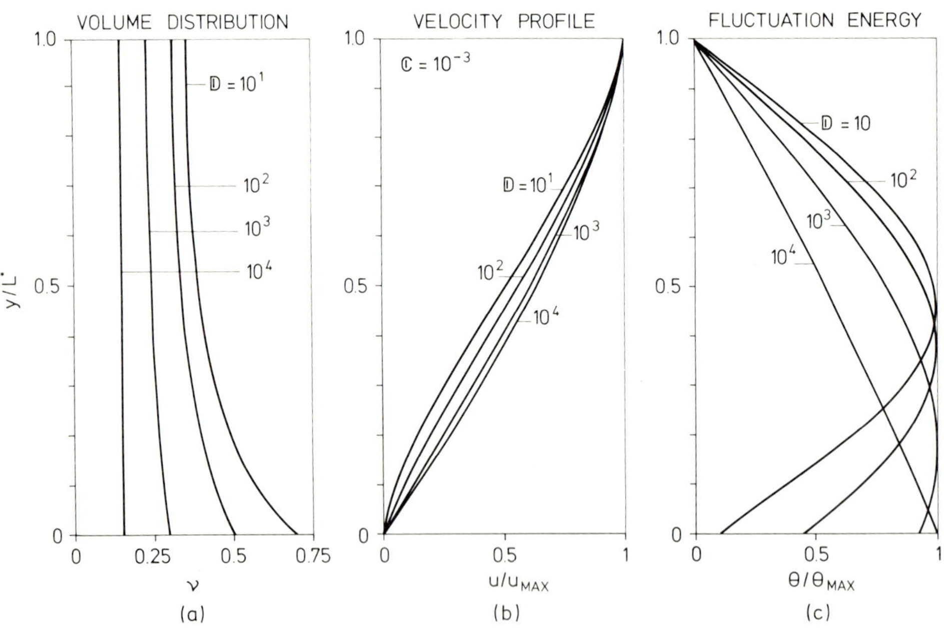

Fig. 5. Same as Fig. 7 but now $\mathbb{D}$ is varied and $\mathbb{C}$ is held fixed.

is held fixed. Both sliding coefficients are significant and apparently may have similar effects. In Fig. 4 C varies between 10^{-3} and 2×10^{-2} and $\mathbb{D} = 10$. In view of equation (6) one may thus be tempted to conclude that basal sliding is small. This is indeed so, as the sliding component in Figs. 4 and 5 is not visible in comparison to the gliding part. However, varying $\mathbb{C}$ and $\mathbb{D}$ in the indicated ranges changes velocity profiles from near Newtonian to the form with inflexion point. Near-Newtonian behavior of the velocity profile is accompanied by both, a uniform density distribution and a linear distribution of the fluctuation energy. Hence, nearly identical density, velocity and fluctuation energy profiles can be obtained with two different choices of the sliding coefficients. Furthermore, it can be shown that surface velocity and the maximal value of the fluctuation energy are increasing functions of $\mathbb{C}$ and $\mathbb{D}$, and both may assume quite large values. To prescribed values of u_{Max} there are at least two pairs with nearly the same density, velocity and fluctuation energy profiles. Comparing the profiles of the volume distribution of Figs. 4 or 5 with those of Fig. 2 shows that uniformity

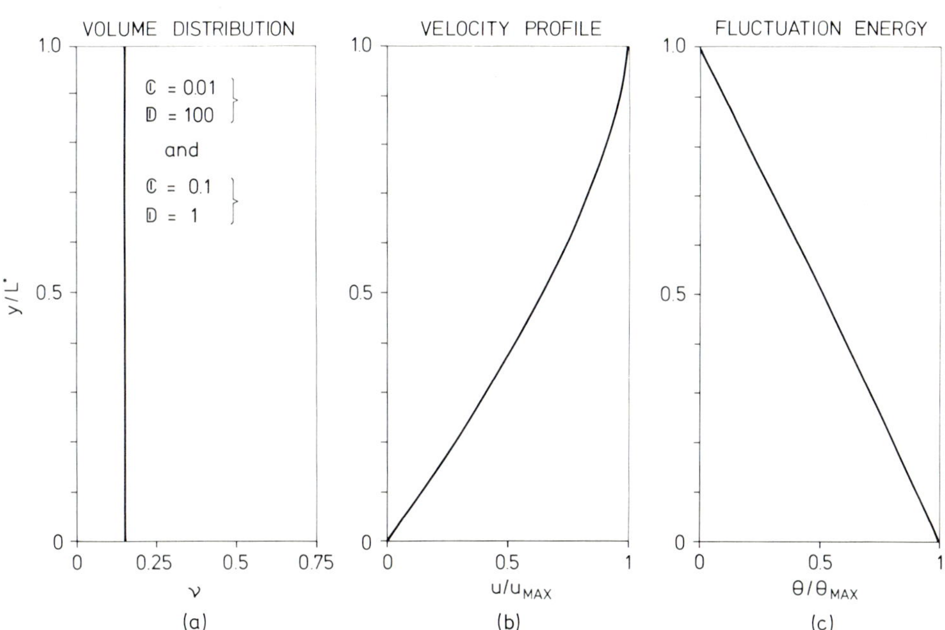

Fig. 6. Profiles of the volume distribution function (a), velocity (b) and fluctuation energy (c) for a chute with $\alpha^* = 45^0$, $L^* = 0.5\,m$, $\mathbb{A} = 0.02$, $\mathbb{R} = 925$ and

	Case I	Case II
(1-e)	10^{-5}	10^{-4}
$\mathbb{C}$	10^{-2}	10^{-1}
$\mathbb{D}$	10^{2}	10^{0}

	Case I	Case II
u_{Max}	195.5 m/s	196.9 m/s
Θ_{Max}	3.73 m^2/s^2	3.60 m^2/s^2

of the profile is obtained for large $\mathbb{C}$ or large $\mathbb{D}$ and, alternatively, for small values of (1-e). This suggests that experimental determination of the volume distribution function for different basal sliding conditions is necessary to separate the effects of inelasticity of collisions and of basal sliding. On the other hand, experimental identification of $\mathbb{C}$ and $\mathbb{D}$ does not seem possible. It requires a theoretical model, perhaps a statistical model of spherical particles bouncing against a rough wall. Proof for the difficulty of the separation of parameters e, C and $\mathbb{D}$ is provided by Fig. 6. It shows the profiles for two different chutes characterized by $\alpha^* = 45^0$, $\mathbb{A} = 2 \times 10^{-2}$, $\mathbb{R} = 925$, and the remaining parameters as indicated in the figure caption. As evident, profiles hardly differ at all and neither is there a difference in u_{Max} and Θ_{Max}. There is no experiment separating the two cases.

Further calculations with power law representations to the sliding law demonstrate that nonlinearities affect profiles and orders of magnitudes substantially. Better knowledge of what happens at the boundary is therefore urgent.

ACKNOWLEDGEMENT

While performing this work Th. Scheiwiller was supported by the Swiss National Science Foundation.

REFERENCES

1 N.F. Carnahan and K.E. Starling, 1969, Equations of state for non-attracting rigid spheres, J. Chem. Phys., Vol. 51, pp. 625-636.
2 M.A. Goodman and S.C. Cowin, 1971, Two problems in the gravity flow of granular materials, J. Fluid Mech., Vol. 45, No. 2, pp. 321-339.
3 M.A. Goodman and S.C. Cowin, 1972, A continuum theory for granular materials, Arch. Rational Mech. Anal., Vol. 44, No. 4, pp. 249-266.
4 K. Hutter, 1982, Glacier flow, American Scientist, Vol. 70, No. 1, pp. 26-34.
5 K. Hutter, 1982, Dynamics of glaciers and large ice masses, Ann. Rev. Fluid Mech., Vol. 14, pp. 87-130.
6 J.T. Jenkins and S.C. Cowin, 1979, Theories for flowing granular materials, The Joint ASME-CSME Appl. Mech. Fluid Engng. and Bioengng, Conf., AMD-Vol.31, pp.79-89.
7 J.T. Jenkins and S.B. Savage, 1981, The mean stress resulting from interparticle collisions in a rapid granular shear flow, In: Proceedings of the Fourth International Conference on Continuum Models of Discrete Systems, (Eds. O. Bruhin and R.K.T. Hsien), North Holland.
8 J.T. Jenkins and S.B. Savage, 1983, A theory for the rapid flow of identical, smooth, nearly elastic spherical particles, J. Fluid Mech. (in press).
9 K.I. Kanatani, 1979, A micropolar continuum theory for the flow of granular materials, Int. J. Engng. Sci., Vol. 17, pp. 419-432.
10 J.W. Nunziato, S.L. Passman and J.P. Thomas Jr., 1980, Gravitational flow of granular materials with incompressible grains, J. Rheology, Vol. 24, No. 4, pp. 395-420.

11 J.R.A. Pearson and C.J.S. Petrie, 1968, On melt flow instability of extruded polymers, In: Polymer Systems, Deformation and Flow. Proceedings of the 1966 Annual Conference of the British Society of Rheology, Ed. R.E. Whetton and R.W. Whorolow, MacMillan, London.
12 S.B. Savage, 1979, Gravity flow of cohesionless granular materials in chutes and channels, J. Fluid Mech., Vol. 92, pp. 53-96.
13 S.B. Savage and D.J. Jeffrey, 1981, The stress tensor in a granular flow at high shear rates, J. Fluid Mech., Vol. 110, pp. 255-272.
14 Th. Scheiwiller and K. Hutter, 1982, Lawinendynamik, Uebersicht über Experimente und theoretische Modelle von Fliess- und Staublawinen, Mitteilung Nr. 56 der Versuchsanstalt für Wasserbau, Hydrologie und Glaziologie an der ETH, Zürich.
15 Th. Scheiwiller and K. Hutter, 1982, On shear flow of cohesionless granular materials down an inclined chute, Advances in the Mechanics and the Flow of Granular Materials (to appear).
16 E. Uhland, 1977, Stratified two-phase flow of molton polymers in circular dies, Polym. Eng. And Sci., Vol. 17, p. 671.
17 E. Uhland, 1979, Das anomale Fliessverhalten von Polyäthylenen hoher Dichte, Rheol. Acta, Vol. 18, 1.

Mechanics of Granular Materials: New Models and Constitutive Relations, edited by
J.T. Jenkins and M. Satake, 1983
Elsevier Science Publishers B.V., Amsterdam — Printed in The Netherlands

RAPID SHEAR FLOW OF DENSELY PACKED GRANULAR SOLIDS

Norbert L. Ackermann and Hayley H. Shen
Department of Civil and Environmental Engineering
Clarkson College, Potsdam, N.Y. 13676 (USA)

ABSTRACT

Present theories cannot yet accurately describe the rapid shear flow of granular solids. Most theoretical formulations of the equations of motion contain undetermined functions or coefficients. Those equations that are fully deterministic predict stresses that differ from experimentally determined results by one or more orders of magnitude. Previous investigations that have provided a quantifiable description of the stress state have adopted a binary collision model to describe the momentum exchange within the granular flow.

A preliminary analysis which incorporates the kinematic constraints of the flow process, indicates that a lateral mass transfer occurs that has a significant effect upon the momentum exchange between colliding solids. These effects have not been previously reported. By incorporating the effects of this lateral mass transfer into the analytical procedure proposed by Shen and Ackermann (1982), theoretically predicted stresses are found which agree well with experimentally determined values.

BACKGROUND

Ogawa, Umemura and Oshima (1980) provided a significant development in the theory describing the flow of granular solids by quantifying the average fluctuation speed of particles within a rapidly sheared granular mixture. They used the following equations previously formulated by Kanatani (1979) and Jenkins and Cowin (1979) as a basis for analysis.

$$\frac{\partial \rho}{\partial t} + \frac{\partial \rho v_j}{\partial x_j} = 0 \tag{1}$$

$$\rho\left(\frac{\partial v_i}{\partial t} + v_j \frac{\partial v_i}{\partial x_j}\right) = \frac{\partial}{\partial x_j} \tau_{ji} + b_i \tag{2}$$

$$\rho\left(\frac{\partial e}{\partial t} + v_j \frac{\partial e}{\partial x_j}\right) = \tau_{ji} \frac{\partial v_j}{\partial x_i} - \frac{\partial q_i}{\partial x_j} - \gamma \tag{3}$$

where $e = v_i' v_i'/2$ represents the fluctuation energy of a particle whose mean velocity is v_j, τ_{ji} = stress tensor, b_i = body force, q_j = energy flux and γ = energy sink. Ogawa et. al. (1980) modeled the microstructure of the granular system and the system's kinematics showing how the collisional transfer of momentum served as the stress generating mechanism. They computed the rate of change of the fluctuation energy in a unit volume and matched terms on the two sides of Eq. 3 to identify τ_{ji} and γ. Since they did not include the gradient of the fluctuation of turbulent energy in their model, q_j was not present. Shen and Ackermann (1982) also used Eqs. 1 through 3 in their analysis of a granular mixture in a rectilinear shear flow. These authors, however, used Eq. 4, first presented by Bagnold (1954), to describe the stress τ_{ji}

$$\tau_{ji} = p_j \, f \, \Delta M_i \tag{4}$$

where p_j = number of particles per unit area whose normal is in the jth coordinate direction, f = frequency with which a particle experiences collisions from adjacent particles, and ΔM_i = the ith component of the momentum change produced by each particle collision. The energy sink γ was expressed by Ackermann and Shen (1982) as

$$\gamma = NFE \tag{5}$$

where N = number particles/unit vol, F = collisions/sec/particle, E = energy loss/particle/collision.

Unfortunately none of the theoretically determined constitutive relationships which have been reported to date accurately predict experimental results without resorting to the use of arbitrary constants or functional relationships. Equations containing no undetermined coefficients such as those developed by McTigue (1979), Ogawa et. al. (1980) and Shen and Ackermann (1982) predict results that were significantly different from experimentally determined values. The relationship proposed by McTigue (1979) did not include the effects of material properties nor the turbulent nature of the flow and in general predicted results that were in error by many orders of magnitude. Results by Ogawa et. al. (1980) erred by a factor of 100 while Shen and Ackermann (1982) differed from experimental results by approximately a factor of 10. Some basic stress generating mechanisms was, therefore, still unidentified!

MASS TRANSFER

In previous investigations the existence of a mass transfer normal to the direction of the mean motion was either neglected or considered not to exist. The trajectory of a particle was considered to follow the path indicated by the mean velocity profile in addition to random excursions of a distance s about this mean position. These random excursions resulted from interparticle collisions.

The local, lateral, mass transfers which occurs normal to the direction of the mean motion could easily be overlooked in view of the fact that within a densely packed granular mixture a particle would appear to have little mobility thus preventing any significant lateral movement. A study of the kinematics of the flow process dictates however that a significant lateral mass transfer must take place. In addition it cannot be expected that a particle immediately acquires a mean velocity component that corresponds to its instantaneous coordinate location.

Calculations, based upon this concept show that the momentum exchange resulting from mass transfer is of vital importance when deriving the constitutive relationships. Consider for example the granular arrangement and flow profile described in Fig. 1. This figure shows a granular mixture in which the solids have a diameter D and intergranular spacing s. The mean velocity profile of the mixture is $\bar{u}_1(x_2)$. Particles A and B however cannot continually maintain both their respective mean coordinate positions x_2 as shown as well as their mean relative velocity $\Delta\bar{u} = (D+s)\cos\phi\, d\bar{u}_1/dx_2$. This conclusion is based upon the geometric constraint that a particle, such as A, cannot overtake or pass a slower moving particle, such as B, if lateral motion of the particle is restricted to excursions of no more than a distance of approximately s where $s << D$.

If particles are translated in the x_1 direction according to distances specified by the mean velocity profile, the particles in rows 1 and 2 would at some later time simultaneously occupy the positions along the inclined lines shown in the figure. However, the particles cannot move solely with their respective mean motions as well as have their centers eventually situated along the inclined lines as shown. In order for the solids to have both their initial and final position as illustrated, they must also move laterally in the x_2 direction; possibly as indicated by the dotted lines.

What is concluded therefore is that in order for a densely packed granular mixture to have a gradient in its mean velocity profile, there must be a significant mass transfer in the x_2 coordinate direction. Consider now that particle A in Fig. 1 has the mean velocity in the x_1 direction as indicated by the mean velocity profile. When undergoing movement in the x_2 direction it is not necessary nor probable that at every instant the particle will maintain a mean velocity exactly consistent with the mean velocity profile corresponding to the particle's instantaneous location. Hence any single particle at level x_2, does not necessarily possess exactly the mean motion of that level as described by the mean velocity profile. However, the average of all of the particles at any given level, x_2, has a value $\langle u_1(x_2)\rangle$ which is equal to the mean value $\bar{u}_1(x_2)$. This distinction enables the mass transfer to be considered in the stress generating mechanism. Fig. 2 describes in terms of a probability distribution, the range of velocities a particle may possess when located at the coordinate level x_2. Particles that approach level x_2 from "above" will,

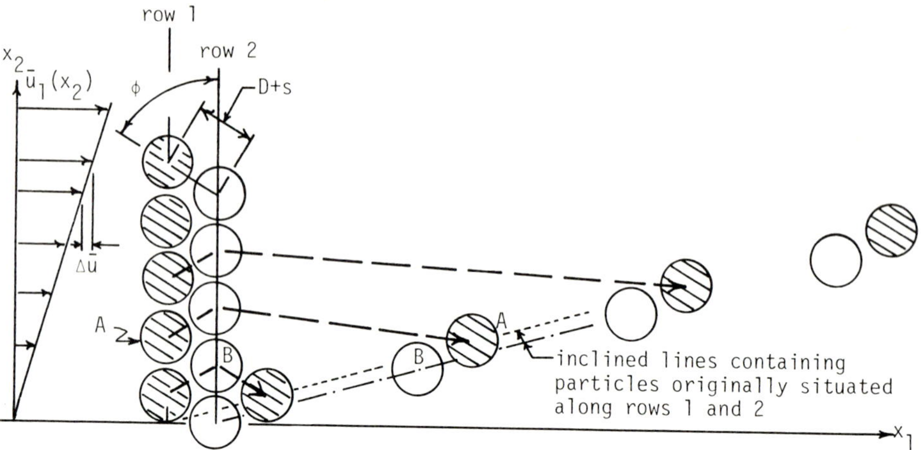

Fig. 1 Two particle configurations consistent with mean flow profile

on the average, have velocities $u_1 > \bar{u}_1(x_2)$ as described by the shaded portion of the probability distribution curve. Particles that approach level x_2 from "below" will, on the average, have velocities $u_1 < \bar{u}_1(x_2)$.

This ability to identify particles as having higher or lower velocities than the mean velocity $\bar{u}_1(x_2)$ depending upon the direction from which these particles arrived, has important consequences when determining the momentum exchange term ΔM_i in Eq. 4. Consider for example the free body diagram of a unit volume shown in Fig. 3. Particles striking particle B from above have, on the average, a velocity greater than $\bar{u}_1(x_2)_B$ where $(x_2)_B$ is the x_2 coordinate of particle B. Particles which strike particle B from below are not stress generating occurrences since such collisions occur between particles interior to the control volume. It is now assumed that particle A, which is exterior to the control volume shown in Fig. 3, has a mean velocity relative to particle B, equal to

$$\vec{V}_A(0) = (D+s)\cos\phi \frac{d\bar{u}_1}{dx_2} + \beta(D+s)\frac{d\bar{u}_1}{dx_2} + \sqrt{2}v' \tag{6}$$

where the first term on the right hand side represents the relative mean motion between particles A and B as indicated by the mean velocity profile, $\beta(D+s)\,d\bar{u}_1/dx_2$ represents the deviation from the mean of all particles striking particle B from "above" and $\sqrt{2}v'$ represents the relative isotropic turbulent velocity fluctuation as derived by Shen and Ackermann (1982). If the relative velocity, $\vec{V}_A(0)$, between colliding particles is known, the model by Shen and Ackermann (1982) is ideally suited to determine the momentum transfer which describes the stress state within a rapidly sheared granular mixture.

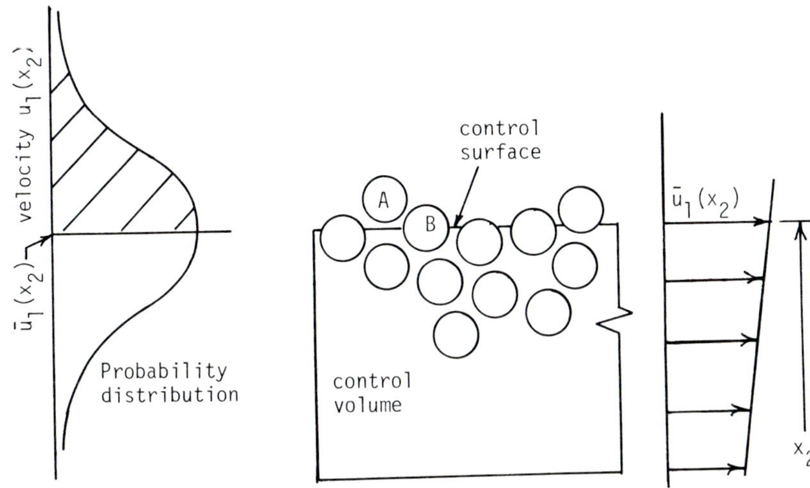

Fig. 2 Probability distribution for velocity component $u_1(x_2)$ of particle B

Fig. 3 Control volume containing particle B on control surface

CONSTITUTIVE RELATIONSHIPS

Consider a pair of particles, A and B, which are about to collide as shown in Fig. 4. The reference frame shown in Fig. 4 moves with the local mean motion $\bar{u}_1(x_2)_B$ where $(x_2)_B$ is the position of particle B just prior to collision. The velocity V_A of particle A relative to particle B is described by Eq. 6. At the point of collision of spheres A and B the normal and parallel components are shown in Fig. 4 and are described as

$$\vec{N} = -\sin\phi\,(\cos\theta\vec{i} - \sin\theta\vec{k}) - \cos\phi\vec{j} \tag{7}$$

$$\vec{P} = \sqrt{\cos^2\phi + \sin^2\phi\,\sin^2\theta}\,\left[\vec{i} - \frac{\sin\phi\,\cos\phi\,\cos\theta\vec{j}}{\cos^2\phi + \sin^2\phi\,\sin^2\theta} + \frac{\sin^2\phi\,\cos\theta\,\sin\theta\vec{k}}{\cos^2\phi + \sin^2\phi\sin^2\theta}\right] \tag{8}$$

where $\vec{i}$, $\vec{j}$, $\vec{k}$ are unit vecotrs in the x_1, x_2, x_3 coordinate directions respectively. The momentum change, $\Delta\vec{M}_B$, of particle B resulting from the collision with particle A is

$$\Delta\vec{M}_B = m_s\,\frac{1+\varepsilon}{2}\,V_{AN}(0)\,[\vec{N} + \mu\vec{P}] \tag{9}$$

where $\vec{V}_{AN}(0)$ is the normal component of $\vec{V}_A$ prior to collision, ε = coefficient of restitution and μ = coefficient of kinetic friction. Since the distribution of v' is considered to be isotropic the contribution of v' to the momentum change component $\Delta\vec{M}_1$ would be negligible. This conclusion is based upon the fact that one-half of the v' components would produce a positive contribution to ΔM_1 while the other half striking the leading half of the reference sphere would produce an equal negative contribution. Therefore the term $\sqrt{2}v'$ of Eq. 6 is neglected when substituting into Eq. 9 to determine ΔM_1.

$$\Delta M_1 = m_s(D+s)\frac{\partial \bar{u}_1}{\partial x_2}\frac{1+\varepsilon}{2}(\sin^2\phi\cos^2\theta-\mu\sqrt{\cos^2\phi+\sin^2\phi\sin^2\theta}\ \sin\phi\cos\theta)(\beta+\cos\phi) \tag{10}$$

By integrating over the range of possible contact points the average momentum change $\overline{\Delta M_1}$ is determined as

$$\overline{\Delta M_1} = \frac{2}{\pi^2}\int_0^{\pi/2} d\phi \int_{\pi/2}^{3\pi/2} \Delta M_1 \ d\theta \tag{11}$$

In the x_2 direction however, the fluctuating components would have a significant effect upon the momentum exchange since all of the velocity fluctuations produce a momentum change having the same sign. For rapidly sheared granular flows the mean relative velocity $(\beta+\cos\phi)(D+s)\ \partial\bar{u}_1/\partial x_2$ is considered to be significantly less than v', making the value of α in Fig. 4 approximately equal to α'. For determining $\overline{\Delta M_2}$, the normal component of Eq. 6 then reduces to

$$V_{AN}(0) = \sqrt{2}v' \cos\alpha \tag{12}$$

The average of the momentum transferred in the $\vec{P}$ direction is zero due to the isotropic distribution of v'. Hence, after substituting Eqs. 7 and 12 into Eq. 9 and dropping the $\vec{P}$ term

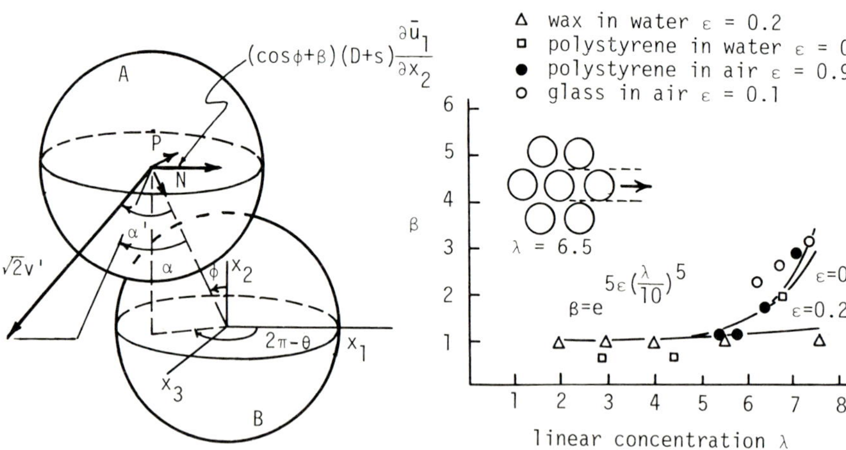

Fig. 4 Spheres at point of collision

Fig. 5 Relationship between β, λ, and ε

$$\overline{\Delta M_2} = - m_s \left(\frac{1+\varepsilon}{2}\right) \sqrt{2}v' \cos\alpha \cos\phi \tag{13}$$

By integrating over the range of all possible contact points

$$\overline{\Delta M_2} = \frac{2}{\pi^3} \int_0^{\pi/2} d\phi \int_0^{2\pi} d\theta \int_0^{\pi/2} (-m_s)\left(\frac{1+\varepsilon}{2}\right) \sqrt{2}v' \cos\alpha \cos\phi \, d\alpha \tag{14a}$$

$$\overline{\Delta M_2} = - m_s \frac{2(1+\varepsilon)}{\pi^2} \sqrt{2}v' \tag{14b}$$

Eqs. 11 and 14b provide the terms ΔM_i which when substituted into Eq. 4 provide the stresses τ_{21} and τ_{22}. However, the term f still remains unidentified. Eq. 3 can now be used to determine f since for a steady rectilinear shear field in which the velocity profile is linear Eqs. 3 and 5 combine to yield

$$\tau_{21} \frac{\partial \bar{u}_1}{\partial x_2} = NFE \tag{15}$$

The frequencies of collision F and f which appear in Eqs. 4 and 5 are related by the equation F = 2f. The term f represents the collisions a particle such as B, shown in Fig. 3, experiences from particles such as A which are exterior to the control volume. The frequency F describes the collisions a particle wholly interior to the control volume experiences with other surrounding interior particles. F can also be described in terms of the fluctuation velocity and integranular spacing as F = v'/s. The energy dissipation E can be expressed in terms of a collision loss produced during grain contact and a loss term caused by viscous dissipation when particles travel thru the interstitial fluid between consecutive collisions. Using the term E as described by Ackermann and Shen (1982)

$$E = \rho_s \frac{\pi D^3}{12} \left[\frac{3}{2} \frac{C_D}{\lambda} \frac{\rho_f}{\rho_s} + \frac{(1-\varepsilon^2)}{4} + \frac{\mu(1+\varepsilon)}{\pi} - \frac{\mu^2(1+\varepsilon)^2}{4}\right] (v')^2 \tag{16}$$

By combining Eqs. 4, 15, and 16 and noting that N = p/D

$$v' = \left\{\frac{(1+\varepsilon)[(0.125+0.144\mu)\beta + (0.053+0.081\mu)] \frac{1+\lambda}{\lambda}}{\frac{3}{2} \frac{C_D}{\lambda} \frac{\rho_f}{\rho_s} + \frac{1-\varepsilon^2}{4} + \frac{\mu(1+\varepsilon)}{\pi} - \frac{\mu^2(1+\varepsilon)^2}{4}}\right\}^{1/2} D \frac{\partial \bar{u}_1}{\partial x_2} \tag{17}$$

All of the terms required to evaluate the shear and normal stresses have now been identified. The shear stress τ_{21} is obtained by substituting Eqs. 11 and 17 into Eq. 4 and noting that f = v'/2s and $p_i = (6C_o/\pi D^2)(\lambda/1+\lambda)^3$ where $\lambda = D/s$ and C_o is the solid concentration at the most dense packing.

$$\tau_{21} = C_o \rho_s D^2 (\frac{\partial \bar{u}_1}{\partial x_2})^2 \frac{\lambda}{2} \{\frac{(1+\varepsilon)^3[(0.125+0.144\mu)\beta+(0.053+0.081\mu)]^3(\frac{\lambda}{1+\lambda})}{\frac{3}{2}\frac{C_D}{\lambda}\frac{\rho_f}{\rho_s} + \frac{1-\varepsilon^2}{4} + \frac{\mu(1+\varepsilon)}{\pi} - \frac{\mu^2(1+\varepsilon)^2}{4}}\}^{1/2} \tag{18}$$

The normal stress τ_{22} is obtained by substituting Eqs. 14 and 17 into Eq. 4

$$\tau_{22} = \frac{-2\sqrt{2}}{\pi^2} C_o \rho_s D^2 (\frac{\partial \bar{u}_1}{\partial x_2})^2 \frac{\lambda}{2} \{\frac{(1+\varepsilon)^2[(0.125+0.144\mu)\beta+(0.053+0.081\mu)](\frac{\lambda}{1+\lambda})^2}{\frac{3}{2}\frac{C_D}{\lambda}\frac{\rho_f}{\rho_s} + \frac{1-\varepsilon^2}{4} + \frac{\mu(1+\varepsilon)}{\pi} - \frac{\mu^2(1+\varepsilon)^2}{4}}\} \tag{19}$$

The quantity β must depend upon the density of the particle packing as well as factors that influence the rate with which a particle would lose its previous mean velocity and acquire that of its new location. The variable most descriptive of the geometric constraints is $\lambda = D/s$ while μ, ε, and C_D would be selected as the variables which dictate how rapidly the "history" of the previous locations is lost. By a process of curve fitting with the experimental data of Bagnold (1954), Savage (1978), and Sayed (1981) the following relationship between β, λ, and ε was found that provides a good agreement between stresses computed by Eq. 7 and measured values.

$$\beta = \exp\{5\varepsilon(\frac{\lambda}{10})^5\} \tag{20}$$

The relationship between β, λ, and ε is shown in Fig. 5 while the relationship between the theoretically determined stresses and measured values are shown in Fig. 6. It is of considerable interest to note that for $\lambda > 6$ the value of β becomes quite sensitive to variations of λ. This would indicate that some physical process becomes markedly more important when the particle packing exceeds the linear density $\lambda \doteq 6$.

The upper portion of Fig. 5 describes spheres having a hexagonal packing with a value of $\lambda = 6.5$. This represents the limiting value where particle motion is still possible without causing a dislocation of continguous spheres. This possible direction of motion is shown by the arrow in the figure. For values of λ larger than 6.5 a particle configuration must occur which facilitates the formation of slip planes by which particles can move laterally to the direction of the mean motion. When λ is less than approximately 6 the value of β is approximately equal to 1. In terms of the concepts of fluid turbulence, the mixing length for such a condition would be said to be approximately one grain diameter. Without such motion it would be impossible that a nonzero value of the velocity gradient could exist. Such a description of particle motion is at this point quite superficial. However, these insights into the mechanism by which the β term originates will provide a basis from which to explore this phenomenon in more detail.

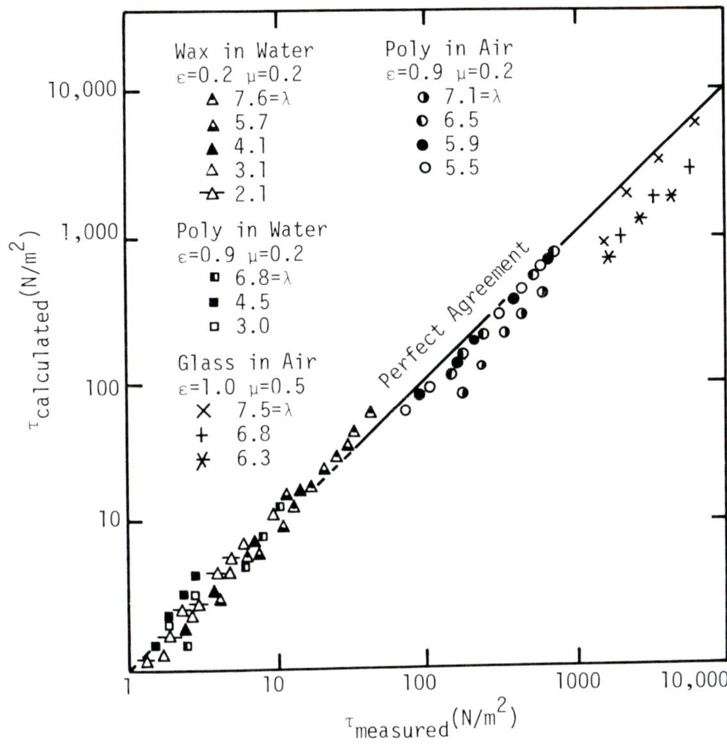

Fig. 6 Relationship between measured and calculated stresses

There exists still another confirmation that the lateral mass transfer and the resulting momentum exchange as described by the β coefficient is an important component of the stress generating mechanism. The ratio of the shear to normal stress, $\tau_r = \tau_{21}/\tau_{22}$ as determined from Eqs. 18 and 19 agrees very closely with the values observed in the experimental results reported by Savage (1978) and Sayed (1981). Table 1 describes the relationship between theoretical and measured values of τ_r. Experimentally determined values of τ_r

TABLE 1

Polystyrene in air (ε=0.9, μ=0.2)

λ	τ_r Measured	Calculated
7.1	.51	.63
6.5	.57	.57
5.9	.58	.53
5.5	.60	.51

Glass in air (ε=1.0, μ=0.5)

λ	Measured	Calculated
7.5	.54	.59
6.8	.54	.49
6.3	.56	.45

obtained by Bagnold for wax spheres in water differed, however, by a factor of 4 from values calculated from Eqs. 18 and 20.

CONCLUSIONS

Constitutive equations have been developed for a rapidly sheared mixture of densely packed spherical solids in a fluid. The theoretically determined stresses agree reasonably well with experimentally determined values. The effect of lateral mass transfer was found to be an essential consideration when deriving the constitutive relationships.

REFERENCES

1 Ackermann, N.L., and Shen, H.H., "Stresses in Rapidly Sheared Fluid-Solid Mixtures," Jour. Engineering Mechanics Div., ASCE, EM1, Feb. 1982, pp. 95-113.
2 Bagnold, R.A., "Experiments on Gravity-Free Dispersion of Large Solid Spheres in a Newtonian Fluid Under Shear," Proc. Roy. Soc. London, Ser. A, V. 225, 1954, pp. 49-63.
3 Jenkins, J.T. and Cowin, S.C., "Theories for Flowing Granular Materials," The Joint ASME-CSME Appl. Mech. Fluid Engrg. & Bioengrg. Conf., AMD, Vol. 31, pp. 79-89, Niagara Falls (1979).
4 Kanatani, K.I., "A Micropolar Continuum Theory for the Flow of Granular Materials," Int. J. Eng. Sco., V. 17, 1979, pp. 419-432.
5 McTigue, D.F., "A Nonlinear Continuum Model for Flowing Granular Materials," Ph.D. Dissert., Stanford Univ., San Francisco, 1979, 165 p.
6 Ogawa, S., Umemura, A., and Oshima, N., "On the Equations of Fully Fluidized Granular Materials," ZAMP, Vol. 31, 1980, pp. 483-493.
7 Savage, S.B., "Experiments of Shear Flows of Cohesionless Granular Materials," Proceedings of the U.S.-Japan Seminar on Continuum Mechanical and Statistical Approaches in the Mechanics of Granular Materials, Sendai, Japan, 1978, pp. 241-254.
8 Sayed, M., "Theoretical and Experimental Studies of the Flow of Cohesionless Granular Materials," Ph.D. Thesis, McGill Univ., 1981.
9 Shen, H. and Ackermann, N.L., "Constitutive Relationships for Fluid-Solid Mixtures," Jour. Engineering Mechanics Div., ASCE (to appear October 1982).

Mechanics of Granular Materials: New Models and Constitutive Relations, edited by
J.T. Jenkins and M. Satake, 1983
Elsevier Science Publishers B.V., Amsterdam — Printed in The Netherlands

SIMULATION OF GRANULAR FLOW

G. W. Hawkins

Bendix Advanced Technology Center, Columbia, Maryland

ABSTRACT

A computer simulation of rapid granular flow has been developed to investigate the properties of granular materials. The system is designed to be very simple and efficient, so that simulated flow problems can be set up easily and run until a steady state flow is achieved. The particles are modeled as two-dimensional rigid disks with realistic coefficients of friction and restitution. Stochastic boundaries are used to eliminate wall effects, for the simulation of uniform flow. An advanced personal computer system is used to perform the computations and to display movies of the flow, for rapid comprehension of the local flow mechanics.

INTRODUCTION

The use of granular materials in advanced material processing systems requires prediction of the flow of dry granular materials, both in general and in specific configurations. Analytical flow models have been developed and flow properties have been measured experimentally in real materials (ref.1), but the numerical simulation of granular flow, based on an analysis of the motions of the particles of the material, has special advantages for investigating this regime. The particles of a granular material can be modified in many ways; the effects on the flow are hard to model analytically, but a direct simulation of the particles can be modified to account for most particle properties. Real experimental flow test configurations have wall effects and flow inhomogeneities that may depend on particle properties. Simulations can have idealized configurations, or can model real configurations to measure wall effects and flow inhomogeneity. The results of a flow simulation can also be displayed as plots or flow movies.

As computations become cheaper, the direct modeling of idealized particles has become common in statistical physics studies (ref.2). Simulation of more realistic macroscopic particles has been done for some time (ref.3), but only recently has simulation become sufficiently inexpensive that extensive simulated flow tests have been carried out by Campbell and Brennen (ref.4). This paper describes the development and use of a granular flow simulation system, using an advanced personal computer to carry out the computations and display the results as movies of the flow. The system is specialized for the measurement of uniform flows, to derive powder flow properties as a function of particle properties.

FLOW SIMULATION SYSTEM

The flow simulation system was designed with three objectives:

1. Ease of use and modification so that different flow configurations and particle properties could be investigated,
2. Sufficient computational efficiency to allow model flows to reach a steady state,
3. Good display and analysis systems to make the flow results meaningful.

To these ends, the main simulation program was made as simple and modular as possible, and the particles were limited initially to two-dimensional rigid disks with the interactions between particles characterized by friction and restitution coefficients. The flow simulation is similar in some parts to the system developed by Campbell and Brennen. Campbell provided consultation and advice, and many of the program design decisions were based on his experience in developing his own system.

The particle simulation system relies on a "ballistic" model for the particles involved. The particles are assumed to travel along straight paths, spinning at constant speed, until disturbed. The program stores the paths of the particles, and computes an ordered list of intersections of particles with other particles and boundaries by extrapolation of the paths. The program advances the time variable to the time of the next intersection and the collision models are used to compute the alteration in the particle paths due to the collision. The new paths are used to compute new intersections, which are merged into the list of intersections (see Fig. 1).

The paths of the particles are stored as the x, y and angular velocities, the x, y and angular position of the last collision and the time of the last collision, all in single precision arrays. The extrapolation of the path from the last collision maintains the accuracy of relative positions for close packed particles, where the mean free path is very short. The use of single precision variables improves the efficiency of the program and can be justified for flows of macroscopic particles where the properties can not be precisely defined. It does require care in the program design to avoid errors or ambiguities.

On a large scale, the particles are rigid, in that the radius is assumed to be exactly constant. The collisions are assumed to be instantaneous. This means that collisions will never interact or interfere, and that only binary collisions need be analyzed. On a smaller scale, the particles are assumed to have energy losses and elastic energy storage at the points of contact,

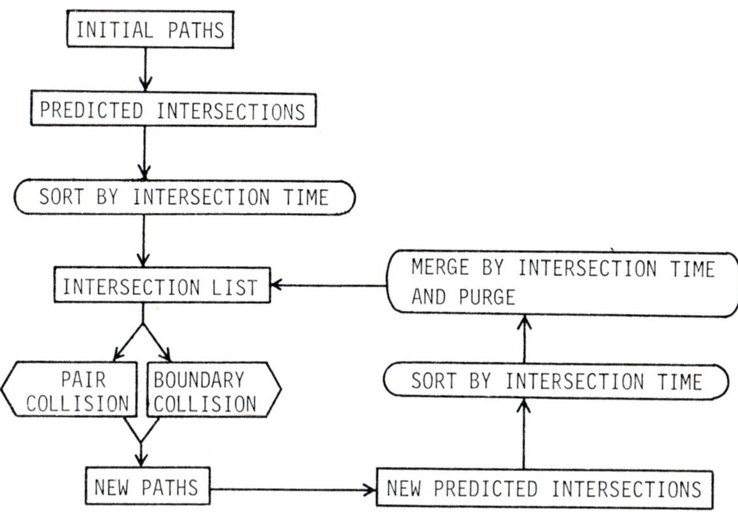

Fig. 1. Flowchart for the Simulation System

expressed in the coefficients of friction and restitution. The algorithm computes the change in relative sliding velocity during the collision, and eliminates the friction force when the particles stop sliding and begin to roll on one another.

COLLISION PREDICTION

The physical realism of the simulation depends heavily on the ability of the program to predict accurately and efficiently whether particles will collide. The collision prediction begins with computation of the time of intersection of each particle with other particles and boundaries. The events that represent real potential collisions in the future are assembled into a intersection list, ordered by intersection time. The next collision to be executed is taken from the top of the list, with no extra checks for particle contact or overlap. If the computations were exact, no difficulty would arise. However, when the round off error of the computation exceeds the time between collisions, the intersection time of a collision that lies a short way into the future may in fact be computed as occurring in the past. If the collision is not executed, the particles will begin to overlap, and the physical realism of the simulation will be lost. The algorithm that decides to accept a potential collision into the intersection list must decide correctly despite the finite error.

For the collision of two particles, solving the linear path equations simultaneously for contact of the two particles produces a quadratic equation with the time of intersection as the smaller root. The larger root can be used to decide whether to accept an intersection. The two solutions define the times of first and last contact of the particles if they continued along their paths and overlapped without collision (see Fig. 2a). If the collision will cause significant changes in the paths of the particles (i.e. not a grazing collision), then the amount of overlap must be significant. If the velocities of the particles are not excessive, then the two contact times must differ by more than the round off error. Even if the first contact time has been moved into the past by the amount of the round off error, the time of last contact must lie in the future. Therefore, the algorithm for selecting intersections accepts those with a last contact time that is in the future. This is easy to compute because most of the work has been done while computing the first contact time.

For boundary intersections, a simpler method suffices. If the particle is moving toward the boundary, then the potential collision must be in the future. The decision can be made more efficiently if the boundaries have a specific direction defining the inside, i.e. the side of the boundary that the particles can hit (see Fig. 2b). Thus if the direction of the particle velocity is opposite that of the boundary direction, the particle is moving toward the boundary and the collision is accepted into the collision list.

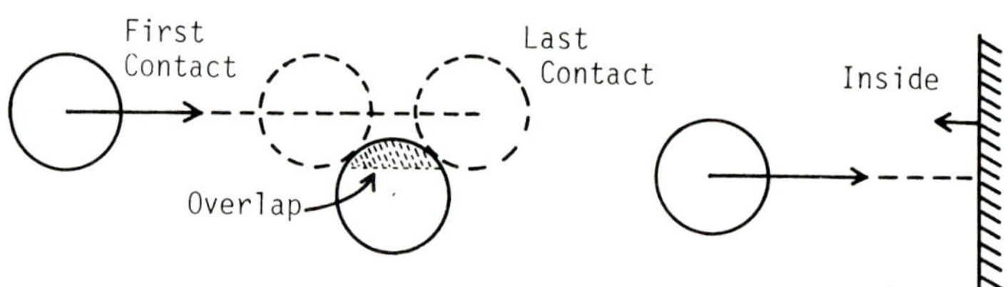

a. collision between particles b. collision with boundary

Fig. 2. Decision Algorithm for Potential Collisions

As a consequence of accepting collisions that roundoff error has moved into the past, the physical reality of the collisions is preserved, but the variable describing the current time may jump backward by an amount equal to the round off error. This does not mean that the simulation runs backwards. Collisions that have already occurred will not be accepted into the collision list because the paths that produced them have changed. Previously nonexistent collisions between other particles will not be accepted, because only the changed paths are examined for new intersections. These methods have successfully avoided particle overlap, even in difficult conditions when the friction is high, the restitution is low and the particles tend to clump together.

The entries that are accepted into the intersection list are stored as an intersection time and the numbers of the particles involved, ordered by intersection time. Boundaries are coded with special numbers so that the boundary collision algorithms can be used. New entries are sorted by intersection time into a new entry list. This list is merged into the intersection list, while the intersection list entries for paths that have been altered are purged. The upkeep of the intersection list consumes a considerable fraction of the total computation time, so efficiency is important. The time involved can be reduced by setting a time horizon, and discarding any intersection that lies beyond it. When the time variable reaches the horizon, the intersection list is empty and the simulation is restarted just like a new problem. The computation required for restarting the simulation occasionally is less than the labor saved in sorting and merging the smaller lists at each collision.

COLLISION MODEL

For the particles involved in a collision, the changes in path are computed from the geometry and mass of the particles, the conservation of linear and angular momentum and the coefficients of friction and restitution. The critical question is whether the relative sliding velocity of the two particles is reduced to zero before the particles separate. Goldsmith (ref.5) solved for the critical relation between the relative sliding velocity and the normal velocity. His plot of the change in normal and tangetial impulse shows clearly when the particles stop sliding and begin to roll on one another.

It is a considerable simplification to describe the properties of real particle collisions by two coefficients, and the details of such collisions need to be analyzed by finite element models, for the low speed, small deformation collisions important in granular flow. Specifically, the use of a coefficient of restitution implies that elastic energy storage occurs in the contact zone for normal displacements. The corresponding energy storage for tangential displacements is neglected. There are varying opinions as to the importance of these terms (ref.6).

BOUNDARY SIMULATION

A special boundary condition was developed to eliminate the wall effect and simulate an infinite homogeneous flow. The boundary reflected the particles elastically and also applied an impulse to the particles, tangential to the boundary. This impulse supplied energy to drive the flow. The collision was assumed to act on the center of mass of the particle, with no effect on the spin. In an infinite uniform flow, much of the shear stress which drives the flow is transferred by normal impacts between particles. If a flat moving wall with friction is used to drive the flow, the stress is transferred by friction on the surface of the particles, with a large input to the spin of the particles nearest to the boundary (ref.4). The special boundary eliminated this wall effect.

The location of the boundary was stochastic, in that the boundary had a finite thickness and the cumulative probability distribution function of collision with the boundary was linear from zero at the inside edge to one at the outside edge of the boundary. If the particle was already inside the boundary thickness, the current position was taken as the zero position. The variation in the location of the boundary helped to avoid "crystallization" of the particles nearest the wall, which can be a severe problem if the particles are all the same size.

FLOW TEST CONFIGURATIONS

The flow configuration used initially was a shear box. A square box with tangential impulse boundaries (see Fig. 3a) was used to generate a pure shear flow in its central region. This configuration was reasonably successful in simulating pure shear, so long as the number of particles was sufficient to limit the boundary effects. An occasional problem was encountered when a particle would become stuck in a corner where the impulses converged. This would stall the computation, with endless little motions of the one particle.

Another configuration was developed to simulate simple shear, using a long narrow rectangular box with tangential impulse boundaries (see Fig. 3b), which has a simple shear flow in its central region. This was an alternative to the use of periodic boundary conditions (ref.4) and also simulated an interesting experimental shear test using a bag as used by Reynolds (ref.7).

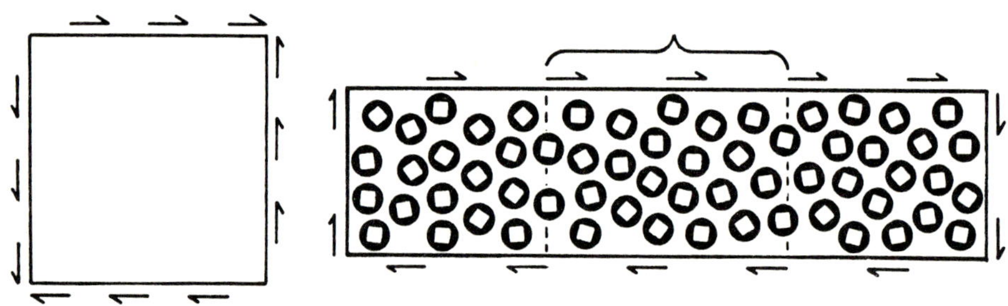

a. Shear box b. Reynolds bag

Fig. 3. Test Configurations for Pure Shear and Simple Shear

ANALYSIS AND DISPLAY OF FLOW RESULTS

The analysis of the results of a flow simulation takes many forms. Large scale parameters like stress and strain components were evaluated by averaging the momentum transfer of the particles. The local statistical parameters like the fluctuation velocity distribution or the angular contact distribution were collected while computing the collision effects. Finally, the actual display of the positions and motions of the particles in the form of flow movies were used to determine the character of the flow.

The ability of an advanced personal computer system to display animated movies of the particle flow is a consequence of the method of displaying text. The character fonts used for the text are stored in the memory used for the display as a set of character images. Text is displayed by copying rectangles including the appropriate character images to the desired location on the screen. The copying operations, called "blits", are performed by special hardware at great speed.

The flow movie generator used the same hardware to move images of the particles to the proper relative location in a work area and then to move the whole group of images to a display area of the screen (see Fig. 4). This reduced the number of operations in the display area and thus limited the flicker of the screen as it is updated. With the overhead involved in computing the desired particle locations, approximately one thousand blits per second were performed, which allowed the animation of about forty particles. The "font" of particle images was generated by using a circle with a square inside to show the angular orientation of the particle. The rotation of the square indicated clearly the angular velocity of the particle.

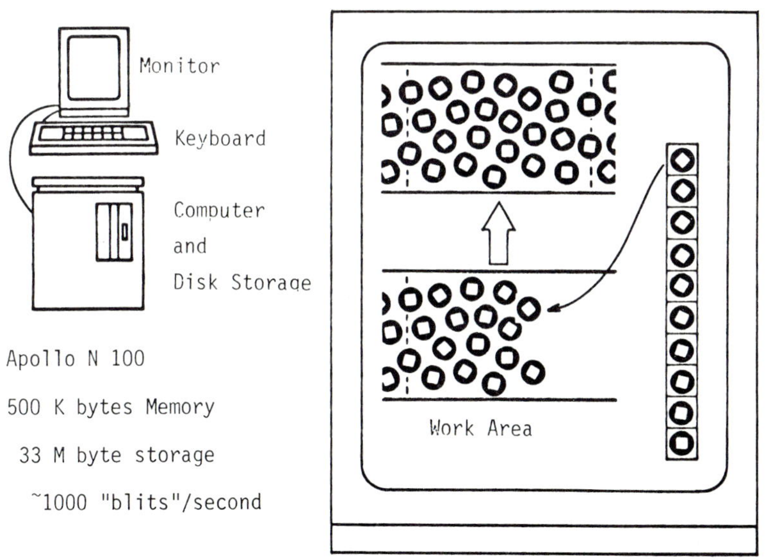

Figure 4. Generation of Flow Movies on an Advanced Personal Computer

The animation provided an immediate picture of the results of the flow simulation. Details of the collisions were examined by zooming in on interesting features. It proved very useful for debugging new features of the flow simulation, and for examining the motions of the particles to get new insight for analytical flow models.

As an example of the benefits of the movie system, initial results for pure shear tests indicated that the influence of the friction and restitution losses on the flow resistance were approximately equal when both effects were moderate However, visual observation of the flow indicated that there are differences in the action of the loss mechanisms. The loss of energy in the coefficient of restitution appears to retard the transmission of energy in pressure waves very strongly compared to sliding friction losses. A series of "acoustic" wave simulations are being performed to clarify and quantify these observations.

CONCLUSIONS

The flow simulation system and movie display system described in this paper provided a very useful means of relating the flow properties of a granular material to the properties of the particles. The collision prediction techniques and intersection list keeping methods made the computation efficient enough to analyze prolonged flows. The boundary conditions simulated a uniform flow and allowed good viscometric measurements. By using the text display hardware on the advanced personal computer, the movie system displayed the flow results very clearly. Using these systems, the direct simulation of granular flow was very productive.

REFERENCES

1. The theories and experiments in rapid granular flow are reviewed in Savage, S.B., Nedderman, R.M., Tuzun, U., and Houlsby, G.T. The Flow of Granular Materials-III (Rapid Shear Flows), Chemical Engineering Science, Review Article Number 9, 1982.
2. for a recent example, see Evans, D.J. Shear induced melting of soft-sphere crystals, Phys. Rev. A, Vol. 25, No. 5, May 1982, pp 2788-2792.
3. Cundall, P.A. and Strack, O.D.L. A discrete numerical model for granular assemblies, Geotechnique, 29:47-65 1979.
   Cundall, P.A. and Strack, O.D.L. Modeling of microscopic mechanisms in granular material, in these proceedings.
4. Campbell, C.S. and Brennen, C.E. Computer simulation of shear flows of granular material, in these proceedings.
   Campbell, C.S. Shear flows of granular material, PhD Thesis and Report No. E-200.7, Div. of Eng. and Appl. Sci., Calif. Inst. of Tech., Pasadena, Ca., 1982.
5. Goldsmith, W. IMPACT, Edward Arnold (pub.) Ltd., London, 1960.
6. Walton, O. Particle dynamics modelling of geological materials, Univ. of Calif. Lawrence Livermore Lab. Report UCRL-52915, 1980.
7. Reynolds, O, On the Dilatancy of Media composed of Rigid Particles in Contact, Phil. Mag. S.5, No. 127, Dec. 1885, pp 473.

Mechanics of Granular Materials: New Models and Constitutive Relations, edited by J.T. Jenkins and M. Satake, 1983
Elsevier Science Publishers B.V., Amsterdam — Printed in The Netherlands

COMPUTER SIMULATION OF SHEAR FLOWS OF GRANULAR MATERIAL

C.S. CAMPBELL AND C.E. BRENNEN
California Institute of Technology, Pasadena, California 91125

ABSTRACT

The purpose of this paper is to present results from computer simulations of Couette flows of granular materials and to examine the detailed rheological behavior inherent in these simulations. Comparison is made with the experimental results of Bagnold (1954) and Savage and Sayed (1980, 1982) as well as with the various theoretical constitutive models (see below).

INTRODUCTION

In recent years there has been a significant increase of interest in the detailed mechanics of granular material flows and considerable progress had been made toward understanding the different regimes of flow. The present state of knowledge has recently been reviewed by Spencer (1981) and Savage (1982). The latter comprehensively reviews the recent experimental and theoretical work in higher speed granular flows which lie in the so-called "grain inertia" regime. It is clear that the experimental work in this area is hindered by lack of well-proven, non-intrusive instrumentation necessary to document precisely the velocity, solid fraction and granular temperature distributions within such flows. On the other hand most of the existing theoretical work is limited to modest solid fractions of the order of 0.3 to 0.4. Considerable work is required to extend theories to solid fractions approaching the critical value. An alternative approach is to utilize the kind of computer simulations which have been quite successful in the analogous problems in molecular gas dynamics (eg. Bird (1976), Barker and Henderson (1976), Wood (1975)).

The purpose of such computer simulations is threefold. First they can provide insights and detailed results which can be used to suggest appropriate assumptions and validate theoretical models particularly at higher solid fractions. At the present time this is best accomplished using simple shear flows or Couette flows. This is the specific objective of the present paper. Two other objectives may, however, be mentioned in passing. One is to compare the simulation results with experimental measurements in order to establish appropriate models of the mechanics of particle/particle and particle/wall interactions. Efforts in this direction are reported elsewhere (Campbell (1982), Campbell and Brennen (1982)) with specific attention to the simulation of chute flows for which some experimental results exist. Parenthetically one might add that a third use of computer simulations might be the exploration of the effects of additional forces (such as those due to the interstitial fluid or electrostatic effects). For simplicity, such effects are not included in the simulations presented here. Discussion of interstitial fluid effects can be found in other recent works (eg. Savage (1982)).

Other computer models of granular material flows include the work of Cundall (1974), Davis and Deresiewicz (1977), Cundall and Strack (1979), Trollope and Berman (1980) and Walton (1980). Most of these are directed toward the simulation of slower flows and smaller deformations. However, in his original work Cundall (1974) did extend his methods to some higher speed but transient flows such as rockfalls and the emptying of a hopper. Also the versatility of Walton's (1980) computer program is readily apparent from the excellent movies which the Lawrence Livermore Laboratory has produced. However to our knowledge none of the existing models have been used to produce continuous "steady" flows which could be used for basic rheological purposes. The intent of the present model was to minimize the complexity of the geometry and the interactions so that steady flows of sufficiently long duration for rheological analysis would be produced. Thus the simulations are just two-dimensional and the particles are circular cylinders. Though extension of the simulation to the third dimension is possible, the necessary computer time would be considerably greater.

Granular material flows in the grain inertia regime have much in common with molecular gas dynamics (Chapman and Cowling (1970), Ferziger and Kaper (1972)) and the recent theoretical work of Blinowski (1978), Kanatani (1979 a,b, 1980), Ogawa and Oshima (1977, 1978a,b, 1980), Savage and Jeffrey (1981), Ackermann and Shen (1982, 1982) and Jenkins and Savage (1982) has drawn extensively from that source. An advantage in granular media flows is that the hard sphere model is appropriate provided the interstitial fluid and electrostatic effects can be neglected. Difficulties do however arise from the dense packing and the inelasticity of the collisions.

Extensive work has, of course, been done on the computer simulation of molecular gas dynamic flows. In that arena, Monte Carlo methods (Bird (1976)) have proved particularly efficient. In the future, such methods may also prove valuable for granular material flow simulations. However the present simulation is entirely mechanistic.

COMPUTER SIMULATION

The present method for the computer simulation of granular material flows has been documented elsewhere (Campbell (1982)) and will only briefly be described in the present paper. Two dimensional unidirectional flows of inelastic circular cylinders are followed mechanistically. The flow solutions sought have no gradients in the flow direction (x-direction, Figure 1). Though both Couette flows and inclined chute flows have been examined the present paper is confined to discussion of Couette flows in the absence of gravity (see Campbell and Brennen (1982) for inclined chute flows). The simulation is initiated by placing a number of cylindrical particles (radius, R; mass, m) in a control volume bounded by the solid boundaries and two perpendicular "periodic" boundaries (see Fig. 1). A particle passing out of the control volume through one of the periodic boundaries immediately re-enters the control volume through the other periodic boundary at the same distance from the solid boundaries and with the same instantaneous velocity. Thus the number of particles initially placed within the control volume remains constant. Simulations with different distances, L, between the periodic boundaries were performed until further doubling of this length had little effect upon the results. In the Couette flow simulations values of L/R of about 10 were found to be acceptable (see Fig. 2).

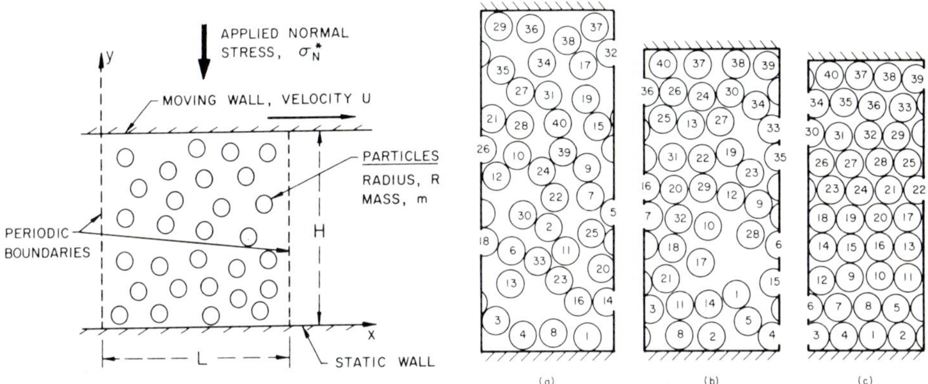

Figure 1. Schematic of computer simulation.

Figure 2. Typical snapshots of the Couette flow simulations for (a)ν=0.56 (b)ν=0.63 (c)ν=0.76.

All simulations were begun with a randomly perturbed square lattice array of particles in the control volume and randomly chosen instantaneous translation velocities (u,v, in the x,y, directions of Fig. 1) and rotational velocity (ω). This entire assemblage of particles is followed mechanistically as the particles collide with each other, the static lower wall and the moving upper wall (velocity, U, in the x direction) until a final asymptotic state is reached in which the flow is steady in the sense of being invariant over long time scales. The first Couette flow simulations attempted maintained a constant wall separation, H, as well as a constant moving wall velocity, U; the intent was to evaluate the asymptotic normal and shear stresses generated at the walls. However it was found that simulations specified in this way were unstable particularly at the lower solid fractions; all the particles tended to migrate to a localized region so that they no longer came into contact with one or both walls. The simulation was then altered to maintain a predetermined normal stress, σ^*_{yy}, on the walls (as well as a given moving wall velocity, U). This required continuous adjustment of the gap, H, according to whether the normal stress exerted by the particles on the walls was greater than or less than the prescribed normal stress (for details see Campbell (1982)). Such a procedure is, of course, analogous to the manner in which the experiments of Savage and Sayed (1980,1982) were performed. These simulations consistently converged, typically requiring 20,000 collisions to reach an asymptotic state.

After convergence a substantial extension of the simulation (of the order of 20,000 collisions) was required to obtain sufficient duration of "steady" flow for meaningful statistical analysis. Since the simulations exhibited fluctuations typical of small thermodynamic systems (Landau and Lifshitz (1958)), this duration had to be longer than one might otherwise anticipate.

SIMULATION DETAILS

Each individual particle/particle or particle/wall collision is assumed to be instantaneous and conventional means are used to solve for the departing linear velocities (u,v in the x,y directions, Fig. 1) and angular velocity, ω, given the incident velocities. In the process, impulses normal and tangential to the contact surfaces are evaluated. In addition to the linear and angular momentum/impulse equations two further relations are required to close a collision solution. First the relative approach and departure velocities normal to the contact surface are related by a coefficient of restitution to represent the inelasticity of the collisions. Different coefficients(denoted by ε_p and ε_w respectively) can be used for particle/particle and particle/wall collisions. Secondly a frictional closure condition pertaining to the relative tangential velocities of the contact points upon departure is necessary. The present simulation assumed that relative tangential motion is destroyed by a collision and hence that the tangential velocities of the contact points are identical upon departure. Parenthetically it should be noted that this choice of the final closure condition is somewhat arbitrary; other choices are possible. However, this second closure condition is used for particle/particle collisions in all the simulations reported here.

On the other hand, for reasons discussed later, two separate types of particle/wall collision (Types (A) and (B)) were investigated. The above mentioned second closure condition is used for particle/wall collisions in the type (A) Couette flows; this choice approximates "smooth" interior surfaces of the walls(see Campbell and Brennen (1982)). In addition simulations (type (B)

Couette flows) were run with an alternative second closure condition intended to approximate "roughened" interior surfaces. In this case a particle colliding with a wall departs with a particle center velocity whose tangential or x component is equal to that of the wall.

ANALYSIS OF THE SIMULATION RESULTS

The results of the Couette flow simulations are presented non-dimensionally by dividing the velocities by the upper wall velocity, U, the stresses by mU^2/R^2, and lengths by either H or R. Consequently the only parameters which affect the results are (i) the wall spacing to particle size ratio, H/R (which is only indirectly specified by fixing the number of particles in the control volume) (ii) the non-dimensional normal stress $\sigma_{yy} = \sigma^*_{yy}R^2/mU^2$ (iii) The coefficients of restitution, ε_w and ε_p and (iv) the ratio of radius of gyration of a particle to its radius, β. This last parameter was maintained at the value for solid cylinders (0.5) throughout.

All flow properties (say, p) were assessed as functions of y/H by dividing the control volume into strips parallel to the x-axis and computing the particle area weighted mean of that property in each strip. The time-averaged value of this is denoted by $<p>$. Mean velocity, $<u>$, and solid fraction profiles are evaluated. In addition, the squares of the components of both the fluctuating translational and rotational components of velocity (as opposed to the velocities, $<u>$, associated with the mean shear flow) were evaluated as

$$<u'^2> = <u^2> - <u>^2 \quad ; \quad <v'^2> = <v^2> \quad ; \quad <\omega'^2> = <\omega^2> - <\omega>^2 \qquad (1)$$

Hence the kinetic energy associated with the random motion of the particles (as opposed to that of the mean shear flow) is given by

$$\tfrac{1}{2}m(<u'^2> + <v'^2> + \beta<\omega'^2>) \qquad (2)$$

and, for convenience, this is termed the "total granular temperature". The part of this associated with fluctuating translational motions (i.e. without the last term of the expression (2)) is termed the "translational granular temperature". Both temperatures are non-dimensionalized by $\tfrac{1}{2}mU^2$.

In the existing theoretical studies (see Savage (1982)) the parameter, S, (Savage's R) is used as a measure of the ratio of the typical velocity difference associated with the mean shear flow, $2R\,d<u>/dy$ and the typical translational fluctuating velocity. Here S values are assessed as

$$S = 2R\frac{d<u>}{dy} / [<u'^2> + <v'^2>]^{\frac{1}{2}} \qquad (3)$$

In the converged state, Couette flows have normal and shear stresses, σ_{yy} and

σ_{xy}, which are uniform throughout the flow and are equal to the stresses exerted by the walls. In the results which follow they are both evaluated simply by summation of the normal and tangential impulses applied to the walls by particle collisions.

VELOCITY, SOLID FRACTION AND TEMPERATURE PROFILES

Three typical instantaneous snapshots of the arrangement of particles in the control volume are presented in Fig. 2. The mean solid fraction in these examples vary from $\nu = 0.56$ to $\nu = 0.76$. For future reference note the increased tendency for the particles to arrange themselves in layers at the higher solid fractions.

Typical velocity and solid fraction distributions for the type (A) (or "smooth wall")Couette flows are presented in Fig. 3 for, $\varepsilon_W = 0.8$, $\varepsilon_p = 0.6$ and various normal stresses and H/R. The velocity distributions are characteristic of all such simulations indicating "slip" at both walls amounting to about 20% of the relative velocity and regions of higher shear close to the walls. The solid fraction profiles exhibit a corresponding decrease in ν in these regions of higher shear.

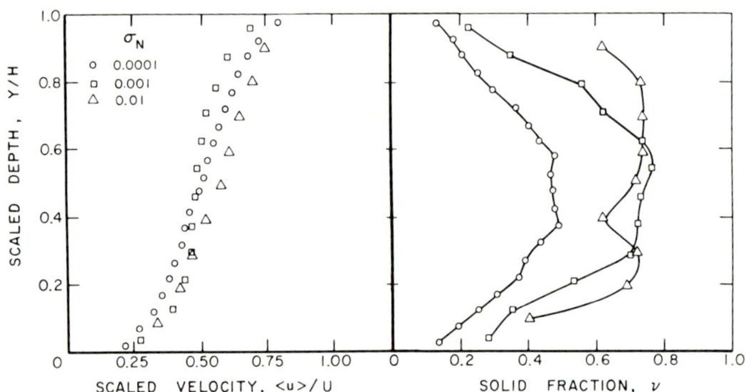

Figure 3. Typical velocity and solid fraction profiles for the type A Couette flow simulations for $\varepsilon_p = 0.6$, $\varepsilon_W = 0.8$. Data is shown for $\sigma_{yy} = 0.0001$ and H/R = 39.6 , $\sigma_{yy} = 0.001$ and H/R = 23.9, $\sigma_{yy} = 0.01$ and H/R = 19.5.

Typical type B (or "rough wall") Couette flow profiles are presented in Fig. 4 for three different normal stresses, σ_{yy} (and H/R though this does not vary much in this set of data). The resulting linear velocity profiles have no slip and the solid fractions are virtually uniform across the flow. The total granular temperature profiles are more scattered but do indicate a temperature which is fairly uniform over the central core of the flow; there is however some indication of a reduced temperature near both of the walls. Globally it is clear that the solid fraction increases and the temperature decreases as the normal stress increases. In fact the type B simulations closely approximate simple shear flow and are therefore well suited for rheological investigations. Therefore the remainder of the paper concentrates on the type B simulations.

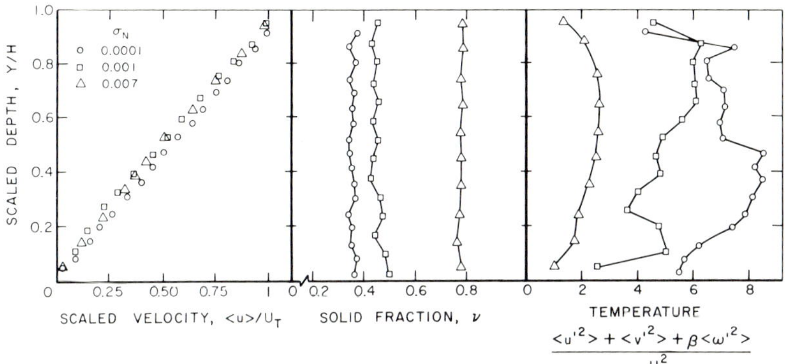

Figure 4. Typical velocity, solid fraction and total granular termperature profiles for the type B Couette flow simulations for $\varepsilon_p = 0.6$, $\varepsilon_w = 0.8$. Data is shown for $\sigma_{yy} = 0.0001$ and $H/R = 39.7$, $\sigma_{yy} = 0.001$ and $H/R = 29.2$, $\sigma_{yy} = 0.007$ and $H/R = 20.8$.

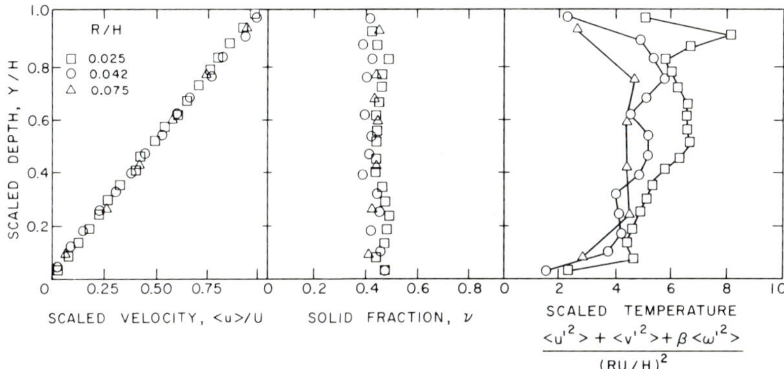

Figure 5. Comparison of type B Couette flow simulation data for different H/R as indicated but all with $\sigma_{yy} = 0.567(R/H)^2$. Note the temperature scale in this figure has been divided by $(R/H)^2$.

Figure 5 presents a comparison of typical results for different values of H/R. Specifically the three cases shown all have a dimensionless applied normal stress equal to $0.567\ (R/H)^2$. According to the existing constitutive models discussed below this should yield the same solid fraction and a temperature proportional to $(R/H)^2$ in all three cases. We note that this indeed appears to be the case.

COMPARISON WITH EXISTING CONSTITUTIVE MODELS

In the absence of interstitial fluid, electrostatic or other additional effects, dimensional analysis requires that the normal and shear stresses (σ^*_{yy} and σ^*_{xy}) in a simple shear flow must be given by

$$\sigma^*_{ij} = \rho_p [R \frac{d<u>}{dy}]^2 f_{ij} \quad (4)$$

where ρ_p is the particle density and the quantities f_{yy} and f_{xy} can only be functions of S, ν and ε_p. This relation agree with Bagnolds (1954) original heuristic constitutive laws. We have already demonstrated in Fig. 5 that the scaling implicit in the relation is indeed manifest by the present simulations.

In non-simple shear flows (such as the type A Couette flows or the chute flows of Campbell and Brennen (1982)) temperature gradients can exist in the y direction; steady flows can be generated in which granular heat is conducted in the y direction and introduced or removed through solid boundaries. Both Ogawa, Oshima et al (1977, 1978a,b , 1980) and Kanatani (1979a,b, 1980) have explored such phenomena theoretically. However since their results for f_{ij} are not in very good agreement with the measurements (Savage (1982)) an appropriate form of the granular energy equation which would be required in order to model non-simple shear flows remains to be established. We have previously shown (Campbell and Brennen (1982)) that such a relation is necessary to understand the development of plugs in chute flows.

Therefore we concentrate here on the less ambitious task of establishing the functions f_{yy} and f_{xy} in simple shear flows. Values of f_{yy} and f_{xy} obtained from the simulations are presented in Figs. 6 and 7 as functions of the solid fraction (and for several values of $\varepsilon_p, \varepsilon_w$). The ratio f_{xy}/f_{yy} which is the

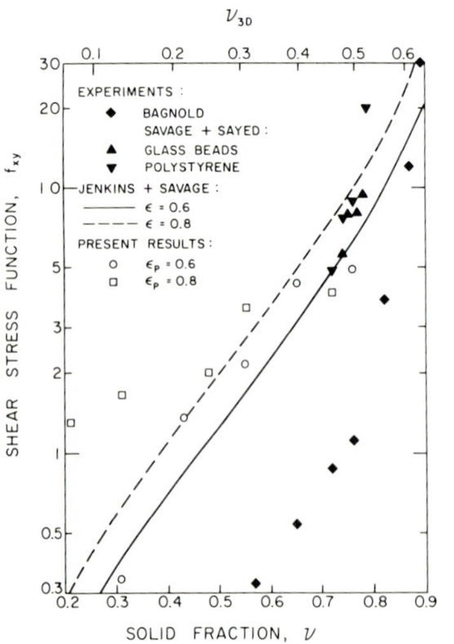

Figure 6. The shear stress function, f_{xy}, as a function of ν.

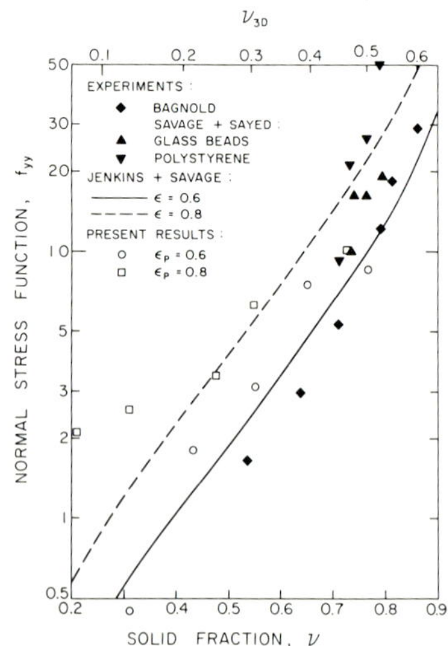

Figure 7. The normal stress function, f_{yy}, as a function of ν.

effective coefficient of friction for the material is presented in Fig. 8. For reference, the corresponding values of S are presented in Fig. 9.

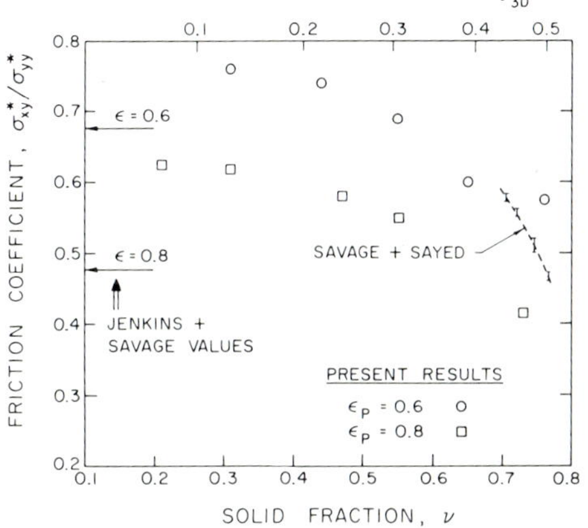

Figure 8. The friction coefficient, $\sigma^*_{xy}/\sigma^*_{yy}$ or f_{xy}/f_{yy}, as a function of ν. Also shown are the theoretical values from Jenkins and Savage which are independent of ν.

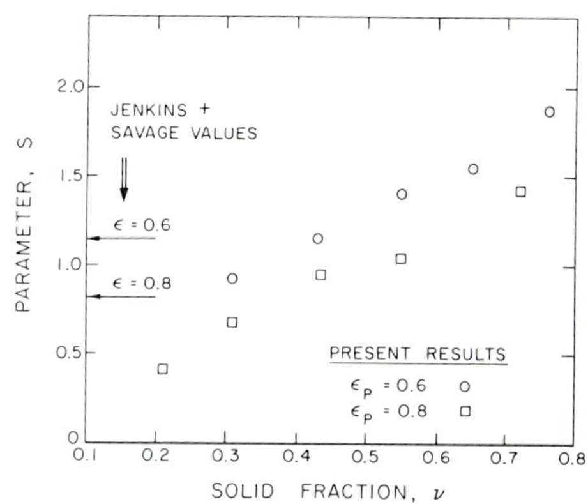

Figure 9. The parameter, S, as a function of ν. Also shown are the theoretical values of Jenkins and Savage which are independent of ν.

In order to make comparisons with the existing experimental and theoretical results for three dimensional near-spherical particles it is necessary to define an approximate equivalent three dimensional solid fraction, ν_{3D}. Here we somewhat arbitrarily choose to define ν_{3D} so that the average interparticle gap is the same as that in the two-dimensional simulation so that $\nu_{3D} = 0.752\nu^{3/2}$. The experimental data of Bagnold (1954) and Savage and Sayed (1980, 1982) as well as the theoretical predictions of Jenkins and Savage (1982) are thus displayed in Figs. 6 to 9.

Though an exact comparison with the three dimensional results is inappropriate it seems clear that the values of f_{xy} and f_{yy} from the simulation (Figs. 6 and 7) are comparable with those from the experiments and from the theory of Jenkins and Savage (1982). The values of these functions increase almost exponentially with the solid fraction and presumeably asymptote to infinity at some maximum packing or critical solids fraction. The values also increase with coefficient of restitution, ε_p.

Though the simulation data of Figs 6 and 7 is somewhat erratic, the ratio of f_{xy}/f_{yy} which is the effective coefficient of friction <u>and</u> the value of the parameter S (which is an inverse measure of the translational temperature for a fixed $Rd\langle u\rangle/dy$ are less erratic and are displayed in Figs. 8 and 9. Contrary to the three dimensional theory of Jenkins and Savage both the friction coefficient and S proved to be functions of ν. As seen in Fig. 8, the values of the friction coefficient are of the same order as those of Jenkins and Savage and, indeed, they decrease as ε_p increases in a similar manner. Moreover they are relatively independent of solid fraction at the lower values of ν where the theory of Jenkins and Savage is most appropriate. However the friction coefficient does appear to decrease as ν reaches higher values. As seen in Fig. 8 this decrease is consistent with the experimental data of Savage and Sayed (1980,1982)

The values of the parameter, S, which can be conveniently envisaged as an inverse measure of the granular translational temperature (for the same $Rd\langle u\rangle/dy$) also exhibit a substantial increase with ν over the whole range of ν examined. Indeed it appears to be a linear function of ν with lower values for a higher coefficient of restitution.

STATISTICAL PROPERTIES OF THE COUETTE FLOWS (TYPE B)

Most of the existing theories assume that the random motions in granular material flows have velocities with Maxwellian distributions. For this reason both translational and rotational velocity distributions were constructed from the Couette flow (type B) simulations (see Campbell (1982)). Since all of these distributions are qualitatively similar we present only one in this paper, namely the distribution of the magnitude of the translation velocity with the mean velocity removed. The abscissa, X, of Fig. 10 is therefore defined as

$$X = \left[\frac{(u-\langle u\rangle)^2 + v^2}{\langle u'^2\rangle + \langle v'^2\rangle}\right]^{\frac{1}{2}}$$

Figure 10 shows that the translation velocity distribution at lower void fractions (and smaller S) is essentially Maxwellian. However the distribution tends to deviate from Maxwellian at higher values of ν (and S).

The collision angle distribution is also of considerable importance in determining the rheology of a granular material flow. The orientation of two particles involved in a collision is non-isotropic due to the fact that the differential velocities inherent in the mean shearing motion are of the same order of magnitude as the random motions. Figure 12 presents the distribution of collisions over the orientation or collision angle, θ, as defined by the inset in that figure. Results are shown for Couette flows (type B) of five different solids fraction. There is clearly a substantial change in the form of this distribution as the solids fraction is increased. At low solids fractions ($\nu\sim 0.35$) or low S where the random motions dominate there is a relatively smooth skewness caused by the shearing motion. This distribution (at $\nu = 0.35$) is very similar in form to that proposed by Savage and Jeffrey (1981) and also used in other theoretical models. Quantitatively these low solids fraction distributions are somewhat different from those proposed by Savage and Jeffrey (1981) and also used in other theoretical models. Campbell (1982) has suggested a modification to their theoretical distribution which results in better quantitative agreement with the simulation results.

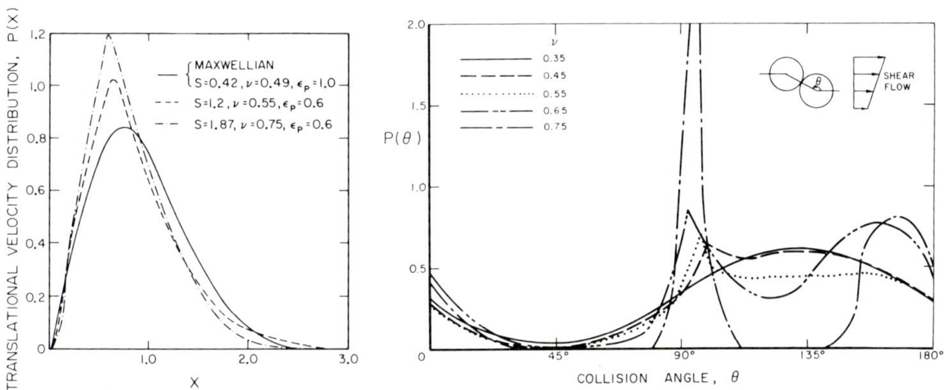

Figure 10. Translational velocity distribution functions from some Couette flow (type B) simulations.

Figure 11. Collision angle distributions for various solid fractions from Couette flow (type B) simulations.

However, the main point of Fig. 12 is to demonstrate that the collision angle distribution undergoes substantial changes as ν increases to higher values. Two peaks around $\theta \approx 90°$ and $\theta \approx 180°$ develop and ultimately dominate the distribution at solids fractions larger that 0.6. These higher solids fraction distributions result from increased geometric order or "microstructure" within the flow. Probability distributions for the relative positions of two particles within the flow (Campbell (1982)) show that as ν increases from 0.35 to 0.75 there is an increased likelihood that particles arrange themselves in layers. Thus the peak at $\theta \approx 90°$ represents glancing collisions between particles in neighboring layers while the peak at $\theta \approx 180°$ represents collisions between particles in the same layer. It seems clear that any rheological model capable of predicting flows over the entire range of solids fraction must incorporate or reflect such substantial variations in the collision angle distribution.

CONCLUSIONS

In summary, this paper has presented results for the computer simulation of Couette flows of a "two-dimensional" granular material consisting of circular cylinders characterized by coefficients of restitution for cylinder/cylinder collisions and for cylinder/wall collisions. Interstitial fluid and gravitational effects are neglected as are other possible forces (eg. electrostatic) that can occur in the rapid flows of granular materials.

Results for two different kinds of wall boundary condition are first presented. The first type corresponds roughly to "smooth walls" and the simulations exhibit slip at both walls. The flow in the neighborhood of the walls has a lower solid fraction and granular temperature than in the core. On the other hand an alternative boundary condition corresponding to "rough walls" yielded linear velocity profiles without slip and uniform solid fractions.

The latter were therefore used to examine the effective constitutive relations for the flowing material. The veracity of a generalized form of Bagnold's constitutive laws was first demonstrated and the problem reduced to establishing the normal and shear stress functions f_{yy} and f_{xy} which may be functions of the solid fraction and coefficient of restitution. It is shown that the simulations yield values of these functions similar to those in the experiments of Savage and Sayed (1980, 1982) and to the theory of Jenkins and Savage (1982). The effective coefficient of friction, f_{xy}/f_{yy}, decreases with increasing solid fraction, ν, and with increasing coefficient of restitution. The former trend is consistent with the experiments of Savage and Sayed.

The statistical properties of the flow are also examined. It is shown, for example, that the velocity distributions tend to deviate from Maxwellian at the higher solid fractions. This is accompanied by an increased microstructure or layering of the particles. This, in turn, radically alters the collision angle distribution function.

ACKNOWLEDGMENTS

The authors wish to express their sincere gratitude to Professor Rolf Sabersky for his help and advice. We also gratefully acknowledge the support of the National Science Foundation (Grant CME 79-15132) and additional support provided by Union Carbide Corporation and Chevron Oil Field Research.

REFERENCES

1 N. L. Ackermann, H. Shen, Stresses in Rapidly Sheared Fluid-Solid Mixtures, Submitted to Proc. ASCE, J. Eng. Mech., 1982. (See also, H. Shen, N. L. Ackermann, Constitutive Relationships for Fluid-Solid Mixtures, Submitted to Proc. ASCE, J. Eng. Mech., 1982.

2 R. A. Bagnold, Experiments on a Gravity Free Dispersion of Large Solid Spheres in a Newtonian Fluid Under Shear, Proc. R. Soc. London, Ser. A255: 49-63, 1954.

3 J. A. Baker, D. Henderson, What is a Liquid? Understanding the States of Matter, Rev. Mod. Phys. 48:587, 1976.

4 G. A. Bird, Molecular Gas Dynamics, Clarendon Press, Oxford, 1976.

5 A. Blinowski, On the Dynamic Flow of Granular Media, Arch. Mech.30:27-34, 1978.

6 C. S. Campbell, Shear Flows of Granular Materials, Ph.D. Thesis and Report No. E-200.7, Div. of Eng. and Appl. Sci., Calif. Inst. of Tech., Pasadena, Ca., 1982.

7 C. S. Campbell, C. E. Brennen, Computer Simulation of Chute Flows of Granular Materials, Proc. IUTAM Symp. on Deformation and Failure of Granular Materials, Delft, Sept. 1982.

8 S. Chapman, T. G. Cowling, The Mathematical Theory of Non-Uniform Gases, Cambridge Univ. Press, 3rd Ed., 1970.

9 P. A. Cundall, A computer Model for Rock-Mass Behavior Using Interactive Graphics for Input and Output of Geometrical Data, U. S. Army Corps of Eng., (Missouri River Div.), Tech. Report MRD-2-74, 1974.

10 P. A. Cundall, O. D. L. Strack, A discrete Numerical Model for Granular Assemblies, Geotech. 29:47-65, 1979.

11 R. A. Davis, H. Deresiewicz, A Discrete Probabilistic Model for Mechanical Response of a Granular Medium, Acta. Mech. 26:69-89, 1977.

12 J. M. Ferziger, H. G. Kaper, Mathematical Theory of Transport Processes in Gases, North Holland Pub. Co., 1972.

13 J. T. Jenkins, S. B. Savage, A Theory for the Rapid Flow of Identical Smooth, Nearly Elastic Particles, Submitted to J. Fluid Mech., 1982.

14 K. Kanatani, A Micropoplar Continuum Theory for the Flow of Granular Materials, Int. J. Engrg. Sci., 17:419-32, 1979a.

15 K. Kanatani, A Continuum Theory for the Flow of Granular Materials, In Theoretical and Applied Mechanics, ed. Japan Nat. Comm. Theor. and Appl. Mech. 27:571-78, 1979b.

16 K. Kanatani, A Continuum Theory for the Flow of Granular Materials (II), In Theoretical and Applied Mechanics, ed. Japan Nat. Comm. for Theor. and Appl. Mech, 28:485-492, 1980.

17 L. D. Landau, E. M. Lifshitz, Statistical Physics, Pergamon Press, 1958.

18 S. Ogawa, Multi-Temperature Theory of Granular Materials, Proc. U. S. Japan Seminar on Continuum-Mechanical and Statistical Approaches in the Mechanics of Granular Materials, pp. 208-17, 1978b.

19 S. Ogawa, N. Oshima, A Thermomechanical Theory of Soil-Like Materials, In Theoretical and Applied Mechanics, ed. Japan Nat. Comm. Theor. and Appl. Mech., 25:229-44, 1977.

20 N. Oshima, Continuum Model of Fluidized Granular Media, Proc. U.S.-Japan Seminar on Continuum-Mechanical and Statistical Approahces in the Mechanics of Granular Materials, pp. 189-202, 1978a.

21 N. Oshima, Dynamics of Fluidized Granular Media, In Theoretical and Applied Mechanics, ed. Japan Nat. Comm. for Theor. and Appl. Mech., 28:475-84, 1980.
22 S. B. Savage, D. J. Jeffrey, The Stress Tensor in a Granular Flow at High Shear Rates, J. Fluid Mech., 110:255-72, 1981.
23 S. B. Savage, M. Sayed, Experiments on Dry Cohesionless Materials in an Annular Shear Cell at High Strain Rates, Presented at EUROMECH 133 - Statics and Dynamics of Granular Materials, Oxford Univ., 1980.
24 S. B. Savage, M. Sayed, Stresses Developed by Dry Cohesionless Granular Materials in an Annular Shear Cell, Submitted to J. Fluid Mech., 1982.
25 S. B. Savage, The Mechanics of Rapid Granular Flows, Appl. Mech. Rev. (to appear), 1982.
26 A. J. M. Spencer, Deformation of an Ideal Granular Material, In Mechanics of Solids, Rodney Hill 60th Anniv. Vol. , eds. H. G. Hopkins, J. J. Sewell, Pergamon Press, 1981.
27 D. H. Trollope, B. C. Burman, Physical and Numerical Experiments with Granular Wedges, Geotech. 30:135-157, 1980.
28 O. Walton, Particle Dynamics Modelling of Geological Materials, Univ. of Calif. Lawrence Livermore Lab. Report UCRL-52915, 1980.
29 W. W. Wood, Computer Studies on Fluid Systems of Hard-Core Particles, In Fundamental Problems of Statistical Mechanics, Vol. 3, ed, E. D. G. Cohen, No. Holland Pub. Co., 1975.

Mechanics of Granular Materials: New Models and Constitutive Relations, edited by
J.T. Jenkins and M. Satake, 1983
Elsevier Science Publishers B.V., Amsterdam — Printed in The Netherlands

PARTICLE-DYNAMICS CALCULATIONS OF SHEAR FLOW*

O. R. WALTON

Earth Sciences Department, Lawrence Livermore National Laboratory, P. O. Box 808, Livermore, CA 94550 USA

ABSTRACT

Two-dimensional discrete particle computer models similar to those of P. Cundall (ref. 1, 2) are described. The "soft-particle" approach used in these models allows them to be applied over a wide range of conditions from static situations through rapid shear conditions. Surface friction between particles and with boundary walls is explicitly modeled. Particular attention is paid to the modeling of dynamic situations wherein the energy losses and momentum transfer during inter-particulate collisions play important roles. Comparisons with analytic solutions have verified the numerical techniques and direct comparison with physical tests involving several particles have verified the models' ability to calculate the motion of real materials. Direct shear tests on oil shale rubble and corresponding calculations indicate qualitatively similar circulation phenomena and both showed large fluctuations in the magnitude of the shearing force. Incline chute flow calculations are providing detailed descriptions of individual particle paths in which shearing and size segregation phenomena can be observed. Initial comparisons with experiments indicate somewhat slower segregation in two-dimensional calculations than in experiments with spherical particles.

INTRODUCTION

Particle-dynamics is a term this author uses to refer to calculational models that are being developed to study the bulk behavior of assemblies of particles by calculating the motion of each individual particle in the assembly. The term especially refers to those calculational models designed to calculate the discrete motion of macroscopic (as opposed to molecular sized) particles where surface friction, collisional energy losses, boundary forces and gravity are important parameters. Peter Cundall did much of the pioneering work in this area (ref. 2, 3), developing what he termed the distinct element method, and several parts of the models described in this paper are patterned after his work. The companion field of molecular dynamics -

*Work performed under the auspices of the U. S. Department of Energy by the Lawrence Livermore National Laboratory under contract No. W-7405-ENG-48.

discrete particle modeling on a molecular scale with no energy loss mechanisms - has undergone extensive development since its origination in the late 1950's, and there are currently some interesting new non-equilibrium techniques in that field that may be adaptable to macroscopic particle models (refs. 4, 5, 6, 7). The success of molecular dynamics calculations in predicting equilibrium equations-of-state and non-equilibrium transport properties in molecular systems is one of the major motivating factors for applying discrete particle models to granular materials.

Soft vs. rigid particles

In addition to Monte Carlo calculations (which to my knowledge are only being applied to thermodynamic equilibrium configurations of molecular systems) there are two techniques in common use for numerically predicting the behavior of systems of particles: rigid-particle and soft-particle models. They are both trajectory generating processes wherein the path of each particle is calculated as it interacts with other particles and boundaries. Both of these techniques are being applied to granular materials and both types of models have been described by others at this Seminar (ref. 2, 9, 10). Rigid particle models assume that interparticle collisions are instantaneous, with the after-collision trajectories determined from the initial trajectories and the "rules" governing two-body collisions (i.e., elastic, inelastic, smooth, rough, frictional, etc.). Smooth trajectories, which may or may not include gravity, are followed between these instantaneous collisions. Such models are particularly well suited for low density situations (like gases). They can also be used in high density situations as long as there is a sufficiently high vibrational energy (or "temperature") in the system. The time step for numerical models based on rigid particles is usually based on the time between collisions for an individual particle - as this time gets shorter so does the time step - so that in the limit of continuous contacts, or simultaneous contacts, this approach cannot usually be used. There have been a few attempts to use rigid particles with continuous contacts (ref. 8) but to my knowledge none of these have been applied to granular materials.

The soft-particle approach requires that collisions are of finite duration with the interaction force between particles usually varying continuously with displacement as particles "overlap" slightly. In order to accurately integrate the equations of motion, the usual explicit schemes require on the order of 10 time steps during a collision. The time step size is also usually kept at one fixed value, whether collisions are occurring or not, so this approach may not be very efficient for dilute systems with only occasional collisions.

However, for dense systems this approach can be more efficient than a rigid particle approach and, perhaps more important, all configurations are inherently soluable - even static assemblies.

Table I lists some of the current researchers calculationally examining discrete macroscopic particles and indicates which type of model (rigid or soft) they are using. In addition to those listed, there are of course a large number of molecular dynamics research efforts underway - involving molecular sized particles and no energy loss mechanisms.

TABLE I

Characteristics of various macroscopic particle models.

	Rigid Particle Models	Soft Particle Models
best suited for...	Dynamic Situations	Static or Dynamic
most efficient with..	significant "granular temperature"	moderate to high densities
Researchers:	C. Brennen/C. Campbell (ref. 9) G. Hawkins G. Dahlquist/P. Lotstedt (ref. 8)	P. Cundall/O. Strack (ref. 2) J. Drake* L. Taylor* G. Hocking M. Voegele* others* O. Walton (ref. 11) T. Kawai (ref. 10)

*Several modeling efforts are using computer programs written by P. Cundall, or revisions of programs originated by him.

MATHEMATICAL MODEL

We have used a "soft-particle" model similar to that of Cundall in two different computer programs. One calculates the motion of 2-dimensional polygonal particles, and the other uses circular particles. Figure 1a is a schematic of the two-dimensional rheologic model used for the interparticulate forces in these computer programs. Figure 1b is a schematic representation of the model used by Cundall for the same type of contact force calculation. The major difference between the two is the inclusion of the damping or dashpot normal force in determining the total normal force, F_n, used for the friction limit, $F_f \leq \mu|F_n|$. The total normal force is just the sum of a linear spring and a linear velocity dependent term (a Kelvin-Voit element),

$$F_n = -Kx - D\dot{x} \tag{1}$$

where K is the spring stiffness, and D is the damping coefficient. Using the total normal force rather than just the spring force for the friction limit results in more realistic dynamic impacts.

For an isolated collision with the normal force acting along the line joining the centers of the two colliding particles the equation-of-motion for either particle is essentially that of a damped harmonic oscillator,

$$m\ddot{x} = -Kx - D\dot{x} \qquad (2)$$

(in this case x would be in center of mass coordinates and m would be a reduced mass). The damping factor, D, is usually chosen based on equation (2) to provide damping that is less than, equal to, or greater than critical damping, as desired.

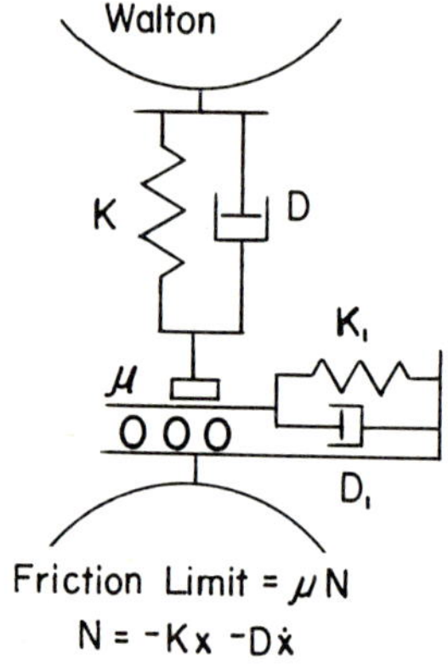

Figure 1a. Schematic of Rheologic model used in these calculations.

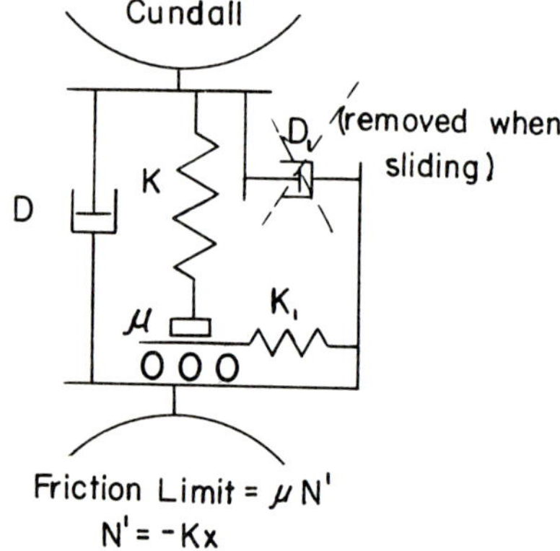

Figure 1b. Schematic of Cundall's contact force model.

The computer model then consists of algorithms to:

1) Detect all current contacts.
2) Based on current positions and velocities, calculate all forces according to equations similar to equation (1) plus a tangential force term.
3) Sum all of the forces and moments acting on each particle from all contacts.
4) Integrate the equations-of-motion for all particles one time step to obtain new positions, then return to step (1).

The details of this calculational procedure have been described elsewhere (refs. 11, 12), suffice it to say that most of the programming effort is in searching and bookkeeping algorithms with only a relatively minor part for calculating forces and integrating equations-of-motion. On the other hand, by only going through step (1) above every 10th time step or so, the overall computer time usage is dominated by calculating forces and integrating equations-of-motion and not by searching or bookkeeping chores.

VERIFICATION OF CALCULATIONAL MODELS

Initial validation of the computer models consisted of comparison with analytic solutions for rocking, sliding, toppling and rolling rigid bodies. These comparisons verified the finite difference algorithms as well as checking out the contact detection, and friction force algorithms. Comparisons with analytic rocking vs. sliding criteria and amplitude vs. frequency formulas for rocking rectangular blocks (ref. 13) established that the calculational models would predict the correct mode of motion for a block on a flat surface. Physical tests of the same configurations confirmed both the analytic predictions and the computer predictions for amplitude vs. frequency and also verified that the computer model changed modes of motion (rocking to sliding) under the same conditions as occur physically. Domino-like series of toppling rectangular blocks were photographed with a high speed motion picture camera and the motion compared to computer calculations. Within the experimental uncertainty, the rate of propagation and the final configurations were the same in both the physical tests and the calculations of toppling fire bricks. The success of these tests and further qualitative comparisons with bin flow tests have given us confidence in the ability of the polygonal particle model to calculate dynamic motion of system or particles.

A different integration scheme is used for the equations of motion in the circular particle model (ref. 14) than in the polygonal particle model. This integration algorithm automatically adjusts time steps as the effective stiffness of the system of equations changes. Several verification calculations were performed to test both the force calculation models and the integration scheme. One rather interesting test of the integration scheme involved a rotating, perfectly elastic, disk bouncing on a horizontal surface. Both normal and tangential springs act in the model to store potential energy during the contact. For a disk with moment of inertia $I_0 = 1/2\ mr^2$, the tangential and normal springs will both return to their rest positions simultaneously if the tangential spring constant, K_t, is given by,

$$K_t = 1/3\ m\ K_n \tag{3}$$

where m is an odd integer and K_n is the normal spring constant. For such a choice of spring constants it can be shown (ref. 15) that a rotating disk, in a uniform gravity field, will bounce back and forth between the same two spots on a flat surface, if the initial angular velocity, $\dot{\theta}_o$, is related to the initial horizontal speed, V_{xo}, according to

$$\dot{\theta}_o = \frac{2}{r} V_{xo} = \frac{2}{r} |\vec{V}| \cos \alpha, \tag{4}$$

where α is the angle of incidence from the horizontal surface. With the appropriate choice of initial conditions according to equations (3) and (4) the circular particle model, using a third order accurate implicit integration of the equation of motion, faithfully calculated a disk (super ball?) bouncing back and forth between the same two points, with cumulative error of less than 1% after many hundreds of time steps. Quantitative comparisons with physical tests have not yet been performed with the circular particle model, however, comparisons of incline plane shearing flow calculations and high speed movies of physical tests, indicate similar qualitative phenomena occurring.

GRANULAR MATERIAL SIMULATION CALCULATIONS

Comprehensive equation-of-state studies of granular assemblies using these two dimensional models have not been performed, however, we have been using these models to aid in design of surface retorts for oil shale. Of interest in such design work is the shearing resistance and gravity flow characteristics of oil shale under various conditions. The following calculations coupled with laboratory tests have contributed to the design of laboratory scale surface retorts, that are testing concepts that may be used in full scale commercial oil shale retorts in the future.

Bin flow calculations and observations

A two dimensional bin flow calculation reported elsewhere (refs. 11, 12) showed formation and collapse of arches near the hopper opening when that opening was only 5 or 6 particle diameters across. Physical tests of two-dimensional polygonal plexiglas particles showed qualitatively similar behavior. Several laboratory tests of rubblized oil shale particles flowing through circular and rectangular openings have shown a "chugging" phenomenon when the hopper opening is only a few particle diameters across (refs. 16, 17). While not quantitatively measured, certain qualitative features of this "chugging" phenomena have been observed that are consistent with the hypothesis that this phenomenon is essentially the same arch formation and collapse phenomenon observed in the calculations. If the opening is less than approximately 6 particle diameters, flow is often stopped by the formation of a stable arch. When the diameter is near this value, apparently, arches form temporarily but subsequently collapse in respone to vibration and loading from the moving material above the temporary arch. The frequency of the "chugging" (forming and then collapse of the arches) increases as the hopper opening is increased in relation to the particle size. Also, the amplitude (or apparent effect) of this "chugging" diminishes as the opening is increased so that for large openings (greater than ten or so particle diameters) the flow appears to be essentially smooth and continuous. The frequency of this action also depends on the absolute size of the particles (lower frequency for larger particles) so that it is relatively easy to observe in particles on the order of a centimeter in diameter but may be quite difficult to observe in fine sand.

Direct Shear

Direct shear tests on oil shale rubble were simulated with the polygonal particle model. Figure 2 shows the configuration used in these calculations. Qualitatively the same circulatory motion was seen in the calculation as was observed in corresponding physical tests. Quantitatively the magnitude of the shearing force in the two-dimensional computer simulation was of the correct order but did not correspond exactly to the measured value on tests of three-dimensional particles. In both the computer simulation and the laboratory tests large fluctuations (±25% or greater) in shearing force were observed when relatively large particles were used (as in Fig. 2). The peaks in the calculated force corresponded to arches or stress networks between the bottom of the lower bin and the lower edges of the upper bin. As these arches

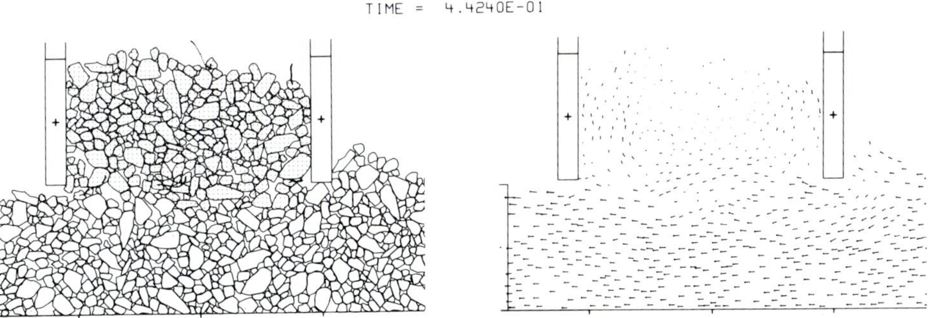

Figure 2. Configuration of direct shear calculations showing both positions and velocity vectors.

or stress networks collapsed (or buckled) the shear force passed through a temporary minimum, then increased again as a new stress network built up.

Calculations were not performed with very small particles, however, in the laboratory tests the magnitude of the fluctuations in the shearing force decreased considerably (to ±10% or so) when fine particles were tested in the same apparatus.

Incline Plane Flow

A limited number of incline plane flow calculations have been performed using both the circular particle and the polygonal particle models. Of special interest for oil shale retort applications are shearing, segregation and mixing phenomena that can occur during incline plane flow. To date only qualitative comparisons have been made with laboratory tests, however the details of the shearing flow seen in the computer simulations are providing insight in interpreting some phenomena observed in laboratory results.

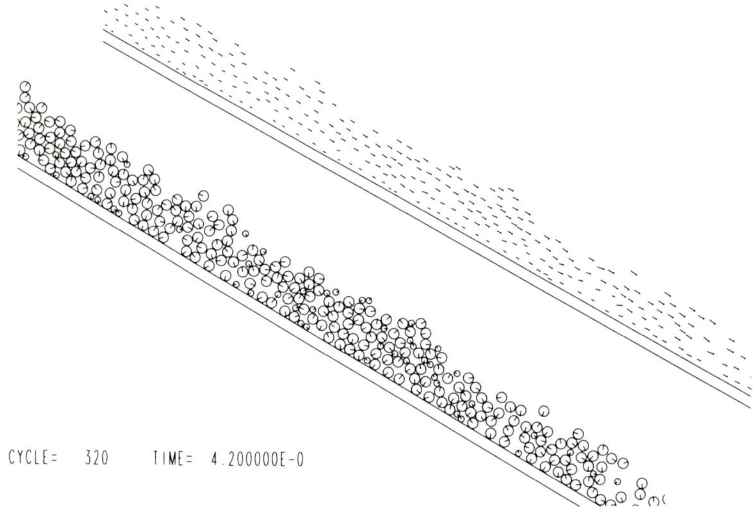

Figure 3. Simulated flow of binary mixture of circular particles on incline.

Both calculations and physical tests of the flow of circular particles on a 30° incline show that a layer of rolling particles develops at the incline surface; the circular particles in the layer immediately above this layer tend to roll on the rough, moving, surface of this bottom layer. The result is a significantly higher velocity for the second layer of particles than the first. The velocity continues to increase as one moves toward the top of the 5 to 6 particle thick layers studied, however, the gradient in the velocity is much smaller once one is past the second particle layer. This phenomena is evident in the snapshot of positions and the velocity vectors shown in Figure 3.

Savage and Lun (ref. 18) have performed particle size segregation tests with binary mixtures of spherical polystyrene beads. Their tests show significantly more rapid segregation than was observed in the two-dimensional calculations or than we obtained in laboratory tests of circular particles on an inclined surface. Savage and Lun used a rough inclined surface composed of a layer of particles stuck to the incline plane. In the calculations and in the two dimensional laboratory tests in this work a planar (but frictional) surface was used for the incline. It is not known at this time if the difference between the results of Savage and Lun and those seen in two dimensions is due to differences between two dimensional and three dimensional particle behavior or if it is primarily due to the nature of the incline surface.

Simulated polygonal particles flowing on a 35° incline (with friction angle of 30° between particles and between surface and particles) are shown in Fig. 4.

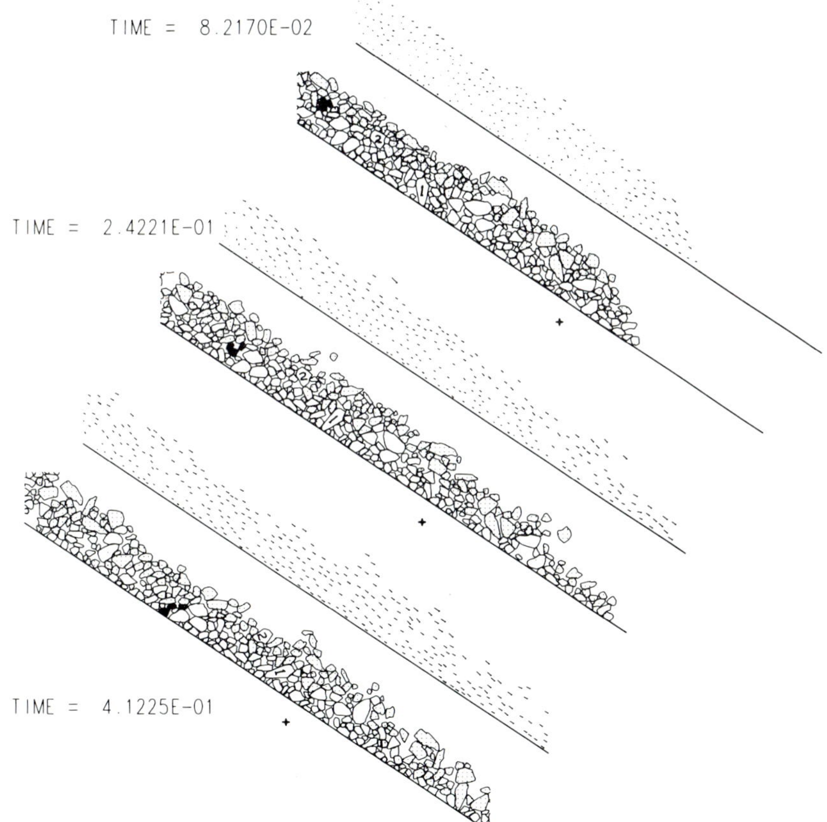

Figure 4. Simulated flow of oil shale rubble down 35° incline showing percolation of small and large particles down and up respectively.

Both positions and velocity vectors are shown for three separate times shortly after the initiation of flow. The particle shapes used in these calculations were obtained by digitizing photographs of oil shale particles. The particle size distribution was chosen to correspond to a truncated log-normal distribution fitting a typical oil shale rubble sample set screened to pass through a 1.2 inch screen and fail to pass through a .25 inch screen. The three pictures shown in this sequence show the percolation of small particles (e.g., heavily shaded in Fig. 4) down through the layer as the flow proceeds. The larger particles also have a tendency to move toward the top of the flowing layer as evidenced by the numbered particles in this figure. This percolation phenomena only occurred during the initial stages of the flow for the particular geometry considered here. As time progressed the amount of shearing occurring gradually decreased (as the sliding velocity of the bottom layer increased) until, after traveling a few hundred particle diameters, there was very little difference between the velocity of the top and bottom particles.

Figure 5 shows the positions and velocity vectors as the flowing layer leaves one 35° incline and impacts another incline in the opposite direction. A vertical wall was placed at the turning point to reduce the number of particles involved in the stationary region that develops. As can be seen

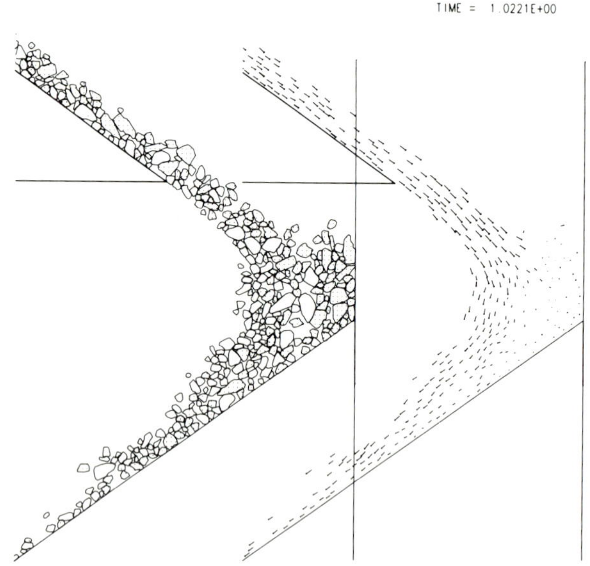

Figure 5. Continuation of flow simulation in Fig. 4.

in the velocity vector plot the material initially on the bottom of the first incline becomes the top layer of the second incline - until shearing and particle segregation rearranges the material on the second incline. Physical tests of particle size segregation in incline chute flow showed somewhat more rapid segregation of sizes than is indicated in the two dimensional simulation calculations (ref. 17). This may be due to the increased shearing caused by the side walls in the relatively narrow chutes used in the laboratory tests ($\sim$6 cm).

CONCLUSION

Several papers presented at this Seminar (Brennen and Campbell, Cundall and Strack, Hawkins, Kawai, and this paper) indicate that discrete particle modeling is becoming a viable means of simulating granular material behavior. Cundall and Strack have shown that quasi-static deformations calculated by discrete particle models produce results very similar to the experimental measurements of photo-elastic discs. Hopefully this work plus the dynamic modeling of Brennen and Campbell establish that dynamic simulations of granular flow can also be performed successfully using discrete particle models. The qualitative results obtained in these very preliminary calculations of granular material flow have already been useful in developing new designs for oil shale retorts. As better diagnostic algorithms are implemented in these models we expect to obtain better information with which to describe the rheologic behavior of the simulated materials.

ACKNOWLEDGMENTS

Thanks are due Professor S. Savage and C. Lun for advance copies of their particle size segregation data. Programming for the circular particle model described in this paper was performed by J. Peterson and R. Drach. Also R. Drach's help in conducting the laboratory tests on circular particles, running circular particle model calculations and in providing the super-ball solution is greatly appreciated, as is R. Thorpe's generation of digitized sample sets for the oil shale simulation calculations. The direct shear tests were done by W. Patrick and C. Sisemore.

REFERENCES

1 P. A. Cundall, J. Marti, P. Beresford, N. Last, M. Asgian (1978) Computer Modeling of Jointed Rock Masses, USA-WES Tech. Rept. N-78-4, Dames and Moore, Los Angeles.

2 P. A. Cundall, O. D. L. Strack (1982) "Modeling of Microscopic Mechanisms in Granular Material", elsewhere in these proceedings.

3 P. A. Cundall (1974) Rational Design of Tunnel Supports, U.S. Army Corps of Engineers Tech. Rept. MRD-2-74, Univ. of Minnesota.

4 W. Ashurst and W. Hoover (1975), Phys. Rev. A, 11, 658.

5 D. J. Evans "The Nonsymmetric Pressure Tensor in Polyatomic Fluids", J. Stat. Phys. Vol. 20, No. 5, 1979.

6 D. J. Evans and S. Murad, "Singularity Free Algorithm for Molecular Dynamics Simulation of Rigid Polyatomics", Mol. Phys. 1977, Vol. 34, No. 2, 327-331.

7 D. J. Evans, "Non-Equilibrium Molecular Dynamic Study of the Rheological Properties of Diatomic Liquids", Mol. Phys. 1981, Vol. 42, No. 6, 1355-1365.

8 G. Dahlquist, P. Lotstedt (1977) "Interactive Simulation of the Progressive Collapse of a House" in Proceedings IAMCS Symposium on Simulation Software and Numerical Methods for Differential Equations, Blacksburg, VA.

9 C. S. Campbell and C. E. Brennen (1982), "Computer Simulation of Shear Flows of Granular Materials", elsewhere in these proceedings.

10 T. Kawai (1982) "New Discrete Models and Their Application to the Mechanics of Granular Materials", elsewhere in these proceedings.

11 O. Walton "Explicit Particle-Dynamics Model for Granular Materials", Fourth Int'l. Conf. on Num. Methods in Geomechanics, Edmonton, May 31-June 4, 1982 (also Lawrence Livermore National Laboratory Rept. No. UCRL-86266).

12 O. Walton (1980) Particle Dynamics Modeling of Geological Materials, Lawrence Livermore National Laboratory Rept. UCRL-52915.

13 G. W. Housner (1963) "The Behavior of Inverted Pendulum Structures During Earthquakes", Bull. Seis. Soc. of Am., Vol. 53, No. 2, pp 403-417.

14 A. Hindmarsh "Stiff System Problems and Solutions at LLNL", Int'l. Conf. on Stiff Computation, Park City, Utah, April 12-14, 1982.

15 R. Drach (1982) private communication.

16 J. Carley "Flow of Crushed Shale Through Orifices" to be presented at: 4th Briefing on Oil Shale Research, Lawrence Livermore National Laboratory, Sept. 22-23, 1982, LLNL-MISC-4059.

17 O. Walton "Rubble Flow and Mixing Experiments Related to Gravity-Mixed Pyrolyzers" to be presented at: Fourth Briefing on Oil Shale Research, Lawrence Livermore National Laboratory, Sept. 22-23, 1982 (LLNL MISC-4059).

18 S. Savage and C. Lun (1982) private communication.

Mechanics of Granular Materials: New Models and Constitutive Relations, edited by
J.T. Jenkins and M. Satake, 1983
Elsevier Science Publishers B.V., Amsterdam — Printed in The Netherlands

MACROSCOPIC PHASE TRANSITIONS IN TWO-DIMENSIONAL GRANULAR MATERIALS

J.T. JENKINS[1] and M. SHAHINPOOR[2]

[1]Department of Theoretical and Applied Mechanics, Cornell University, Ithaca, New York, 14853 (U.S.A.)

[2]Department of Mechanical and Industrial Engineering, Clarkson College of Technology, Potsdam, New York, 13676 (U.S.A.)

ABSTRACT

We review a theory that is successful in predicting a first order phase transition between solid and fluid states of identical rigid disks and examine its relevance to the mechanics of granular materials.

INTRODUCTION

Recent theories for rapid deformations of granular materials have exploited the analogy between the irregular motion of the grains, resulting from collisions driven by the rapid deformation, and the thermal motion of the molecules of a dense gas or liquid (ref. 1-4). Then the kinetic theory of gases provides a method with well developed techniques for deriving a continuum theory for rapid deformations of granular materials from a consideration of the details of a collision between a pair of grains and a statistical characterization of the likelihood of such collisions. Although the predictions of such theories have not yet been thoroughly tested and it is possible that the analogy does not strictly apply (the characterization of the interaction between the particles as instantaneous, binary collisions may well be too simple, for example), it seems worthwhile to investigate what other results from statistical studies of liquids and dense gases may have some applicability to the mechanics of granular materials.

The two phenomena of interest to us here are the vibrational fluidization of granular materials as occurs, for example, in the neighborhood of an oscillating boundary, and the onset of granular jams (local regions experiencing relatively slow deformations in an otherwise rapidly deforming granular material). We regard these two phenomena as phase changes between ordered and disordered states of the granular material; then fluidization and jamming correspond, respectively, to melting and solidification. With these interpretations, we review the analyses of Collins (ref. 5) and Kawamura (ref. 6) of the melting and solidification of a simple system consisting of identical circular discs, elaborate upon it, and discuss its relevance to granular materials.

THEORY

Collins and Kawamura focus attention on plane close packed arrangements of identical disks. The disks are assumed to have a diameter σ and the centers of neighboring disks are assumed, on average, to be a distance b apart. The exact location of a center fluctuates in the plane because of the thermal energy of the disk. If the average locations of the centers of neighboring disks are joined by straight lines, the plane is covered by regular or quasi-regular polygons with the length of their edges equal to b.

Collins and Kawamura adapt Bernal's characterization of the solid and fluid states of a simple monatomic substance to this two dimensional system. In Bernel's view (ref. 7-10) a liquid may possess short range order, in which, for example, the centers of neighboring molecules are, on average, the same distance apart; but long range order (a lattice) is present only in a solid.

In terms of the polygons, an example of a solid state is the hexagonal lattice formed by identical triangles. Fluid states are disordered arrays of the polygons. If the relative numbers of the different polygons are taken to specify the state of the solids or fluid; then, because there are many possible configurations in each state, a configurational entropy may be defined in terms of the probability of such states.

Adopting a free energy incorporating this configurational entropy, Collins and Kawamura ask whether there exist first order phase transitions between irregular close packed arrays and lattices. At such a transition the free energy is minimized in both an ordered and a disordered state having equal free energies. While the average distance between nearest neighbors in the solid and fluid states is the same, if, at the transition, the average number of nearest neighbors in the fluid is less than that in the solid, the configuration of the fluid will be expanded over that of the solid and the phase transition will be of first order.

In a further simplification Collins and Kawamura consider arrays consisting only of triangles and squares. Collins argues that, because of their relatively large areas, near the solid-liquid transition the pentagons and higher polygons should be far less common than the triangles and squares. Because triangles and square can close pack in any proportion, little is lost by this assumption. Typical fluid and solid states of triangles and squares are indicated in Fig. 1.

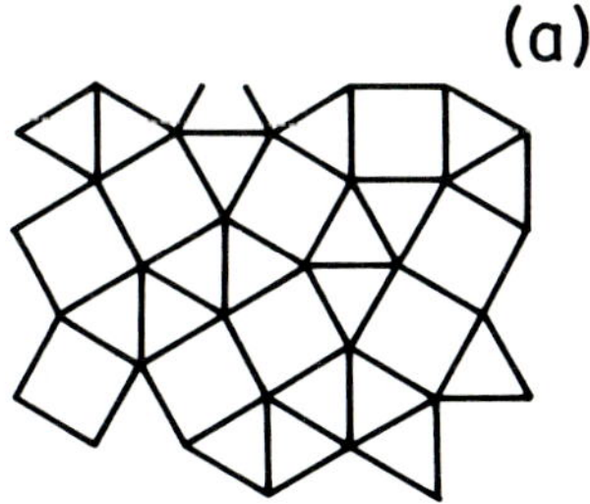

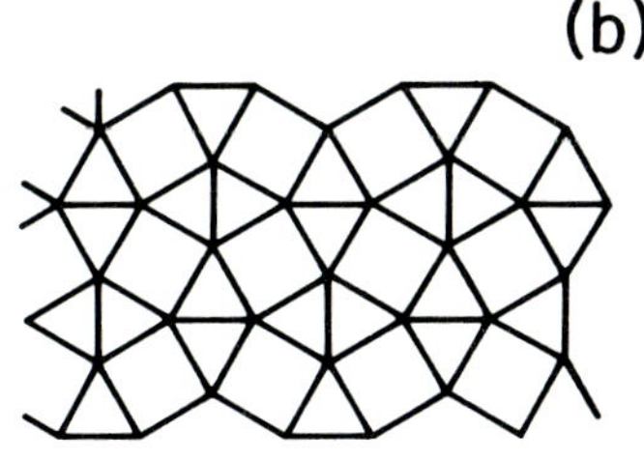

Fig. 1. (a) A typical fluid state. (b) A plane tesselation of triangles and squares.

In the neighborhood of a disk, then, there are four possible local configurations. Three of these are indicated in Fig. 2.

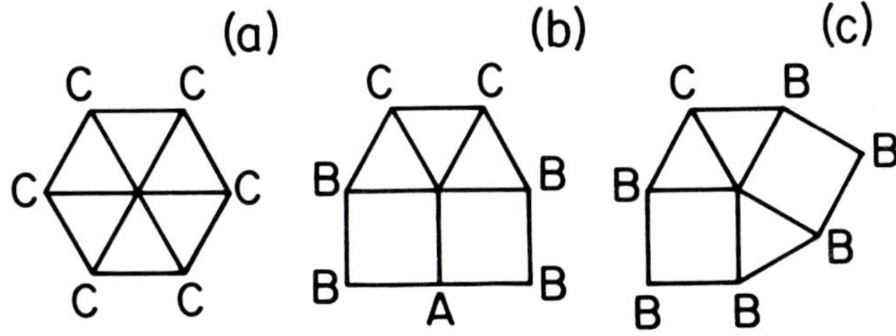

Fig. 2. (a) 6-disk, (b) 5α-disk, (c) 5β-disk.

A 6-disk is surrounded by six triangles and has six nearest neighbors. There are two 5-disks, 5α and 5β, consisting of three triangles and two squares and involving five nearest neighbors. A 4-disk with four nearest neighbors is surrounded by four squares. Kawamura discards this configuration as involving too much area to be common near a transition.

If there are a total of N disks of which N_r are r disks (r = 5α, 5β, 6), then

$$N = N_{5\alpha} + N_{5\beta} + N_6 \ . \tag{1}$$

The Wigner-Seitz cell associated with an r-disk is the region contained within the perpendicular bisectors of lines joining the center of the disk to the centers of its neighbors. The areas a_r of the Wigner-Seitz cells associated with the r-disks are:

$$a_5 \equiv a_{5\alpha} = a_{5\beta} = b^2 \ ; \quad a_6 = \frac{\sqrt{3}}{2} b^2 \ . \tag{2}$$

Then the total area A of the system is given by

$$A = (N_{5\alpha} + N_{5\beta})a_5 + N_6 a_6 \tag{3}$$

In terms of the fraction $n_r \equiv N_r/N$ of r-disks, equations (1) and (3) may be written as

$$n_{5\alpha} + n_{5\beta} + n_6 = 1 \ , \tag{4}$$

and

$$(n_{5\alpha} + n_{5\beta})a_5 + n_6 a_6 = A/N \ . \tag{5}$$

Equations (4) and (5) are constraints on the relative fractions of the types of disk.

In order to obtain an expression for the configuration entropy it is necessary to determine the number of possible ways to arrange the N_6, $N_{5\alpha}$, and $N_{5\beta}$ disks in the plane. If there were no geometrical constraints this number would be $N!/N_6!N_{5\alpha}!N_{5\beta}!$; this is the number of ways to choose N of the distinguishable r-disks completely at random. However these r-disks must be arranged in such a way that their local configurations fit together to cover the plane. Unfortunately, the number of ways to arrange the disks so that their local configurations satisfy this global geometric constraint is not yet known. This is an example of the type of problem in statistical geometry that Bernal's characterization of liquids gives rise to.

Kawamura provides a local solution to this problem that insures that at least the local configurations of nearest neighbors fit together. From an inspection of the local configurations in Fig. 2 it is clear that an A-site must be the center of a 5α-disk while a B-site may be the center of either a 5α- or a 5β-disk. A C-site imposes no restriction on neighboring configurations. Consequently, for each 5α-disk there must be one 5α-disk and four 5α- or 5β-disks and for each 5β-disk there must be six 5α- or 5β-disks. The number W of possible ways to choose the r-disks in order to satisfy these constraints is given by

$$W = \frac{N!}{N_6!N_{5\alpha}!N_{5\beta}!} (n_{5\alpha} + n_{5\beta})^{(4N_{5\alpha} + 6N_{5\beta})/2} (n_{5\alpha})^{(N_{5\alpha})/2} \ . \tag{6}$$

The configurational entropy S_c is defined in terms of W by

$$S_c = \mathrm{Ln}\ W \ . \tag{7}$$

Collins did not take into account any geometric constraints on the arrangements of the disks and came to the conclusion that no phase change was possible.

After introducing the variables x and t so that

$$n_{5\alpha} + n_{5\beta} = x \ , \quad n_6 = 1 - x \ , \tag{8}$$

$$n_{5\alpha} = tx \ , \quad n_{5\beta} = (1 - t)x \ , \tag{9}$$

with $0 \leq x \leq 1$ and $0 \leq t \leq 1$, the configurational entropy may be written as

$$S_c/N = -(1-x)\mathrm{Ln}(1-x) - \frac{tx}{2}\mathrm{Ln}(tx) - (1-t)x\mathrm{Ln}[(1-t)x] + (3-t)x\mathrm{Ln}x \ . \tag{10}$$

Kawamura also introduces a vibrational entropy S_v associated with the thermal fluctuations of the centers of the disks about their mean position. To calculate this he assumes that an r-disk can wander freely in the area available to it in its local configuration when the centers of its nearest neighbors are fixed at their mean positions. When the value of $\alpha \equiv (b/\sigma) - 1$ is small, the free areas v_r of the r-disks are

$$v_5 \equiv v_{5\alpha} = v_{5\beta} = (2 + \sqrt{3})\alpha^2\sigma^2 \ , \tag{11}$$

and

$$v_6 = 2\sqrt{3}\alpha^2\sigma^2 \ . \tag{12}$$

The mean free area $\bar{v}$ is, then,

$$\bar{v} = (1 - x)v_6 + xv_5 \ , \tag{13}$$

$$= v_6(1 + Cx) \ , \tag{14}$$

where $C = (2\sqrt{3} - 3)/6$. Then

$$S_v/N \equiv - \mathrm{Ln}\ \bar{v} \tag{15}$$

$$= - \mathrm{Ln}(2\sqrt{3}\alpha^2\sigma^2) - \mathrm{Ln}(1 + CX) \ . \tag{16}$$

In order to determine the possible equilibrium values for x and t for conditions of constant pressure and temperature, the Gibbs free energy G must be minimized. The Gibbs free energy is given in terms of the internal energy E , the entropy, the temperature T , the pressure P , and the area A of the system by

$$G = E - ST + PA \ . \tag{17}$$

The temperature T is the mean kinetic energy associated with the fluctuations in position of the centers of the disks. This is the same interpretation of the temperature as in the recent theories for granular materials (ref. 1-4). In the system of hard disks the only internal energy is the total kinetic energy. Because it is proportional to T and independent of x and t, it does not influence the possibility of a phase transition; so we ignore it. The pressure P has its usual interpretation in terms of the mean momentum transfer in collisions. The area A of the system is given through (3) in terms of x, α, and the closed packed area $A_o \equiv \sqrt{3}\sigma^2 N/2$ by

$$A = (1 + \alpha)^2 (1 + Cx) A_o \ . \tag{18}$$

With this G/NT may be expressed as a function of x, t, α, and the parameter $\xi = PA_o/NT$ as

$$G/NT = (1-x)\mathrm{Ln}(1-x) + \frac{tx}{2}\mathrm{Ln}(tx) + (1-t)x\mathrm{Ln}[(1-t)x] - (3-t)x\mathrm{Ln}x - \mathrm{Ln}(2\sqrt{3}\alpha^2) - \mathrm{Ln}(1+Cx) + \xi(1+\alpha)^2(1+Cx) \ . \tag{19}$$

If the first partials of this function with respect to t and α are calculated and set equal to zero, t and α may be expressed as functions of x as

$$t(x) = [2 + xe^{-1} - \sqrt{(2 + xe^{-1}) - 4}\,]/2 \ , \tag{20}$$

and

$$\alpha(x) = [\sqrt{1 + 4\xi^{-1}(1 + Cx)^{-1}} - 1]/2 \ , \tag{21}$$

Kawamura shows that these correspond to local minima of G with respect to t and α. If these are used in G, it becomes a function of x and the parameter ξ alone. To examine the x dependence of G, Kawamura introduces a function g that depends upon x in the same way as G but takes the value zero at $x = 0$ for all values of ξ:

$$g(x) = (1-x)\mathrm{Ln}(1-x) + \frac{tx}{2}\mathrm{Ln}(tx) + (1-t)x\mathrm{Ln}[(1-t)x] - (3-t)x\mathrm{Ln}x - 2\mathrm{Ln}(\alpha/\alpha_o) - \mathrm{Ln}(1+Cx) + \xi[(1+\alpha)^2(1+Cx) - (1+\alpha_o)^2] \ , \tag{22}$$

where $\alpha_o \equiv \alpha(0)$. A sketch of $g(x)$ versus x for several values of ξ is shown in Fig. 3.

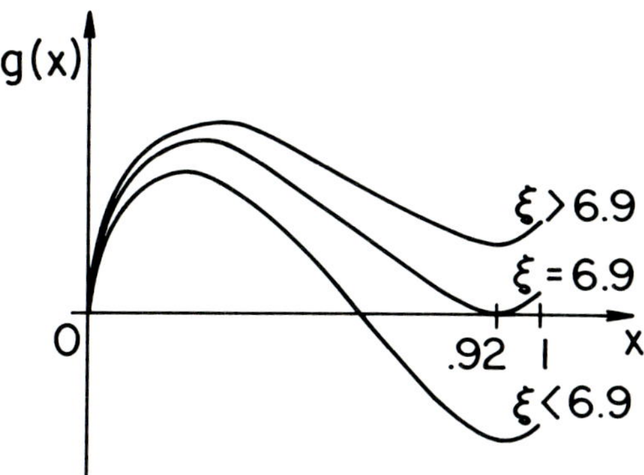

Fig. 3. $g(x)$ versus x for several values of ξ .

A phase transition occurs at $\xi = 6.9$. There the hexagonal close packed configuration at $x = 0$ and the configuration at $x = .92$ that contains triangles and squares have the same free energy. When ξ is greater than 6.9 the solid state has lower free energy. When ξ is less than 6.9 the fluid state has a lower free energy. At the transition the values of α and $\bar{v}$ in the solid and fluid are

$$\alpha_s = 0.128 \ , \quad \alpha_f = 0.121 \ , \tag{23}$$

and

$$\bar{v}_s = 0.578\sigma^2 \ , \quad \bar{v}_f = 0.054\sigma^2 \ . \tag{24}$$

The change in total area across the transition, $(A_s - A_f)/A_o$, is 0.072 .

The existence of a phase transition between equilibrium states corresponding to solid and liquid configurations of identical rigid disks may indicate the existence of a similar phase transition in the analogous macroscopic granular system. An important difference between the two systems is that in the macroscopic system the fluctuation energy that is the equivalent to the temperature exists only if there is an energy input to balance the energy dissipated in collisions. That is, the macroscopic system may be spatially homogeneous and steady, but it is not in equilibrium. An example of such a state of a three dimensional granular material consisting of identical imperfectly elastic spheres, is the simple shearing motion between perfectly elastic parallel plates considered by Jenkins and Savage (ref. 4). Here the ratio of the pressure to the temperature is determined in terms of the mean shear rate, the solid volume fraction, and the diameter and coefficient of restitution of the spheres. In the corresponding flow involving disks a phase transition from the liquid to solid state may take place when the ratio of the pressure to the temperature exceeds the value of 6.9 N/A_o predicted by Kawamura. Then such a transition should also be observed in the numerical simulations of the simple shear of identical, imperfectly elastic disks.

Interpretations of the critical void ratio important in critical state soil mechanics (ref. 11) have been made in a statistical mechanical context by Shahinpoor (ref. 12, 13). However the possibility that the critical void ratio corresponds to a parameter governing a phase transition in what are typically quasi-static deformation is far less certain, largely due to the uncertainty in the identification of the analogs of the pressure and the temperature in these slow deformations. Here the appropriate approach to phase transitions may involve the confining pressure and a consideration of ensemble averages.

REFERENCES

1 S. Ogawa, U. Umemura, and N. Oshima, Z.A.M.P., 31 (1980) 483-493.
2 S.B. Savage and D.J. Jeffrey, J. Fluid Mech., 110 (1981) 255-272.
3 H. Shen and N.L. Ackermann, J. Eng. Mech. Div., ASCE, 108 (1982) 748-763.
4 J.T. Jenkins and S.B. Savage, (submitted for publication).
5 R. Collins, Proc. Phys. Soc., 83 (1964) 553-564.
6 H. Kawamura, Prog. Theor. Phys., 61 (1979) 1584-1596.
7 J.D. Bernal, Trans. Faraday Soc., 33 (1937) 29-32.
8 J.D. Bernal, Nature, 183 (1959) 141-147.
9 J.D. Bernal, Nature, 195 (1960) 68-70.
10 J.D. Bernal, Proc. Roy. Soc. Lond., A280 (1964) 299-236.
11 K.H. Roscoe, A.N. Schofield, and C.P. Wroth, Geotechnique, 22 (1958) 1-
12 M. Shahinpoor, Powder Tech., 25 (1980) 163-176.
13 M. Shahinpoor, Bulk Solids Handling, 1 (1981) 31-36.

Mechanics of Granular Materials: New Models and Constitutive Relations, edited by
J.T. Jenkins and M. Satake, 1983
Elsevier Science Publishers B.V., Amsterdam — Printed in The Netherlands

KINEMATIC WAVES IN VERTICAL SAND COLUMNS

S.C. COWIN

Biomedical Engineering Department, Tulane University, New Orleans, LA 70118 (U.S.A.)

ABSTRACT

Waves of bulk density change in vertical columns of sand are modeled as kinematic waves. A flow-concentration curve for sand movement in a vertical column is developed and shown to be similar to flow concentration curves for traffic flow on long crowded roads. As an application of kinematic wave theory, the speed of a strong rarefaction wave in a sand column is obtained.

INTRODUCTION

The problem of certain types of waves in vertical sand columns is considered here. A sand column is a volume of sand contained within a rigid cylinder of arbitrary cross section. The long axis of the cylinder is coincident with the direction of gravity, and the internal surface of the cylinder is rough. We use sand in a generic sense and we thereby include all dry, cohesionless granular materials in which pneumatic effects are not significant. Examples of sand columns satisfying this definition include hopper sections of bulk material handling equipment, silos for the storage of bulk agricultural products, and standpipes that supply granular materials to certain chemical production processes.

This study is motivated by observations of waves in a sand column which do not reflect at interfaces. Attention is thus directed here to kinematic rather than dynamic waves because for kinematic waves, as pointed out by Lighthill and Whitham (ref.1), "no reflexion of any kind is possible". Dynamic waves are obtained from Newton's second law and they satisfy a characteristic hyperbolic differential equation of the form

$$\left(\frac{\partial}{\partial t} - c\frac{\partial}{\partial z}\right)\left(\frac{\partial}{\partial t} + c\frac{\partial}{\partial z}\right)q = 0 \ , \tag{1}$$

where c is the wave velocity, z is the spatial coordinate and q the physical quantity. Kinematic waves, on the other hand, are governed by a first order partial differential equation

$$\left(\frac{\partial}{\partial z} + c\frac{\partial}{\partial z}\right)q = 0 \ . \tag{2}$$

Therefore, while a physical quantity q satisfying (1) will have information projected along two characteristics $dz = cdt$ and $dz = -cdt$, the kinematic wave will only have information projected along the characteristic $dz = cdt$. Thus, at an interface that would give rise to a reflected dynamic wave, there is no reflection of the kinematic wave because of the non-existence of the second characteristic. In mechanics, dynamic waves generally arise from the second law of Newton whereas kinematic waves arise from mass conservation requirements. In general these two effects will be coupled and both kinematic and dynamic waves will exist. The dynamic waves should, in general, have a higher wave velocity and attenuate more rapidly.

Kinematic wave theory is surveyed and applied in two long papers by Lighthill and Whitham, (ref.1 and ref.2). In the first paper Lighthill and Whitham develop the theory as it applies to flood movements in long rivers. In the second paper they apply it to traffic flow on long crowded roads. It is suggested in the present study that sand movement in vertical columns bears many similarities to traffic flow on long crowded roads.

The theory of kinematic waves, described in the following section, is based on the assumed existence of a flow-concentration curve for the phenomenon. Arguments are presented in the section after next that suggest the form of the flow-concentration curve for sand movement in vertical columns. The slope of the flow-concentration curve is the wave velocity and the velocity of a wave carrying a concentration discontinuity is the slope of the straight line between the two points on the flow-concentration curve associated with values of the concentration at the leading and trailing edges of the wave. This last result is discussed in a section on concentration discontinuity waves. As an example of a concentration discontinuity wave, a rarefaction wave induced in a column of sand at rest by removing the support at the bottom of the column is considered. The predicted speed of propagation of the rarefaction wave corresponded well with speed measurements obtained in a preliminary experiment. The rarefaction wave results were obtained by Cowin and Comfort (ref.3), but were not placed in the context of kinematic waves by them.

KINEMATIC WAVES

Consider sand flowing down a long cylinder whose axis is coincident with the direction of the gravity field. The cylinder is a right cylinder of arbitrary cross section. It has rough walls. The cross section of the cylinder is in the x-y plane and z is taken to be positive in the direction opposite to that of the gravity field. The bulk density ρ of the sand is written as a product of the density of the sand grains γ and the solid volume fraction ν of the grains,

$$\rho = \gamma\nu . \tag{3}$$

It is assumed here that the density of the sand grains is constant in space and time, thus the continuity equation can be written

$$\frac{\partial \nu}{\partial t} + \frac{\partial}{\partial x}(\nu v_x) + \frac{\partial}{\partial y}(\nu v_y) + \frac{\partial}{\partial z}(\nu v_z) = 0 , \tag{4}$$

where v_x, v_y and v_z are the three components of the velocity vector.

The equation of continuity (4) will now be reduced by averaging it over the cross sectional area A of the cylinder. The average value of the quantity f(x, y, z, t) over the cross section z = constant is denoted by $\overline{f}(z,t)$ and is defined by

$$\overline{f}(z,t) = \frac{1}{A} \iint_A f(x,y,z,t)\,dxdy \tag{5}$$

Integration of the continuity equation (4) over the area A, with subsequent division by A, and application of the divergence theorem, yields

$$\frac{\partial \overline{\nu}}{\partial t} + \frac{\partial}{\partial z}(\overline{\nu v_z}) + \frac{1}{A} \int_C (\nu v_x n_x + \nu v_y n_y)\,ds = 0 \tag{6}$$

where the line integral is around the perimeter C of the cross section. Since there is no flux through the wall the last term is zero and we have that

$$\frac{\partial \overline{\nu}}{\partial t} + \frac{\partial}{\partial z}(\overline{\nu v_z}) = 0 \quad . \tag{7}$$

Shifting now to the terminology of kinematic wave theory (ref.1) the <u>flow</u> q of the sand and the <u>concentration</u> k of the sand are defined by

$$q \equiv \overline{\nu v_z} \;, \quad k = \overline{\nu} \tag{8}$$

and equation (7) may be rewritten as

$$\frac{\partial k}{\partial t} + \frac{\partial q}{\partial z} = 0 \; . \tag{9}$$

Kinematic waves are said to exist if, to sufficient approximation, there exists a functional relationship of the form

$$q = q(k,z) \; , \tag{10}$$

because, multiplication of (9) by

$$c = \left(\frac{\partial q}{\partial k}\right)_{z=\text{const.}} = c(k,z) \ , \tag{11}$$

yields

$$\frac{\partial q}{\partial t} + c\frac{\partial q}{\partial z} = 0 \ . \tag{12}$$

This condition requires that q be constant on waves moving past the point z with the velocity c given by (11). Waves satisfying the differential equations (2) and (12) are said to be the kinematic waves to distinguish them from the dynamic waves satisfying equation (1).

The wave velocity c defined by (11) is the slope of the flow concentration curve q (k,z). In terms of the mean velocity v at the station z,

$$v \equiv \frac{1}{k}q(k,z) \ , \tag{13}$$

the wave velocity c is

$$c = \frac{d}{dk}(vk) = v + k\frac{dv}{dk} \ . \tag{14}$$

When $c > v$ the mean velocity increases with concentration (as in flood movements down rivers) and when $c < v$ it decreases with concentration (as in traffic flow and sand movement in vertical columns).

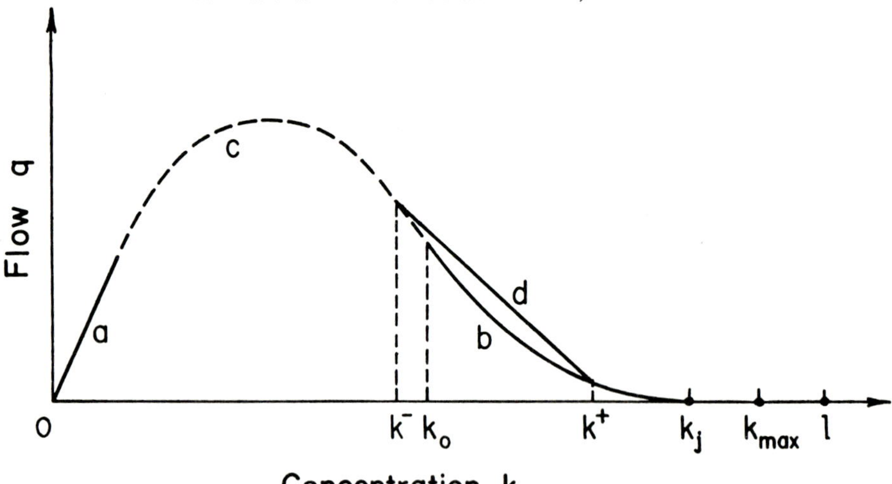

Fig.1. The flow-concentration curve for sand movement in a vertical cylinder. The initial slope of the curve, a, is the terminal velocity of a typical grain falling in air. The right leg of the curve, b, follows roughly the relation (15) of the text. The dashed portion of the curve, c, is a guess. The slope of the line d between the points + and − of the curve is the velocity of a concentration discontinuity wave carrying a change of concentration from k^+ to k^-.

THE FLOW-CONCENTRATION CURVE FOR A VERTICAL SAND COLUMN

The properties of kinematic waves for particular systems are determined by their particular flow-concentration curves. For flood movements in long rivers q is proportional to $k^{3/2}$. For sand movement in vertical cylinders it will be shown here that the flow concentration curve has a shape similar to that characterizing traffic flow on a long crowded road. The flow concentration curve in these cases will rise sharply after from the origin, levels to a plateau, then drops to zero flow at a finite concentration, called the jam concentration. A hypothetical flow concentration curve for a sand column is shown in Fig. 1.

The straight line slope of left leg of the flow-concentration curve is simply the terminal velocity of a grain of sand in air. At low concentrations the grains do not interact and their descent will then be determined by their weight and their drag in the air. The shape of the more sharply curved right leg of the flow concentration curve, the high concentration leg, is determined by a more lengthy argument given in the next paragraph. In between these two legs the curve is conjectured to follow the dashed portion of Fig. 1.

Between the concentration at which the flow jams, k_j, and the concentration at which the internal friction between particles vanish, k_o, it is argued that the flow q is related to the concentration k by

$$q = Vk^2(k_j - k)^4 \text{ for } k_o \leq k \leq k_j \,, \tag{15}$$

where V is a parameter of the system that can vary with z. A flow-concentration relation of the form (15) is implied by the work of Bagnold (ref.4); see also Savage (ref.5). Bagnold experiments suggested that the shear stress T_{zx} in a granular material sheared in the z-x plane is proportional to square of the velocity gradient and inversely proportional to eighth power of the difference between the concentration and the jam-concentration,

$$T_{zx} = \mu(k_j - k)^{-8}\left|\frac{\partial v_z}{\partial x}\right|\left(\frac{\partial v_z}{\partial x}\right) . \tag{16}$$

In the present situation the velocity gradient is approximated by the mean velocity v divided by some length L_1,

$$\frac{\partial v_z}{\partial x} \sim \frac{v}{L_1} \tag{17}$$

To obtain the result (15) the weight of sand in a unit length of the vertical sand column, $k\gamma gA$, is equated to the shear force per unit length acting on sand $T_{zx}L$ where L is the perimeter of the cross section. This relation yields

$$v = kV(k_j - k)^4 \tag{18}$$

where all factors not proportional to k have been lumped into V. The result (15) follows from substitution of (18) into (13).

The argument leading to (15) is admittedly loose. It does however lend credence to flow-concentration curve of the general form illustrated in Figure 1. This curve is similar in shape to those measured for traffic on long roads, except the flow-concentration curve for traffic does not have the sharply curved toe at the right. It comes down more directly to k_j.

WAVES OF CONCENTRATION DISCONTINUITY

In this section statements of the conservation of mass and momentum across concentration discontinuities are recorded. Waves with concentration discontinuities include both rarefaction waves and condensation waves. Consider the

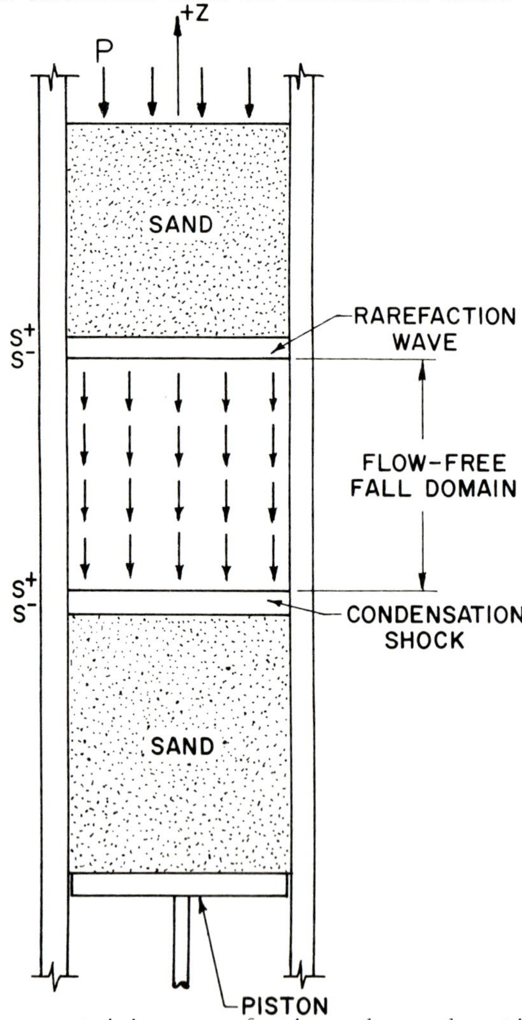

Fig. 2. The sand column containing a rarefaction and a condensation wave.

rarefaction wave illustrated in Fig. 2. The leading surface of this wave is denoted by s^+ and the trailing surface by s^-. In front of the wave the flow is q^+ and the concentration is k^+ and behind the wave they are q^- and k^-, respectively. Across the wave there is a steep change in both q and k. If the wave has a spatial velocity u, then the quantity crossing it per unit time may be written either as $q^+ - uk^+$ or as $q^- - uk^-$. This gives the velocity of the wave as

$$u = \frac{[q]}{[k]} \tag{19}$$

where $[f]$ of a function f means the difference between f^+ and f^-, $f^+ - f^-$. The velocity u of the concentration discontinuity is illustrated in Fig. 1 as a slope of the straight line from the point + on the flow-concentration curve to the point -.

The material velocity of surfaces s^+ and s^- are denoted by U^+ and U^-, respectively and related to the concentration discontinuity wave velocity u by

$$U^+ = u - v^+ \quad , \quad U^- = u - v^- \tag{20}$$

where v^+ and v^- are the mean velocities in front and behind the wave, respectively. Multiplication of the two formulas (20) by k^+ and k^- respectively and subsequent application of the definition (13) both in front and behind the wave yields

$$k^+U^+ = k^+u - q^+ \quad , \quad k^-U^- = k^-u - q^- . \tag{21}$$

It is easy to see that (21) and (19) imply that

$$[kU] = 0 \tag{22}$$

This is a statement of mass balance across a concentration discontinuity wave.

To obtain a statement of momentum conservation across the concentration discontinuity wave the mass velocity M, or the mass per unit area entering or leaving the wave front per unit time, is introduced,

$$M \equiv \gamma k^+U^+ = \gamma k^-U^- \tag{23}$$

The change in normal stress $[T_{zz}]$ is equal to rate at the mass in the discontinuity in gaining momentum, $[T_{zz}] = M(u - v^+) - M(u - v^-) = -M[v]$, or, from (20),

$$[T_{zz}] = M[U] \ . \tag{24}$$

An expression for $(U^+)^2$ is obtained from (24) using (22) and (23),

$$(U^+)^2 = -\frac{k^-}{\gamma k^+}\frac{[T_{zz}]}{[k]} \tag{25}$$

Thus, using (23) and (25), the speeds U^+ and U^- of concentration discontinuity wave are determined by the normal stress jump $[T_{zz}]$ and the values of the concentrations k^+ and k^- at the leading and trailing edges of the wave, respectively, and are independent of other factors.

THE RAREFACTION WAVE

We consider now what happens when the piston in Fig. 2 is suddenly dropped at an acceleration greater than that of gravity. Such an action leaves the bottom level of the sand unsupported. The sand will become less densely packed and begin to fall freely, creating a rarefaction wave. The surface s^+ is the lowest surface that is still in static equilibrium. Below the surface s^+, the shear stress sustained by friction between the sand and the wall diminishes in a short distance to zero. The surface s^- is defined to be the sand plane where the sand has become sufficiently disperse that it can no longer sustain intergranular stresses.

In order to determine the velocity of the rarefaction wave the jump in the normal stress T_{zz} across the wave must be determined. Above the rarefaction wave the sand is in equilibrium and the mean stress is given by

$$-T_{zz} = \gamma k_r g \ell_o + (P - \gamma k_r g \ell_o)e^{z/\ell_o} \tag{26}$$

where

$$\ell_o = \frac{A}{\mu LK} \tag{27}$$

and where g is the acceleration of gravity, the static coefficient of friction between the cylinder and the sand, k_r is the concentration in the static equilibrium situation in front of the wave, P is the surcharge stress at the free surface z = 0 of the sand, and K is the ratio of horizontal to vertical stress called Janssen's coefficient. The coefficient K is constant for most bulk materials; a summary of K values is given by Sundaram and Cowin (ref.6). A derivation of the formula (26) is given by Cowin (ref.7). It should be kept in mind that the last term in (26) represents an exponential decay because the admissible values of z are all negative.

The formula (26) gives the normal stress at the leading edge of the wave where the concentration is k_r. At the trailing edge of the wave the normal stress T^-_{zz} is assumed to vanish and the concentration is k_o, the value of the concentration at which intergranular stress disappears. When these values for T^+_{zz}, T^-_{zz}, $[k]$, k^+ and k^- are substituted in (25) we obtain

$$(U^+)^2 = v_o^2 + (v_1^2 - v_o^2)e^{z/\ell_o} \tag{28}$$

where v_o and v_1 are velocities defined by

$$v_o^2 = \frac{g\ell_o k_o}{k_r - k_o}, \quad v_1^2 = \frac{k_o P}{\gamma k_r(k_r - k_o)} \quad . \tag{29}$$

Since the mean velocity in front of the wave, v^+, is zero the speed of the rarefaction wave u coincides with the speed of its leading edge and

$$u^2 = v_o^2 + (v_1^2 - v_o^2)e^{z/\ell_o} \tag{30}$$

This result shows that far away from the free surface u tends to v_o (since z tends to negative infinity) and that near the free surface it tends to v_1. In particular, if the free surface has no surcharge (i.e., P = 0), the wave speed u will go to zero as the free surface is approached.

The time for a rarefaction shock to travel from a location $-z_o$ to the free surface can be calculated using the formula (30). From (30) it follows that

$$u = \frac{dz}{dt} = [v_o^2 + (v_1^2 - v_o^2)e^{z/\ell_o}]^{1/2} \tag{31}$$

Thus,

$$t = \int_{-z_o}^{0} [v_o^2 + (v_1^2 - v_o^2)e^{z/\ell_o}]^{1/2} dz \quad . \tag{32}$$

Integration yields

$$t = \frac{\ell_o}{v_o} \ell n\left(\frac{2v_o^2}{(v_o + v_1)^2} e^{z_o/\ell_o} + \frac{v_1 - v_o}{v_1 + v_o}\right) , \tag{33}$$

and, in the special case when the free surface is unstressed (P = 0, hence $v_1 = 0$),

$$t = \frac{\ell_o}{v_o} \ell n(2e^{z_o/\ell_o} - 1). \tag{34}$$

Equation (33) is a general expression for the time t that it takes a rarefaction wave to travel from a depth z_o to the free surface. Equation (34) is an expression for the same time t in the special case of an unstressed free surface. Cowin and Comfort (ref.3) indicate that this predicted speed of propagation for a rarefaction wave corresponds well with the speed measurements in a preliminary experiment.

CONCLUSION

The general conclusion of this study is that the loose framework of kinematic wave theory is applicable to sand movement in vertical columns and can be successfully applied to explain observed phenomena. The similar shapes of the flow-concentration curves for traffic flow on long crowded roads and concentration waves in vertical columns suggests that many of the results of the study of Lighthill and Whitham (ref.2) for traffic flow might be transferred to sand columns. For example, the analysis of a traffic bottleneck (ref.2) can easily be transferred to a bottleneck* in a sand column. The results of Cowin and Comfort (ref.3) on rarefaction and condensation waves in sand columns have been shown to be consistent with kinematic wave theory.

REFERENCES

1 M.J. Lighthill and G.B. Whitham, Proc. Roy. Soc. A, 229 (1955) 281-316.
2 M.J. Lighthill and G.B. Whitham, Proc. Roy. Soc. A, 229 (1955) 317-345.
3 S.C. Cowin and W.J. Comfort, III, J. Appl. Mech., 49 (1982) 497-500.
4 R.A. Bagnold, Proc. Roy. Soc. A, 225 (1954) 49-63.
5 S.B. Savage, J. Fluid Mech., 92 (1979) 53-96.
6 V. Sundaram and S.C. Cowin, Powder Technology, 22 (1979) 23-32.
7 S.C. Cowin, J. Appl. Mech., 44 (1977) 409-412.
8 G.B. Wallis, One Dimensional Two-phase Flow, McGraw Hill (1969).

POSTSCRIPTS

C.E. Brennen and J.W. Nunziato have pointed out that kinematic wave theory is used extensively in two phase flow; c.f., e.g. the book by Wallis (ref.8).

In an unpublished study, A.J.M. Spencer and V.G. Hart have considered kinematic wave models of certain cyclic phenomena in materials handling equipment.

---

*The sand column bottleneck in the present situation would be a region of increased wall friction rather than a change in column cross-sectional area.

Mechanics of Granular Materials: New Models and Constitutive Relations, edited by
J.T. Jenkins and M. Satake, 1983
Elsevier Science Publishers B.V., Amsterdam — Printed in The Netherlands

NONLINEAR INSTABILITIES IN FLUIDIZED BEDS

J. T. C. LIU
The Division of Engineering, Brown University
Providence, Rhode Island 02912 (U.S.A.)

ABSTRACT

The interpretation of linear instabilities of voidage disturbances in fluidized beds (ref. 12) is reviewed and the natural extension to nonlinear instabilities presented in the light of interpretation of observations (refs. 9, 10). The role of the constitutive relations in the analysis is also discussed.

INTRODUCTION

In the technologically important problems of pneumatic transport of granular materials, fluidized beds and sedimentation, for instance, regimes corresponding to fluidization and granular flows take place simultaneously in certain regions of such multiphase flow systems (ref. 1-5). While the difficult but important constitutive relations bridging the entire "spectrum" from granular material flows to fluidized flows have as yet to be considered from first principles, progress has been made in the limiting situations as evidenced by the papers given at the present Seminar on the mechanics of granular materials and by the literature on fluidization (ref. 6).

Although not entirely free from criticisms, the stress and rate-of-strain relationship for the solid phase in the fluidization limit is a linear one (refs. 6-8), modeled after the Navier-Stokes relationship. The interactions between the suspended solid phase and the interstitial fluid require the postulation of an interaction force which consists of drag and forces due to virtual mass effect and pressure-gradients. These must be supplemented by a number of relations. For instance, an equation of state relating the changes of the solid phase pressure contributions due to changes in the void fraction, leading to a sound speed owing to compressibility effects in the solid phase. How the drag coefficient changes with the void fraction is another relation that is needed. We refer to Jackson (ref. 6) for a discussion of these issues as well as the continuum formulation of the fluidization problem.

The state of uniform fluidization, in which the force exerted by the upward flowing fluid on the solid phase balances the downward force it experiences due to gravity, is an exact solution to the continuum equations. Ideally, the prospect of uniform fluidization renders the fluidized bed as an attractive device for bringing fluids into contact with particulate solids for a variety of technical applications. Both theoretically and experimentally, it is well known that the state of uniform fluidization is unstable to voidage wave disturbances. What is more important is that there is observational evidence that bubble formation can be traced to the nonlinear stages of voidage wave development (ref. 9), at least in a deep, liquid-solid bed.

The intent of the present paper is to provide the outline of an elucidation of the development of voidage wave disturbances from the linear into finite-amplitude region while the waves are still planar but nevertheless have reached what appear to be (momentary) equilibrium (ref. 9). Further cross-stream or secondary instabilities of the finite planar disturbances lead to bubble-like voidage disturbances. The analysis of nonlinear planar disturbances discussed in this paper would provide the necessary basis for further work on the secondary instabilities. In the present discussion, which also brings out the weak dependence of the analysis upon the constitutive relations, it will be shown how observations can be successfully interpreted in terms of ideas from nonlinear wave-hierarchies and hydrodynamic stability.

WAVE-HIERARCHY ANALYSIS

The experiments of Anderson and Jackson (ref. 10) and El-Kaissy and Homsy (ref. 9) in deep, liquid-solid fluidized beds indicate that the propagating voidage wave disturbances are planar and grow exponentially with vertical distance above the distributor. This is in accordance with the linearized theory (refs. 10, 11). The experiments also indicate that the rate of growth is eventually modified and the amplitude of the disturbances eventually equilibrates.

The particular features of the observations which also emerges from the present analysis is that disturbances with an initially large growth rate equilibrates to larger amplitude. The behavior is the same for the phase velocity in that it increases towards a higher equilibrium level for initially larger disturbance growth rates.

It has been shown (ref. 12) that the linearized stability problem of a uniformly fluidized bed (ref. 6) can be recast into the wave-hierarchy form, written now for planar disturbances

$$\tau\left(\frac{\partial}{\partial t} + c_1\frac{\partial}{\partial x}\right)\left(\frac{\partial}{\partial t} + c_2\frac{\partial}{\partial x}\right)\varepsilon + \left(\frac{\partial}{\partial t} + a_1\frac{\partial}{\partial x}\right)\varepsilon = \nu_e\tau\frac{\partial^3\varepsilon}{\partial t\partial x^2}, \tag{1}$$

where ε is the perturbation void fraction, x the vertical distance from the distributor plate, t is the time. The detailed relationship between the constants that emerges and a particular fluidization model (ref. 6) is given in ref. 12. It would be more illuminating to give the constant coefficients their physical interpretation here. A relaxation time is denoted by τ, which reflects the time scale for velocity equilibration between the two phases brought about by the drag force. The higher-order wave speeds are c_1 and c_2 and their interpretation is made simple for a gas-solid system where the fluid to solid density ratio, ρ_f/ρ_s, is small, in which case $c_1 = a_0^s$ and $c_2 = -a_0^s$ are just the sound speed in the solid phase due to compressibility effects. For ρ_f/ρ_s finite, $c_1 > c_2$ and they are modified by inertia such as the virtual mass effect and fluid pressure gradient force. The lower order wave speed a_1 arises from the restoring force due to the drag force and the (competing) effect of gravity. The effective viscosity contributed by the solids is ν_e.

The presence of ν_e is a direct consequence of a postulated Navier-Stokes stress and rate of strain relation for the solid phase, including the contribution of the solids to a pressure p_s. An equation of state for this pressure, as function of the void fraction, has been implied, with the interpretation that $a_0^{s^2} = [\partial p_s/\partial \rho_s(1-\varepsilon)]_0$.

The wave-hierarchy stability condition (here $c_1 > c_2$) is $c_1 > a_1$ according to Lighthill and Whitham (refs. 13, 14). Thus, for the uniformly fluidized bed to be unstable, one must have $c_1 < a_1$. This is, indeed, the case from estimates (ref. 12) of such wave speeds from experiments (ref. 11). In the unstable situation, the higher-order waves, which correspond to higher-frequency components closer to the distributor plate, have wave speeds that are slower than those of the lower-order waves. The latter correspond to lower-frequency waves in the upper regions of the bed. If viewed as an "initial value" problem, the voidage disturbances would grow exponentially following the c_1 waves but these will eventually be left behind by the faster a_1 waves following which the disturbances will be "focused" by an effective-negative diffusion coefficient according to wave-hierarchy interpretations (ref. 12). Observations of voidage disturbances (ref. 9) show that slower traveling, higher frequency waves are being surpassed by faster traveling, lower frequency waves. This bears strong resemblence to the present interpretations. Although in the experimental situation the disturbance waves are already nonlinear in the upper regions of the bed, their "kinematics" are still susceptible to the present linearized wave-hierarchy interpretations according to (1).

The nonlinear dynamical features, concerning amplitude and phase velocity development and saturation, however, requires the following consideration. We note that according to estimates (ref. 12) from observations (ref. 9), the respective wave-hierarchy speeds are such that $(c_1-a_1)/c_1 \ll 1$. For the initial value problem this implies that in the x-t plane the a_1 and c_1 waves make a small angle so that charges in voidage disturbances are much larger across than along the waves. From a normal mode analysis of linearized hydrodynamic stability (e.g., ref. 6), the condition $(c_1-a_1)/c_1 << 1$ corresponds to small amplification rates. In this situation a weak nonlinear extension of (1) suffices, and this is obtained from the same set of nonlinear equations that give rise to (1). The technique is similar to that used in works on weak waves in gas dynamics for which nonequilibrium or relaxation effects are important (refs. 15, 16). The weak nonlinear extension of (1) is thus a hierarchy of Burgers-like equation,

$$\tau[(\frac{\partial}{\partial t} + c_1 \frac{\partial}{\partial x})(\frac{\partial}{\partial t} + c_2 \frac{\partial}{\partial x}) + c_3^2 \frac{\partial}{\partial x}(\varepsilon \frac{\partial}{\partial x})]\varepsilon +$$

$$+ [\frac{\partial}{\partial t} + (a_1 + a_3\varepsilon)\frac{\partial}{\partial x}]\varepsilon = \nu_e \tau \frac{\partial^3 \varepsilon}{\partial t \partial x^2} . \qquad (2)$$

The nonlinear modification of the viscosity, which could be included, has been omitted. The nonlinear corrections enter through the modification of the propagation speeds through $c_3^2\varepsilon$ and $a_3\varepsilon$. The constants c_3^2 and a_3 are functions of the fluidization parameters. It suffices to mention that the leading contribution to c_3^2 is $(c_3/c_1)^2 \sim 2\varepsilon_0^2/(1-\varepsilon_0) > 0$. The effect of this correction is that the higher-order propagation speed would be increased for increasing amplitude of the disturbance ε. The dominant contribution (for the linearly unstable situation) towards the lower-order wave speed correction is $a_3/c_1 \sim -2\varepsilon_0^2/(1-\varepsilon_0) < 0$. This directly decreases the propagation speed of the lower-order waves for increasing amplitude of ε. Thus, the effect of nonlinearity is the tendency to restore the stability condition in the linearized sense. The nonlinear effects from the wave-hierarchy point of view is directly related to amplitude equilibration in the nonlinear hydrodynamic stability analysis.

NONLINEAR HYDRODYNAMIC INSTABILITY

Jackson (ref. 5) has given an excellent summary of the work on the linear, normal mode analysis of the hydrodynamic stability of a uniform fluidized bed. The temporal problem is considered, where the upward propagating instabilities amplify or decay in time. The nonlinear analysis considers the temporal problem in which the disturbance amplitude becomes finite and equilibrates in time. The technique here follows closely the nonlinear hydrodynamic stability work on shear flows (ref. 17). However, the present problem is considerably simpler in that the basic flow is uniform.

The planar wavy disturbances are expandable into the form

$$\varepsilon(x,t) = [A_1(t)\psi_1(x) + A_1(t)|A_1(t)|^2\psi_{11}(x)]\exp(-i\alpha c_r t)$$

$$+ A_1^2(t)\psi_2(x)\exp(-2i\alpha c_r t) + c.c., \qquad (3)$$

where $A_1(t)$ is the time-dependent amplitude that grows exponentially for initially amplified infinitesimal disturbances and becomes finite with time, $\psi_1 \sim exp(i\alpha x)$ is the fundamental disturbance of the linear theory and α is the (real) wave number, $\psi_{11}(x)$ is the function which modifies the fundamental as the amplitude becomes finite, c_r is the phase velocity, $\psi_2(x)$ is the first harmonic component and c.c. denotes the complex conjugate. Insertion of (3) into (2) results in a series of problems. The linear problem $\psi_1(x)$ yields a characteristic equation for α and $c = c_r + ic_i$, where c_i is the complex phase velocity and αc_i is the amplification rate, with τ, c_1, c_2, a_1 and ν_e as parameters. A neutral curve could be found and expressed in terms of the dimensionless wave number $\alpha^* = \alpha u_0 \tau$, dimensionless propagation speeds $a_1^* = a_1/u_0$, $c_1^* = c_1/u_0$ and a Reynolds number $R = u_0^2\tau/\nu_e$,

$$\alpha_c^{*2} = R_c a_1^*/(c_1^*-1), \tag{4}$$

where the subscript c denotes critical conditions along the neutral curve. For a given α_c^*, $R - R_c > 0$ corresponds to amplified disturbances for which $c_i > 0$ whereas $R - R_c < 0$, $c_i < 0$. The weakly nonlinear theory concerns small departures from the neutral curve for which the amplification rates are small and the truncation of higher order effects result in a system which is adequate for a full nonlinear description.

A solvability condition similar to Stuart (1960) leads to the determination of a nonlinear amplitude equation for $A_1(t)$ as well as the coefficient k of the nonlinear term in that equation. The complex constant k being known as the Landau-Stuart constant. Thus, we have

$$\tau\{\frac{d}{dt} + \alpha c_i + i\alpha[(c_1-c_r) + (c_2-c_r)]\}(\frac{dA_1}{dt} - \alpha c_i A_1)$$
$$+ (1+\nu_e\tau\alpha^2)(\frac{dA_1}{dt} - \alpha c_i A_1) - \alpha k A_1|A_1|^2 = 0. \tag{5}$$

The equation for $A_1(t)$ is second order, however, the second and third terms in (5) dominate over most of the time history of the disturbance development in view of the smallness of the relaxation time τ compared to the advective time x/u_0 estimated previously (ref. 12). Thus,

$$\frac{dA_1}{dt} - \alpha c_i A_1 - \alpha k(c_1/a_1)A_1|A_1|^2 \simeq 0, \tag{6}$$

which is of the same form as the shear flow problem (ref. 17). The coefficient of the nonlinear term is given by

$$k = \frac{\int_a^b \Phi[(-\tau c_3^2 \frac{d}{dx} + a_3)\frac{d\tilde{\psi}_1\psi_2}{dx}]dx}{\int_a^b \Phi\ \alpha\psi_1 dx}, \tag{7}$$

where $(b-a)$ is one wavelength, $\Phi \sim exp(-i\alpha x)$ is the adjoint of the fundamental $\psi_1 \sim exp(i\alpha x)$, $\tilde{\psi}_1$ is the complex conjugate of ψ_1 and $\psi_2 \sim exp(2i\alpha x)$ is the first harmonic of ψ_1.

The equilibrium amplitude is obtained by setting all time derivatives in (5) to zero (the same result holds true for (6)),

$$|A_1|_e^2 = c_i \frac{a_1/c_1}{-k_{rc}}, \tag{8}$$

where k_{rc} is the real part of k_c, provided that for $c_i > 0$ (supercritical disturbances) $-k_{rc} > 0$. Appropriate evaluation of the integrals in (7) and taking the value of k on the neutral curve, we obtain

$$\frac{k_{rc}}{|B_1|^2} = \alpha_c \tau a_3 c_3^2/2(a_1-c_1), \tag{9}$$

where B_1 is the undetermined constant corresponding to the linear theory for ψ_1. The solution for $|A_1|^2$ is

$$|A_1|^2 = \frac{c_i\ C\ exp(2\alpha c_i t)}{1 + \frac{c_i C}{|A_1|_e^2} exp(2\alpha c_i t)}$$

obtainable from (6), where C is a real, arbitrary constant. Now, the supercritical case $c_i > 0$ corresponds to $a_1 > c_1$ and $a_3 < 0$ from wave-hierarchy considerations. In this case $-k_{rc} > 0$ for $c_i > 0$ and an equilibrium amplitude does exist as $t \rightarrow \infty$.

For the subcritical case $c_i < 0$ and $a_1 < c_1$ ($a_3 < 0$, regardless) an equilibrium amplitude exists for $t \rightarrow -\infty$. As long as the initial amplitude is less than the threshold $|A_1|_{et}^2 = |c_i|(a_1/c_1)/|k_{rc}|$, the disturbance will decay to zero with time for the subcritical case; the disturbance will grow to unbounded levels if the initial amplitude is larger than the threshold amplitude $|A_1|_{et}^2$, in which case the weakly nonlinear theory would no longer be valid. The agreement with observations here is that the observed (ref. 9) equilibrium amplitudes for supercritical disturbances increase and that the equilibration process is achieved earlier with increasing initial amplification rates.

The phase velocity of the disturbance can be estimated from an equation for the phase of the disturbance, $A_1/\tilde{A}_1$, where $\tilde{A}_1$ is the complex conjugate of A_1, and again, obtainable from (6). In the supercritical case it can be shown that the amplitude develops from a non-oscillatory one from $t \to -\infty$ to an oscillatory one as $t \to \infty$. The nonlinear correction to the equilibrium phase velocity c_{re} as $t \to \infty$ is obtained as

$$c_{re} = c_r - c_i k_{ic}/(-k_{rc}), \tag{11}$$

where k_{ic} is the imaginary part of k_c. Now

$$\frac{k_{ic}}{|B_1|^2} = -\frac{2(\alpha\tau^2 c_3^2)^2 + a_3^2}{6(a_1 - c_1)}$$

and $k_{ic} < 0$ for supercritical disturbances for which $a_1 > c_1$. With $-k_{rc} > 0$ as already shown, then the nonlinear correction is positive

$$-c_i k_{ic}/(-k_{rc}) > 0$$

for $c_i > 0$. Thus, the observed (ref. 9) increase in the phase velocity for increasing supercriticality is again consistent with the present theory.

In concluding this paper, it would be worthwhile to point out how the constitutive relations (see, for instance, Jackson, ref. 6) affect the present analysis. First of all, the instability mechanism is a "dynamical" one in that the effective viscosity of the solid phase has at most a stabilizing influence. The competition between higher-order compressibility-like waves and lower-order waves, whose restoring force comes from the drag and gravity, gives rise to instability. Such a wave-hierarchy interpretation depends upon the wave speeds obtained via (1) the contribution to an effective pressure from the solid phase self-interaction or particle-particle collisions and (2) the interaction forces between the solid and fluid phases, such as drag, pressure-gradient force, the virtual mass effect and the gravitational force. The effective viscosity of the solid phase is needed to obtain a neutral curve, the perturbation from which forms the basis of the weakly nonlinear theory.

In spite of the uncertainties concerning both the constitutive relations and the values of the "constants" therein, it is encouraging that the present analysis is capable, in a limited sense, of explaining some of the fascinating features of instabilities in fluidized beds when the disturbances have become finite and nonlinear. The consideration of the dynamics of cross-stream instability modes leading to bubble-like structures appear to be the forthcoming topic for study, from which the present work serves as a springboard.

This work was initiated while the author was on sabbatical leave at the Department of Mathematics, Imperial College of Science and Technology, London. The hospitality of and helpful discussions with J. T. Stuart, F.R.S. and the partial support from the United Kingdom Science Research Council through its Senior Visiting Fellowship Programme are gratefully acknowledged.

REFERENCES

1 R. Jackson, in H. Littman (Ed.), Proceedings of the NSF Workshop on Fluidization and Fluid-Particle Systems Research Needs and Priorities, Rensselaer Polytechnic Institute, Troy, 1979, pp. 49-55.
2 J. C. Ginestra, S. Rangachari and R. Jackson, in J. R. Grace and J. M. Matsen (Eds.), Fluidization, Plenum, New York, 1980, pp. 477-484.
3 K. Konrad, D. Harrison, R. M. Nedderman and J. F. Davidson, Prediction of the Pressure Drop for Horizontal Dense Phase Pneumatic Conveyance of Particles, Proceedings of the Fifth International Conference on the Pneumatic Transport of Solids in Pipes, BHRA Fluid Engineering, Cranfield, 1980, pp. 225-244.
4 L. S. Leung, in J. R. Grace and J. M. Matsen (Eds.), Fluidization, Plenum, New York, 1980, pp. 25-68.
5 G. K. Batchelor, The Mechanics of Suspensions of Small Particles, Fourth International Conference on Physico-Chemical Hydrodynamics, New York, 1982.
6 R. Jackson, in J. F. Davidson and D. Harrison (Eds.), Fluidization, Academic Press, London, 1971, pp. 65-119.
7 S. B. Savage, J. Fluid Mech., 92 (1979) 53-96.
8 S. B. Savage and D. J. Jeffrey, J. Fluid Mech., 110 (1981) 255-272, 117 (1982) 531.
9 M. M. El-Kaissy and G. M. Homsy, Int. J. Multiphase Flow, 7 (1976) 379-395.
10 T. B. Anderson and R. Jackson, Ind. Engng. Chem. Fundam., 8 (1969) 137-144.
11 G. M. Homsy, M. M. El-Kaissy and A. Didwania, Int. J. Multiphase Flow, 6 (1980) 305-318.
12 J. T. C. Liu, Proc. R. Soc. Lond., A380 (1982) 229-239.
13 M. J. Lighthill and G. B. Whitham, Proc. R. Soc. Lond., A229 (1955) 281-345.
14 G. B. Whitham, Linear and Nonlinear Waves, John Wiley and Sons, New York, 1974, 636 pp.
15 P. A. Blythe, J. Fluid Mech., 37 (1969) 31-50.
16 H. Ockendon and D. A. Spence, J. Fluid Mech., 39 (1969) 329-345.
17 J. T. Stuart, J. Fluid Mech., 9 (1960) 353-370.